#2

To Marx,
My good friend & colleague.

Barry

GEOSYNTHETIC ENGINEERING

Robert D. Holtz, PhD., P.E.

University of Washington
Department of Civil Engineering
Seattle, Washington

Barry R. Christopher, PhD., P.E.

Consultant
Atlanta, Georgia

Ryan R. Berg, P.E.

Consultant
St. Paul, Minnesota

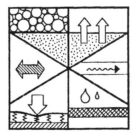

Published by
BiTech Publishers Ltd.
173 - 11860 Hammersmith Way
Richmond, British Columbia, Canada V7A 5G1

Printed and bound in Canada

First Printing, June 1997

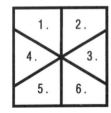

COVER KEY:
1. Separation Function
2. Filtration Function
3. Drainage Function
4. Reinforcement Function
5. Cushioning Function
6. Fluid Barrier Function

ISBN 0-921095-20-1

Preface

This Geosynthetic Engineering book has evolved from the following FHWA manuals:

- Holtz, R.D., Christopher, B.R. and Berg, R.R. (1995), *Geosynthetic Design and Construction Guidelines*, U.S. Department of Transportation, Federal Highway Administration, Washington, D.C., FHWA-HI-95-038, May, 398 p.
- Berg, Ryan R. (1993), *Guidelines for Design, Specification, and Contracting of Geosynthetic Mechanically Stabilized Earth Slopes on Firm Foundations*; U.S. Department of Transportation, Federal Highway Administration, Washington, D.C., FHWA-SA-93-025, January, 88 p.
- Christopher, B.R. and Holtz, R.D. (1992), *Geotextile Design & Construction Guidelines - Participant Notebook,* U.S. Department of Transportation, Federal Highway Administration, Washington, D.C., FHWA-HI-90-001, 398 p.; October 1988 and selectively updated to April 1992.
- Christopher, B.R., Gill, S.A., Giroud, J.P., Mitchell, J.K., Schlosser, F. and Dunnicliff, J. (1990), *Reinforced Soil Structures - Volume I, Design and Construction Guidelines, and Volume II Summary of Research and Systems Information*; U.S. Department of Transportation, Federal Highway Administration, Washington, D.C., FHWA-RD-89-043, November, 445 p.
- Christopher, B.R. and Holtz, R.D. (1985), *Geotextile Engineering Manual*, U.S. Department of Transportation, Federal Highway Administration, Washington, D.C., 917 p.
- Haliburton, T.A., Lawmaster, J.D. and McGuffey, V.C. (1981), *Use of Engineering Fabrics in Transportation Type Related Applications* by U.S. Department of Transportation, Federal Highway Administration, Washington, D.C.

Acknowledgement

Jerry A. DiMaggio, P.E. of the FHWA was the Technical Consultant for most of the manuals noted above, and his guidance and input was invaluable.

SI CONVERSION FACTORS

APPROXIMATE CONVERSIONS FROM SI UNITS

Symbol	When You Know	Multiply By	To Find	Symbol
LENGTH				
mm	millimeters	0.039	inches	in
m	meters	3.28	feet	ft
m	meters	1.09	yards	yd
km	kilometers	0.621	miles	mi
AREA				
mm^2	square millimeters	0.0016	square inches	in^2
m^2	square meters	10.764	square feet	ft^2
m^2	square meters	1.195	square yards	yd^2
ha	hectares	2.47	acres	ac
km^2	square kilometers	0.386	square miles	mi^2
VOLUME				
ml	millimeters	0.034	fluid ounces	fl oz
l	liters	0.264	gallons	gal
m^3	cubic meters	35.	cubic feet	ft^3
m^3	cubic meters	71	cubic yards	yd^3
		1.307		
MASS				
g	grams	0.035	ounces	oz
kg	kilograms	2.202	pounds	lb
TEMPERATURE				
°C	Celsius	1.8 C + 32	Fahrenheit	°F
WEIGHT DENSITY				
kN/m^3	kilonewton/cubic meter	6.36	poundforce/cubic foot	pcf
FORCE and PRESSURE or STRESS				
N	newtons	0.225	poundforce	lbf
kN	kilonewtons	225	poundforce	lbf
kPa	kilopascals	0.145	poundforce/square inch	psi
kPa	kilopascals	20.9	poundforce/square foot	psf

TABLE OF CONTENTS

LIST OF FIGURES

LIST OF TABLES

1.0 INTRODUCTION, IDENTIFICATION and EVALUATION

1.1 BACKGROUND

This manual was prepared to assist design engineers, specification writers, estimators, construction inspectors, and maintenance personnel with the design, selection, and installation of geosynthetics. In addition to providing a general overview of these materials and their applications, step-by-step procedures are given for the cost-effective use of geosynthetics in drainage and erosion control systems, roadways, reinforced soil structures, and in containment applications. Although the title refers to the general term *geosynthetic*, specific applications address the appropriate use of subfamilies of geotextiles, geogrids, geocomposites, and geomembranes.

The basis for much of this manual is the FHWA *Geotextile Engineering Manual* (Christopher and Holtz, 1985). Other sources of information include the books by Koerner (1994), John (1987), and Veldhuijzen van Zanten (1986). If you are not already familiar with geosynthetics, you are encouraged to read Richardson and Koerner (1990) and Ingold and Miller (1988), especially if you are attempting to use geosynthetics for the first time. A listing of other geosynthetics literature can be found in Cazzuffi and Anzani (1992) and Holtz and Paulson (1988), both of which are reproduced in Appendix A. Comprehensive geosynthetic bibliographies have recently been prepared by Giroud (1993, 1994). If you are unfamiliar with geosynthetics terminology, see ASTM (1994) D 4439 Standard Terminology for Geosynthetics. Basic terms are defined in Appendix B. The authors assume that you are already familiar with the engineering basics of geotechnical, highway, hydraulic, retaining wall, and pavement design. Common notation and symbols are used throughout this manual, and a list is provided in Appendix C for easy reference. These notations and symbols are generally consistent with the International Geosynthetic Society's (IGS) Recommended Mathematical and Graphical Symbols (1993).

Sample specifications included in this manual were developed in several cases by Task Force 25 Subcommittee of the American Association of State Highway and Transportation Officials (AASHTO, 1990), the Association General Contractors (AGC), and the American Road and Transportation Builders Associations (ARTBA) Joint Committee, along with representatives from the geosynthetic industry. Important input has also been obtained from the AASHTO-AGC-ARTBA Task Force 27 Subcommittee (1990). Specifications from the FHWA Guidelines for Design, Specification, and Contracting of Geosynthetic Mechanically Stabilized Earth Slopes on Firm Foundations (Berg, 1993) are also used with this manual. Finally, sample specifications were obtained from some state Departments of Transportation. **These specifications are meant to serve only as guidelines and should be modified as required by engineering judgment and experience, based upon project specific design and performance criteria.**

Chapter 1 introduces you to the functions and applications of geosynthetics, to the identification of the materials, and to the methods used to evaluate their properties. The remaining nine chapters give specific details about important application categories of geosynthetics, such as drainage and roadways. Each chapter provides a systematic approach to applying geosynthetics so that successful cost-effective designs and installations can be achieved.

1.2   DESIGN APPROACH

We recommend the following approach to designing with geosynthetics:
1. Define the purpose and establish the scope of the project.
2. Investigate and establish the geotechnical conditions at the site (geology, subsurface exploration, laboratory and field testing, etc.).
3. Establish application criticality, severity, and performance criteria. Determine external factors that may influence the geosynthetic's performance.
4. Formulate trial designs and compare several alternatives.
5. Establish the models to be analyzed, determine the parameters, and carry out the analysis.
6. Compare results and select the most appropriate design; consider alternatives versus cost, construction feasibility, etc. Modify the design if necessary.
7. Prepare detailed plans and specifications including: a) specific property requirements for the geosynthetic; and b) detailed installation procedures.
8. Hold preconstruction meeting with contractor and inspectors.
9. Approve geosynthetic on the basis of specimens' laboratory test results and/or manufacturer's certification.
10. Monitor construction.
11. Inspect after events (e.g., 100 year rainfall) that may tax structure performance.

By following this systematic approach to designing with geosynthetics, cost-effective designs can be achieved, along with improved performance, increased service life, and reduced maintenance costs. Good communication and interaction between all concerned parties is imperative throughout the design and selection process.

1.3   DEFINITIONS, MANUFACTURING PROCESSES and IDENTIFICATION

ASTM (1994) has defined a *geosynthetic* as a planar product manufactured from a polymeric material used with soil, rock, earth, or other geotechnical-related material as an integral part of a civil engineering project, structure, or system. A *geotextile* is a permeable geosynthetic made of textile materials. There are a number of other materials available today that technically are not textiles -- including webs, grids, nets, meshes, and composites -- that are used in combination with or in place of geotextiles. These are sometimes referred to as geotextile-related materials. Geotextiles and related materials all fall under the principal category of geosynthetics. *Geogrids*, geosynthetics primarily used for reinforcement, are formed by a regular network of tensile elements with apertures of sufficient size to interlock with surrounding fill material. *Geomembranes* are low-permeability geosynthetics used as fluid barriers. Geotextiles and related products, such as nets and grids, can be combined with geomembranes and other synthetics to complement the best attributes of each material. These products are called *geocomposites*, and they may be geotextile-geonets, geotextile-geogrids, geotextile-geomembranes, geomembrane-geonets, geotextile-polymeric cores, and even three-dimensional polymeric cell structures. There is almost no limit to the combinations of geocomposites.

A convenient classification scheme for geosynthetics is provided in Figure 1-1. For details on the composition, materials, and manufacturing processes, see Koerner (1994), Ingold and Miller (1988), Veldhuijzen van Zanten (1986), Christopher and Holtz (1985), Giroud and Carroll (1983), Rankilor (1981), and Koerner and Welsh (1980). Most geosynthetics are made from synthetic polymers of polypropylene, polyester, or polyethylene. These polymer materials are highly resistant to biological and chemical degradation. Less-frequently-used polymers include polyamides (*nylon*) and glass fibers. Natural fibers, such as cotton, jute, etc., could also be used as geotextiles, especially for temporary applications, but they have not been researched or utilized in the U.S. as widely as polymeric geotextiles.

In manufacturing geotextiles, elements such as fibers or yarns are combined into planar textile structures. The fibers can be continuous *filaments*, which are very long thin strands of a polymer, or *staple fibers,* which are short filaments, typically 20 to 150 mm long. The fibers may also be produced by slitting an extruded plastic sheet or film to form thin flat tapes. In both filaments and slit films, the extrusion or drawing process elongates the polymers in the direction of the draw and increases the fiber strength.

Geotextile type is determined by the method used to combine the filaments or tapes into the planar structure. The vast majority of geotextiles are either *woven* or *nonwoven*. Woven geotextiles are made of *monofilament*, *multifilament*, or *fibrillated* yarns, or of slit films and tapes. The weaving process is as old as Homo sapiens' textile cloth-making. Nonwoven textile manufacture is a modern development, a *high-tech* process industry, in which synthetic polymer fibers or filaments are continuously extruded and spun, blown or otherwise laid onto a moving belt. Then the mass of filaments or fibers are either *needlepunched*, in which the filaments are mechanically entangled by a bed of needles, or *heat bonded*, in which the fibers are *welded* together by heat and/or pressure at their points of contact in the nonwoven mass.

The manufacture of geotextile-related products is as varied as the products themselves. Geonets, geosynthetic erosion mats, geogrids, etc., can be made from large and often rather stiff filaments formed into a mesh with integral junctions or which are welded or glued at the crossover points. Geogrids with integral junctions are manufactured by extruding and orienting sheets of polyolefins. These types of geogrids are usually called stiff geogrids. Geogrids are also manufactured of polyester yarns, joined at the crossover points by a knitting or weaving process, and encased with a polymer-based, plasticized coating. These types of geogrids are generally called flexible geogrids. Manufacture of geomembranes and other geosynthetic barriers is discussed in Chapter 10 - Geomembranes and Other Geosynthetic Barriers.

Geocomposites result when two or more materials are combined in the geosynthetic manufacturing process. Most are used in highway drainage applications and waste containment. A common example of a geocomposite is a prefabricated drain formed by wrapping a fluted or dimpled polymeric sheet, which acts as a conduit for water, with a geotextile which acts as a filter.

Geosynthetics are generically identified by:

1. polymer (descriptive terms, e.g., high density, low density, etc., should be included);
2. type of element (e.g., fiber, yarn, strand, rib, coated rib), if appropriate;
3. distinctive manufacturing process (woven, needlepunched nonwoven, heatbonded nonwoven, stitchbonded, extruded, knitted, roughened sheet, smooth sheet), if appropriate;
4. primary type of geosynthetic (geotextile, geogrid, geomembrane, etc.);
5. mass per unit area, if appropriate (e.g., for geotextiles, geogrids, GCL's, erosion control blankets) and/or thickness, if appropriate (e.g., for geomembranes); and,
6. any additional information or physical properties necessary to describe the material in relation to specific applications.

Four examples are:

- polypropylene staple filament needlepunched nonwoven, 350 g/m^2;
- polyethylene geonet, 440 g/m^2 with 8 mm openings;
- polypropylene extruded biaxial geogrid, with 25 mm x 25 mm openings; and
- high-density polyethylene roughened sheet geomembrane, 1.5 mm thick.

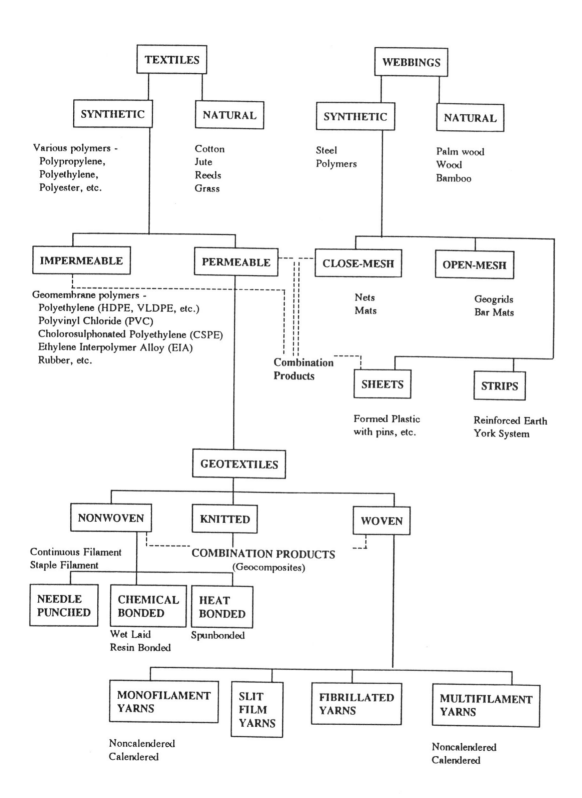

Figure 1-1   *Classification of geosynthetics and other soil inclusions (after Rankilor, 1981).*

1.4   FUNCTIONS and APPLICATIONS

Geosynthetics have six primary functions:
1. filtration
2. drainage
3. separation
4. reinforcement
5. fluid barrier, and
6. protection

Geosynthetic applications are usually defined by their primary, or principal, function. For example, geotextiles are used as filters to prevent soils from migrating into drainage aggregate or pipes, while maintaining water flow through the system. They are similarly used below riprap and other armor materials in coastal and stream bank protection systems to prevent soil erosion.

Geotextiles and geocomposites can also be used as drainage, or lateral transmission media, by allowing water to drain from or through soils of lower permeability. Geotextile applications include dissipation of pore water pressures at the base of roadway embankments. For situations with higher flow requirements, geocomposite drains have been developed. These materials are used as pavement edge drains, slope interceptor drains, and abutments and retaining wall drains. Filtration and drainage are addressed in Chapter 2 - Geosynthetics in Subsurface Drainage Systems.

Geotextiles are often used as separators to prevent road base materials from penetrating into the underlying soft subgrade, thus maintaining the design thickness and roadway integrity. Separators also prevent fine-grained subgrade soils from being pumped into permeable, granular road bases. Separators are discussed in Chapter 5 - Geosynthetics in Roadways and Pavements.

Geogrids and geotextiles can also be used as reinforcement to add tensile strength to a soil matrix, thereby providing a more competent structural material. Reinforcement enables embankments to be constructed over very soft foundations and permits the construction of steep slopes and retaining walls. Reinforcement applications are presented in Chapter 7 - Reinforced Embankments on Soft Foundations, Chapter 8 - Reinforced Slopes, and Chapter 9 - Reinforced Soil Retaining Walls and Abutments.

Geomembranes, thin-film geotextile composites, geosynthetic clay liners, and field-coated geotextiles are used as fluid barriers to impede the flow of a liquid or gas from one location to another. This geosynthetic function has wide application in asphalt pavement overlays, encapsulation of swelling soils, and waste containment. Barrier applications are summarized in Chapters 6 - Pavement Overlays, 10 - Geomembranes and Other Geosynthetic Barriers, and 11 - Geotextiles in Waste Containment Systems.

In the sixth function, protection, the geosynthetic acts as a stress relief layer. Temporary geosynthetic blankets and permanent geosynthetic mats are placed over the soil to reduce erosion caused by rainfall impact and water flow shear stress. A protective cushion of nonwoven geotextiles is often used to prevent puncture of geomembranes (by reducing point stresses) from stones in the adjacent soil or drainage aggregate during installation and while in service. Erosion control is presented in Chapter 3 - Geotextiles in Riprap Revetments and Other Permanent Erosion Control Systems, and Chapter 4 - Temporary Runoff and Sediment Control.

In addition to the primary function, geosynthetics usually perform one or more secondary functions. The primary and secondary functions make up the total contribution of the geosynthetic to a particular application. A listing of common applications according to primary and secondary functions is presented in Table 1-1. It is important to consider both the primary and secondary functions in the design computations and specifications.

Table 1-1
Representative Applications and Controlling Functions of Geosynthetics

PRIMARY FUNCTION	APPLICATION	SECONDARY FUNCTION(S)
Separation	Unpaved Roads (temporary & permanent)	Filter, drains, reinforcement
	Paved Roads (secondary & primary)	Filter, drains
	Construction Access Roads	Filter, drains, reinforcement
	Working Platforms	Filter, drains, reinforcement
	Railroads (new construction)	Filter, drains, reinforcement
	Railroads (rehabilitation)	Filter, drains, reinforcement
	Landfill Covers	Reinforcement, drains, protection
	Preloading (stabilization)	Reinforcement, drains
	Marine Causeways	Filter, drains, reinforcement
	General Fill Areas	Filter, drains, reinforcement
	Paved & Unpaved Parking Facilities	Filter, drains, reinforcement
	Cattle Corrals	Filter, drains, reinforcement
	Coastal & River Protection	Filter, drains, reinforcement
	Sports Fields	Filter, drains, protection
Drainage-Transmission	Retaining Walls	Separation, filter
	Vertical Drains	Separation, filter
	Horizontal Drains	Reinforcement
	Below Membranes (drainage of gas & water)	Reinforcement, protection
	Earth Dams	Filter
	Below Concrete (decking & slabs)	Protection
Protection	Geomembrane Cushion	Drains
	Temporary Erosion Control	Fluid barrier
	Permanent Erosion Control	Reinforcement, fluid barrier

Table 1-1 (continued)

Representative Applications and Controlling Functions of Geosynthetics

PRIMARY FUNCTION	APPLICATION	SECONDARY FUNCTION(S)
Filter	Trench Drains	Separation, drains
	Pipe Wrapping	Separation, drains, protection
	Base Course Drains	Separation, drains
	Frost Protection	Separation, drainage, reinforcement
	Structural Drains	Separation, drains
	Toe Drains in Dams	Separation, drains
	High Embankments	Drains
	Filter Below Fabric-Form	Separation, drains
	Silt Fences	Separation, drains
	Silt Screens	Separation
	Culvert Outlets	Separation
	Reverse Filters for Erosion Control:	Separation
	Seeding and Mulching	
	Beneath Gabions	
	Ditch Armoring	
	Embankment Protection, Coastal	
	Embankment Protection, Rivers & Streams	
	Embankment Protection, Lakes	
	Vertical Drains (wicks)	
Reinforcement	Pavement Overlays	----------
	Subbase Reinforcement in Roadways & Railways	Filter
	Retaining Structures	Drains
	Membrane Support	Separation, drains, filter, protection
	Embankment Reinforcement	Drains
	Fill Reinforcement	Drains
	Foundation Support	Drains
	Soil Encapsulation	Drains, filter, separation
	Net Against Rockfalls	Drains
	Fabric Retention Systems	Drains
	Sand Bags	----------
	Reinforcement of Membranes	Protection
	Load Redistribution	Separation
	Bridging Nonuniformity Soft Soil Areas	Separation
	Encapsulated Hydraulic Fills	Separation
	Bridge Piles for Fill Placement	----------
Fluid Barrier	Asphalt Pavement Overlays	----------
	Liners for Canals and Reservoirs	----------
	Liners for Landfills and Waste Repositories	----------
	Covers for Landfill and Waste Repositories	----------
	Cutoff Walls for Seepage Control	----------
	Waterproofing Tunnels	----------
	Facing for Dams	----------
	Membrane Encapsulated Soil Layers	----------
	Expansive Soils	----------
	Flexible Formwork	----------

1.5   EVALUATION of PROPERTIES

Today, there are more than 600 different geosynthetic products available in North America. Because of the wide variety of products available, with their different polymers, filaments, weaving (or nonwoven) patterns, bonding mechanisms, thicknesses, masses, etc., they have a considerable range of physical and mechanical properties. Thus, the process of comparison and selection of geosynthetics is not easy. Geosynthetic testing has progressed significantly since the first Geotextile Engineering Manual (Christopher and Holtz, 1985) was published. Specific test procedures are given in ASTM (1994), AASHTO (1990a), and GRI (1994). Testing procedures developed by the Geosynthetics Research Institute of Drexel University are considered only interim standards until an equivalent ASTM standard is adopted.

The particular, required design properties of the geosynthetic will depend on the specific application and the associated function(s) the geosynthetic is to provide. The properties listed in Table 1-2 cover the range of important criteria and properties required to evaluate geosynthetic suitability for most applications. It should be noted that not all of the listed requirements will be necessary for all applications. For a specific application requirements, refer to the subsequent chapter covering that application.

All geosynthetic properties and parameters to be considered for specific projects are listed in Table 1-3. Again, see ASTM (1994), AASHTO (1990a), and GRI (1994) for test procedures for each specific property. Manufacturers can provide information on general properties. The December issue of Geotechnical Fabrics Report magazine, published by the Industrial Fabrics Association International (IFAI), is formatted as a Specifier's Guide. General and some index properties are listed according to product type and manufacturer. The Specifier's Guide also contains a directory of product manufacturers, product distributors, geosynthetic installers, design engineers and testing laboratories, with contact person, address, telephone and facsimile numbers noted.

The tests listed in Table 1-3 include index tests and performance tests. Index tests do not produce an actual design property in most cases, but they do provide a general value from which the property of interest can be qualitatively assessed. Index tests are primarily used by manufacturers for quality control purposes. When determined using identical test procedures, index tests can be used for product comparison, specifications, and quality control evaluation.

On the other hand, performance tests require testing of geosynthetics with its companion material (e.g., soil) to obtain a direct assessment of the property of interest. Since performance tests should be performed under specific design conditions with soils from the site, manufacturers should no be expected to have the capability or the responsibility to perform such tests. These tests should be performed under the direction of the design

Table 1-2
Important Criteria and Principal Properties Required for Geosynthetic Evaluation

CRITERIA AND PARAMETER	PROPERTY[1]	FUNCTION					
		Filtration	Drainage	Separation	Reinforcement	Barrier	Protection
Design Requirements:							
Mechanical Strength							
Tensile Strength	Wide Width Strength	—	—	—	✓	✓	—
Tensile Modulus	Wide Width Modulus	—	—	—	✓	✓	—
Seam Strength	Wide Width Strength	—	—	—	✓	✓	—
Tension Creep	Creep Resistance	—	—	—	✓	✓	—
Compression Creep	Creep Resistance	—	✓ [2]	—	—	—	—
Soil-Geosynthetic Friction	Shear Strength	—	—	—	✓	✓	✓
Hydraulic							
Flow Capacity	Permeability	✓	✓	✓	✓	✓	—
	Transmissivity	—	✓	—	—	—	✓
Piping Resistance	Apparent Opening Size	✓	—	✓	✓	—	✓
	Porimetry	✓	—	—	—	—	✓
Clogging Resistance	Gradient Ratio or Long-Term Flow	✓	—	—	—	—	✓
Constructability Requirements:							
Tensile Strength	Grab Strength	✓	✓	✓	✓	✓	✓
Seam Strength	Grab Strength	✓	✓	✓	—	✓	—
Bursting Resistance	Burst Strength	✓	✓	✓	✓	✓	✓
Puncture Resistance	Rod or Pyramid Puncture	✓	✓	✓	✓	✓	✓
Tear Resistance	Trapezoidal Tear	✓	✓		✓	✓	✓
Longevity (Durability):							
Abrasion Resistance[3]	Reciprocating Block Abrasion	✓	—	—	—	—	—
UV Stability[4]	UV Resistance	✓	—	—	✓	✓	✓
Soil Environment[5]	Chemical	✓	✓	?	✓	✓	?
	Biological	✓	✓	?	✓	✓	?
	Wet-Dry		✓	—	—	—	✓
	Freeze-Thaw	✓	✓	—	—	✓	—

NOTES
1. See Table 1-3 for specific procedures.
2. Compression creep is applicable to some geocomposites.
3. Erosion control applications where armor stone may move.
4. Exposed geosynthetics only.
5. Where required.

Table 1-3

Geosynthetic Properties and Parameters

PROPERTY	TEST METHOD	UNITS OF MEASUREMENT
I. GENERAL PROPERTIES (from manufacturers)		
Type and Construction	N/A	-----
Polymer	N/A	-----
Mass per Unit Area	ASTM D 5261	g/m^2
Thickness (geotextiles & geomembranes)	ASTM D 5199	mm
Roll Length	Measure	m
Roll Widths	Measure	m
Roll Weight	Measure	kg
Roll Diameter	Measure	m
Specific Gravity & Density	ASTM D 792 and D 1505	g/m^3
Surface Characteristics	N/A	-----
II. INDEX PROPERTIES		
MECHANICAL STRENGTH - UNIAXIAL LOADING		
a) Tensile Strength (Quality Control)		
1) Grab Strength (geotextiles & CSPE reinforced geomembranes)	ASTM D 4632	N
2) Single Rib Strength (geogrids)	GRI:GG1	N
3) Narrow Strip (geomembranes)		
- EDPM, CO, IR, CR	ASTM D 412	N
- HDPE	ASTM D 638	N
- PVC, VLDPE	ASTM D 882	N
b) Tensile Strength (Load-Strain Characteristics)		
1) Wide Strip (geotextiles)	ASTM D 4595	N
2) Wide Strip (geogrid)	no standard	N
3) Wide Strip Strength (geomembranes)	ASTM D 4885	N
4) 2% Secant Modulus (PE geomembranes)	ASTM D 882	N
c) Junction Strength (geogrids)	GRI:GG2	%
d) Dynamic Loading	no standard	
e) Creep Resistance	ASTM D 5262 (Note: interpretation required)	creep strain: $\%\varepsilon/ s$ creep rupture: kN/m
f) Index Friction	GRI:GS7	dimensionless
g) Seam Strength		
1) Sewn (geotextiles)	ASTM D 4884	% efficiency
2) Factory Peel and Shear (geomembranes)	ASTM D 4545	kg/mm
3) Field Peel and Shear (geomembranes)	ASTM D 4437	kg/mm
h) Tear Strength		
1) Trapezoid Tearing (geotextile)	ASTM D 4533	N
2) Tear Resistance (geomembranes)	ASTM D 1004	N

Table 1-3 (continued)
Geosynthetic Properties and Parameters

PROPERTY	TEST METHOD	UNITS OF MEASUREMENT
II. INDEX PROPERTIES (continued)		
MECHANICAL STRENGTH - RUPTURE RESISTANCE		
a) Burst Strength		
1) Mullen Burst (geotextiles)	ASTM D 3786	Pa
2) CBR (geotextiles, geonets, geomembranes)	GRI:GS1	Pa or N
3) Large Scale Hydrostatic (geomembranes and geotextiles)	ASTM D 5514	Pa
b) Puncture Resistance		
1) Index (geotextiles and geomembranes)	ASTM D 4833	N
2) Pyramid Puncture (geomembranes)	ASTM D 5494	N
3) CBR (geotextile, geonets, and geomembranes)	GRI:GS1	N
c) Penetration Resistance (Dimensional Stability)	no standard	
d) Geosynthetic Cutting Resistance	no standard	
e) Flexibility (Stiffness)	ASTM D 1388	mg/cm^2
ENDURANCE PROPERTIES		
a) Abrasion Resistance (geotextile)	ASTM D 4886	%
b) Ultraviolet (UV) Radiation Stability		
1) Xenon-Arc Apparatus (geotextile)	ASTM D 4355	%
2) Outdoor Exposure	GRI:GT3 (geotextiles)	%
c) Chemical Resistance		
1) Chemical Immersion	ASTM D 5322	N/A
2) Oxidative Induction Time	GRI:GS9	minutes
3) Environmental Exposure	EPA 9090	% change
d) Biological Resistance		
1) Biological Clogging (geotextile)	ASTM D 1987	m^3/s
2) Biological Degradation	ASTM G 21 and G 22	
3) Soil Burial	ASTM D 3083	% change
e) Wet and Dry Stability	no standard	
f) Temperature Stability		
1) Temperature Stability (geotextile)	ASTM D 4594	% change
2) Dimensional Stability (geomembrane)	ASTM D 1204	% change

Table 1-3 (continued)

Geosynthetic Properties and Parameters

PROPERTY	TEST METHOD	UNITS OF MEASUREMENT
II. INDEX PROPERTIES (continued)		
HYDRAULIC		
a) Opening Characteristics (geotextiles)		
1) Apparent Opening Size (AOS)	ASTM D 4751	mm
2) Porimetry (pore size distribution)	Use AOS for O95, O85, O50, O15, and O5	mm
3) Percent Open Area (POA)	(see Christopher & Holtz, 1985)	%
4) Porosity (n)	(V_{voids}/V_{total}) 100	%
b) Permeability (k) and Permittivity (Ψ)	ASTM D 4491	m/s and s^{-1}
c) Soil Retention Ability	Empirically Related to Opening Characteristics	
d) Clogging Resistance	ASTM D 5101 and GRI:GT8	
e) In-Plane Flow Capacity (Transmissivity, q)	ASTM D 4716	m^2/s
III. PERFORMANCE PROPERTIES		
Stress-Strain Characteristics:		kN/m and % strain
a) Tension Test in Soil	(see McGown, et al., 1982)	
b) Triaxial Test Method	(see Holtz, et al., 1982)	
c) CBR on Soil Fabric System	(see Christopher & Holtz, 1985)	
d) Tension Test in Shear Box	(see Christopher & Holtz, 1985)	
Creep Tests:		kN/m and % strain
a) Extension Test in Soil	(see McGown, et al., 1982)	
b) Triaxial Test Method	(see Holtz, et al., 1982)	
c) Extension Test in Shear Box	(see Christopher & Holtz, 1985)	
d) Pullout Method	(see Christopher, et al., 1990)	
Friction/Adhesion:		
a) Direct Shear (soil-geosynthetic)	ASTM D 5321	degrees (°)
b) Direct Shear (geosynthetic-geosynthetic)	ASTM D 5321	degrees (°)
c) Pullout (geogrids)	GRI:GG5	dimensionless
d) Pullout (geotextiles)	GRI:GT7	dimensionless
e) Anchorage Embedment (geomembranes)	GRI:GM2	kN/m

Table 1-3 (continued)
Geosynthetic Properties and Parameters

PROPERTY	TEST METHOD	UNITS OF MEASUREMENT
III. PERFORMANCE PROPERTIES (continued)		
Dynamic and Cyclic Loading Resistance:	no standard procedures	N/A
Puncture a) Gravel, truncated cone or pyramid	ASTM D 5494	kPa
Chemical Resistance: a) In Situ Immersion Testing	ASTM D 5496	N/A
Soil Retention and Filtration Properties: a) Gradient Ration Method - for noncohesive sand and silt type soils	ASTM D 5101	dimensionless
b) Hydraulic Conductivity Ratio (HCR) - for fine-grained soils	ASTM D 5567	dimensionless
c) Slurry Method - for silt fence applications	ASTM D 5141	%

engineer. Performance tests are not normally used in specifications; rather, geosynthetics should be preselected for performance testing based on index values, or performance test results should be correlated to index values for use in specifications.

Brief descriptions of some of the basic properties of geosynthetics (after Christopher and Dahlstrand, 1989) are presented below.

Mass per Unit Area: The unit weight of a geosynthetic is measured in terms of area as opposed to volume due to variations in thickness under normal stress. This property is mainly used to identify materials.

Thickness: Thickness is not usually required information for geotextiles except in permeability-flow calculations. It is used as a primary identifier for geomembranes. When needed, it can be simply obtained using the procedure in Table 1-3, but it must be measured under a specified normal stress. The nominal thickness used for product comparison is measured under a normal stress of 2 kPa for geotextiles and 20 kPa for geogrids and geomembranes.

Tensile Strength: To understand the load-strain characteristics, it is important to consider the complete load-strain curve. It is also important to consider the nature of the test and the testing environment. With most materials, it is usual to use stress in strength and modulus determination. However, because of the thin, two-dimensional nature of geosynthetics, it is awkward to use stress. Therefore, it is conventional with geosynthetics to use force per unit length along the edge of the material. Then, strength and modulus have units of FL^{-1} (i.e., kN/m).

There are several types of tensile strength tests. Specific geosynthetic specimen shapes and loadings are indicated by the referenced procedures in Table 1-3. These tests all give different results.

The plane-strain test represents the loading for many applications, but because it is complicated to perform, it is not a practical test for many routine applications. Therefore, it is approximated by a strip tensile test. Since many narrow strip geosynthetic specimens neck when strained, most applications use wide, short specimens. This is called a wide strip tensile test.

Geosynthetics may have different strengths in different directions. Therefore, tests should be conducted in both principal directions.

The *grab tensile test* is typically used in the specification of geotextiles and is an unusual test. It is widely used and almost universally misused. The grab test may be useful in some applications, but it is difficult, if not impossible, to relate to actual strength without direct correlation tests. The grab tensile test normally uses 25 mm jaws to grip a 100 mm specimen. The strength is reported as the total force needed to cause failure -- not the force per unit width. It is not clear how the force is distributed across the sample. The effects of the specimen being wider than the grips depend on the geotextile filament interaction. In nonwoven geotextiles, these effects are large. In woven geotextiles, they are small.

The *burst test* is performed by applying a normal pressure (usually by air pressure) against a geosynthetic specimen clamped in a ring. The burst strength is given in pascals. This is not the stress in the specimen - it is the normal stress against the geosynthetic at failure. The burst strength depends on the strength in all directions and is controlled by the minimum value. Burst strength is a function of the diameter of the test specimen; therefore, care must be used in comparing tests.

Creep is a time-dependent mechanical property. It is strain at constant load. Creep tests can be run for any of the tensile test types, but are most frequently performed on a wide strip specimen by applying a constant load for a sustained period. Creep tests are influenced by the same factors as tensile load-strain tests - specimen length to width ratio, temperature, moisture, lateral restraint, and confinement.

Short-term strain is strongly influenced by the geosynthetic structure. Geogrids and woven geotextiles have the least; heat-bonded geotextiles have intermediate; and needled geotextiles have the most. Longer-term creep strain are controlled by structure and polymer type. Of the most common polymers, polyester has lower creep rates than polypropylene. The creep limit is the most important creep characteristic. It is the load per unit width above which the geosynthetic will creep to rupture. The creep limit is controlled by the polymer and ranges from 20% to 60% of the material's ultimate strength.

Friction: Soil-geosynthetic and geosynthetic-geosynthetic friction are important properties. It is common to assume a soil-geotextile friction value between 2/3 and 1 of the soil angle of friction. For geogrid materials, the value approaches the full friction angle. Caution is advised for geomembranes where soil-geosynthetic friction angle may be much lower than the soil angle of friction. For important applications, tests are justified.

The direct friction test is simple in principle, but numerous details must be considered for accurate results. Recent procedures proposed by ASTM indicate a minimum shear box size of 300 mm by 300 mm to reduce boundary effects. For many geosynthetics, the friction angle is a function of the soils on each side of the geosynthetic and the normal stress; therefore, test conditions must model the actual field conditions.

Durability Properties: Other properties that require consideration are related to durability and longevity. Exposure to ultraviolet light can degrade some geosynthetic properties. The geosynthetic polymer must be compatible with the environment chemistry. The environment should be checked for such items as high and low pH, chlorides, organics and oxidation agents such as ferruginous soils which contain Fe_2SO_3, calcareous soils, and acid sulfate soils that may deteriorate the geosynthetic in time. Other possible detrimental environmental factors include chemical solvents, diesel, and other fuels. Each geosynthetic is different in its resistance to aging and attack by chemical and biological agents. Therefore, each product must be investigated individually to determine the effects of these durability factors. The geosynthetic manufacturer should supply the results of product exposure studies, including, but not limited to, strength reduction due to aging, deterioration in ultraviolet light, chemical attack, microbiological attack, environmental stress cracking, hydrolysis, and any possible synergism between individual factors. Unless otherwise provided, AASHTO (1992) recommends that the allowable geosynthetic strength be determined with a factor of 2.0 to account for durability. AASHTO also recommends a minimum reduction factor of 1.1 be used in all cases.

Hydraulic Properties: Hydraulic properties relate to the pore size of the geosynthetic and correspondingly its ability to retain soil particles over the life of the project while allowing water to pass. Hydraulic properties may also be affected by chemical and biological agents. Ionic deposits as well as slime growth have been known to clog filter systems (granular filters as well as geotextiles).

The ability of a geotextile to retain soil particles is directly related to its apparent opening size (AOS) which is the apparent largest hole in the geotextile. The AOS value is equal to the size of the largest particle that can effectively pass through the geotextile in a dry sieving test.

The ability of water to pass through a geotextile is determined from its hydraulic conductivity (coefficient of permeability, k), as measured in a permeability test. The flow capacity of the material can then be determined from Darcy's law. Due to the compressibility of geotextiles, the permittivity, ψ (permeability divided by thickness), is often determined from the test and used to directly evaluate flow capacity.

The ability of water to pass through a geotextile over the life of the project is dependent on its filtration potential or its ability not to clog with soil particles. Essentially, if the finer particles of soil can pass through the geotextile, it should not clog. Effective filtration can be evaluated through relations between the geotextile's pore size distribution and the soil's grain size distribution; however, such formulations are still in the development phase. For a precise evaluation, laboratory performance testing of the proposed soil and candidate geotextile should be conducted.

One popular filtration test is the gradient ratio test (ASTM D 5101). This test is primarily suitable for sandy and silty soils (k $\leq$ 10^{-7} m/s). In this test, a rigid wall permeameter, with strategically located piezometer ports, is used to obtain a ratio of the head loss in the soil to the head loss at the soil-geotextile interface under different hydraulic gradients. Although the procedure indicates that the test may be terminated after 24 hours, to obtain meaningful results, the test should be continued until stabilization of the flow has clearly occurred. This may occur within 24 hours, but could require several weeks. A gradient ratio of 1 or less is preferred. Less than 1 is an indication that a more open *filter bridge* has developed in the soil adjacent to the geotextile. However, a continued decrease in the gradient ratio indicates piping, and an alternate geotextile should be evaluated. A high gradient ratio indicates a flow reduction at the geotextile. If the gradient ratio approaches 3 (the recommended maximum by the U.S. Army Corps of Engineers, 1977), the flow rate through the system should be carefully evaluated with respect to design requirements. A continued increase in the gradient ratio indicates clogging, and the geotextile is unacceptable.

For fine-grained soils, the hydraulic conductivity ratio (HCR) test (ASTM D 5567) should be considered. This test uses a flexible wall permeameter and evaluates the long-term permeability under increasing gradients with respect to the short-term permeability of the system at the lowest hydraulic gradient. A decrease in HCR indicates a flow reduction in the system. Since measurements are not taken near the geotextile-soil interface and soil permeability is not measured, it is questionable whether an HCR decrease is the result of flow reduction at the geotextile or blinding within the soil matrix itself. An improvement to this method would be to include piezometer or transducers within these zones (after the gradient ratio method) to aid in interpretation of the results.

Other filtration tests for clogging potential include the Caltrans slurry filtration test (Hoover, 1982), which was developed by Legge (1990) into the Fine Fraction Filtration (F3) test (Sansone and Koerner, 1992), and the Long-Term Flow (LTF) test (Koerner and Ko, 1982; GRI Test Method GT1). According to Fischer (1994), all of these tests have serious disadvantages that make them less suitable than the Gradient Ratio (GR) test for determining the filtration behavior of the soil-geotextile system. The GR test must be run longer than the previous ASTM-specified 24 hours, and proper attention must be paid to the test details (Maré, 1994) to get reproducible results.

Some additional hydraulic properties often required in filtration design are the Percent Open Area (POA) and the porosity. As noted in Table 1-3, there are no standard tests for these properties, although there is a suggested procedure for POA given by Christopher and Holtz (1985), which follows Corps of Engineers procedures. Basically, POA is determined on a light table or by projection enlargement. Porosity is readily calculated just as it is with soils; that is, porosity is the volume of the voids divided by the total volume. The total volume is, for example, 1 m^2, times the nominal thickness of the geotextile. The volume of voids is the total volume minus the volume of the fibers and filaments (*solids*), or the mass of 1 m^2 divided by the specific gravity of the polymer.

1.6   SPECIFICATIONS

Specifications should be based on the specific geosynthetic properties required for design and installation. Standard geosynthetics may result in uneconomical or unsafe designs. To specify a particular type of geosynthetic or its equivalent can also be very misleading. As a result, the contractor may select a product that has completely different properties than intended by the designer. In almost every chapter of this manual, guide specifications are given for the particular application discussed in the chapter.   See Richardson and Koerner (1990) and Koerner and Wayne (1989) for additional guide specifications.

For small projects, the cost of ASTM acceptance/rejection criterion testing is often a significant portion of the total project cost and may even exceed the cost of the geosynthetic itself. In such cases, a manufacturer's product certification specification requirement or an approved product list type specification may be satisfactory.

All geosynthetic specifications should include:
- general requirements
- specific geosynthetic properties
- seams and overlaps
- placement procedures

- repairs, and
- acceptance and rejection criteria

General requirements include the types of geosynthetics, acceptable polymeric materials, and comments related to the stability of the material. Geosynthetic manufacturers and representatives are good sources of information on these characteristics. Other items that should be specified in this section are instructions on storage and handling so products can be protected from ultraviolet exposure, dust, mud, or any other elements that may affect performance. Guidelines concerning on-site storage and handling of geotextiles are contained in ASTM D 4873, Standard Guide for Identification, Storage, and Handling of Geotextiles. If pertinent, roll weight and dimensions may also be specified. Finally, certification requirements should be included in this section.

Specific geosynthetic physical, index, and performance properties as required by the design must be listed. Properties should be given in terms of minimum (or maximum) average roll values (MARVs), along with the required test methods. MARVs are simply the smallest (or largest) anticipated average value that would be obtained for any roll tested (Koerner, 1994). This average property value must exceed the minimum (or be less than the maximum) value specified for that property based on a particular test. Ordinarily it is possible to obtain a manufacturer's certification for MARVs.

If performance tests have been conducted as part of the design, a list of approved products could be provided. **The language *or equal* and *or equivalent* should be avoided within the specification, unless equivalency is spelled out in terms of the index properties and the performance criteria that were required to be included on the approved list.** Approved lists can also be developed based on experience with recurring application conditions. Once an approved list has been established, new geosynthetics can be added as they are approved. Manufacturers' samples should be periodically obtained so they can be examined alongside the original tested specimens to verify whether the manufacturing process has changed since the product was approved. Development of an approved list program will take considerable initial effort, but once established, it provides a simple, convenient method of specifying geosynthetics with confidence.

Seam and overlap requirements should be specified along with the design properties for both factory and field seams, as applicable. A minimum overlap of 0.3 m is recommended for all geotextile applications, but overlaps may be increased due to specific site and construction requirements. Sewing of seams, discussed in Section 1.8, may be required for special conditions. Also, certain geotextiles may have factory seams. The seam strengths specified should equal the required strength of the geosynthetic, in the direction perpendicular to the seam length, using the same test procedures. For designs where wide width tests are used (e.g., reinforced embankments on soft foundations), the required seam strength is a calculated design value. Therefore, seam strengths should not be specified as a percent of the geosynthetic strength.

Geogrids and geonets may be connected by mechanical fasteners, though the connection may be either structural or a construction aid (i.e., strength perpendicular to the seam length is not required by design). Geomembranes are normally thermally bonded and specified in terms of peel and shear seam strengths, as discussed in Chapter 10.

For sewn geotextiles, geomembranes, and structurally connected geogrids, the seaming material (thread, extrudate, or fastener) should consist of polymeric materials that have the same or greater durability as the geosynthetic being seamed. For example, nylon thread, unless treated, which is often used for geotextile seams may weaken in time as it absorbs water.

Placement procedures should be given in detail within the specification and on the construction drawings. These procedures should include grading and ground-clearing requirements, aggregate specifications, aggregate lift thickness, and equipment requirements. These requirements are especially important if the geosynthetic was selected on the basis of survivability. Detailed placement procedures are presented in each application chapter.

Repair procedures for damaged sections of geosynthetics (i.e., rips and tears) should be detailed. Such repairs should include requirements for overlaps, sewn seams, fused seams, or replacement requirements. For overlap repairs, the geosynthetic should extend the minimum of the overlap length requirement from all edges of the tear or rip (i.e., if a 0.3 m overlap is required, the patch should extend at least 0.3 m from all edges of the tear).

Acceptance and rejection criteria for the geosynthetic materials should be clearly and concisely stated in the specifications. It is very important that all installations be observed by a designer representative who is knowledgeable in geotextile placement procedures and who is aware of design requirements. Sampling (e.g., ASTM D 4354, Standard Practice for Sampling of Geosynthetics for Testing) and testing requirements during construction should also be specified. Guidelines for acceptance and rejection of geosynthetic shipments are contained in ASTM D 4759, Standard Practice for Determining the Specification Conformance of Geosynthetics.

1.7   FIELD INSPECTION

Problems with geosynthetic applications are often attributed to poor product acceptance and construction monitoring procedures on the part of the owner, and/or inappropriate installation methods on the part of the contractor. A checklist for field personnel responsible for observing a geosynthetic installation is presented in Table 1-4. Recommended installation methods are presented in the application chapters.

Table 1-4
Geosynthetic Field Inspection Checklist

☐ 1. Read the specifications; determine if geosynthetic is specified by (a) specific properties or (b) an approved products list.

☐ 2. Review the construction plans.

☐ 3. (a) For specification by specific properties, check listed material properties of supplied geosynthetic, from published literature, against the specific property values specified; OR
(b) obtain the geosynthetic name(s), type, and style, along with a small sample(s) of approved material(s) from the design engineer. Check supplied geosynthetic type and style for conformance to approved material(s). If the geosynthetic is not listed, contact the designer with a description of the material and request evaluation and approval or rejection.

☐ 4. On site, check the rolls of geosynthetics to see that they are properly stored; check for any damage.

☐ 5. Check roll and lot numbers to verify whether they match certification documents.

☐ 6. Cut two 100 mm to 150 mm square samples from a roll. Staple one to your copy of the specifications for comparison with future shipments and send one to the design engineer for approval or information.

☐ 7. Observe materials in each roll to make sure they are the same. Observe rolls for flaws and nonuniformity.

☐ 8. Obtain test samples according to specification requirements from randomly selected rolls. Mark the machine direction on each sample and note the roll number.

☐ 9. Observe construction to see that the contractor complies with specification requirements for installation.

☐ 10. Check all seams, both factory and field, for any flaws (*e.g.*, missed stitches in geotextile). If necessary, either reseam or reject materials.

☐ 11. If possible, check geosynthetic after aggregate or riprap placement for possible damage. This can be done either by constructing a trial installation, or by removing a small section of aggregate or riprap and observing the geosynthetic after placement and compaction of the aggregate, at the beginning of the project. If perforations, tears, or other damage has occurred, contact the design engineer.

☐ 12. Check future shipments against the initial approved shipment and collect additional test samples. Collect samples of seams, both factory and field, for testing. For field seams, have the contractor sew several meters of a *dummy* seam(s) for testing and evaluation.

1.8 FIELD SEAMING

Some form of geosynthetic seaming will be utilized in those applications that require continuity between adjacent rolls. Seaming techniques include overlapping, sewing, stapling, tying, heat bonding, welding and gluing. Some of these techniques are more suitable for certain types of geosynthetics than others. For example, the most efficient and widely used methods for geotextiles are overlapping and sewing, and these techniques are discussed first.

The first technique, the *simple overlap*, will be suitable for most geotextile and biaxial geogrid projects. The minimum overlap is 0.3 m. Greater overlaps are required for specific applications. If stress transfer is required between adjacent rolls, the only strength provided by an overlap is the friction between adjacent sheets of geotextiles, and by friction and fill *strike-through* of substantial apertures of biaxial geogrids. Unless overburden pressures are large and the overlap substantial, very little stress can actually be transferred through the overlap.

The second technique, *sewing*, offers a practical and economical alternative for geotextiles when overlaps become excessive or stress transfer is required between two adjacent rolls of fabric. For typical projects and conditions, sewing is generally more economical when overlaps of 1 m or greater are required. To obtain good-quality, effective seams, the user should be aware of the following sewing variables (Koerner, 1994; Diaz and Myles, 1990; Ko, 1987):

- Thread type: Kevlar, aramid, polyethylene, polyester, or polypropylene (in approximate order of decreasing strength and cost). Thread durability must be consistent with project requirements.
- Thread tension: Usually adjusted in the field to be sufficiently tight; but not cut the geotextile.
- Stitch density: Typically, 200 to 400 stitches per meter are used for lighter-weight geotextiles, while heavier geotextiles usually allow only 150 to 200 stitches per meter.
- Stitch type: Single- or double-thread chainstitch, Types 101 or 401; with double-thread chain- or *lock*-stitch preferred because it is less likely to unravel (Figure 1-2(a)).
- Number of rows: Usually two or more parallel rows are preferred for increased safety.
- Seam type: Flat or *prayer* seams, J- or Double J-type seams, or *butterfly* seams are the most widely used (Figure 1-2(b)).

When constructed correctly, sewn seams can provide reliable stress transfer between adjacent sheets of geotextile. However, there are several points with regard to seam strength that should be understood, as follows.

1. Due to needle damage and stress concentrations at the stitch, sewn seams are weaker than the geotextile (good, high-quality seams have only about 50% to 80% of the intact geotextile strength based on wide width tests).
2. Grab strength results are influenced by the stitches, so the test yields artificially high seam strengths.
3. The maximum seam strengths achievable at this time are on the order of 200 kN/m under factory conditions, using 330 kN/m geotextiles.
4. Field seam strengths will most likely be lower than laboratory or factory seam strengths.
5. All stitches can unravel, although lock-type stitches are less likely to.

6. Unraveling can be avoided by utilizing high-quality equipment and proper selection of needles, thread, seam and stitch type, and by using two or more rows of stitches.
7. Careful inspection of all stitches is essential.

Field sewing is relatively simple and usually requires two or three laborers, depending on the geotextile, seam type, and sewing machine. Good seams require careful control of the operation, cleanliness, and protection from the elements. However, adverse field conditions can easily complicate sewing operations. Although most portable sewing machines are electric, pneumatic equipment is available for operating in wet environments.

Since the seam is the weakest link in the geotextile, all seams, including factory seams, should be carefully inspected. To facilitate inspection and repair, the geotextile should be placed (or at least inspected prior to placement) with all seams up (Figure. 1-2(c)). Using a contrasting thread color can facilitate inspection. Procedures for testing sewn seams are given in ASTM D 4884, Standard Test Method for Seam Strength of Sewn Geotextiles.

Seaming of biaxial geogrids and geocomposites is most commonly achieved by overlaps, and the remarks above on overlap of geotextiles are generally appropriate to these products. Uniaxial geogrids are normally butted in the along-the-roll direction. Seams in the roll direction of uniaxial geogrids are made with a *bodkin* joint for HDPE geogrids, as illustrated in Figure 1-3, and may be made with overlaps for coated PET geogrids.

Seaming of geomembranes and other geosynthetic barriers is much more varied. The method of seaming is dependent upon the geosynthetic material being used and the project design. Overlaps of a designated length are typically used for thin-film geotextile composites and geosynthetic clay liners. Geomembranes are seamed with thermal methods or with solvents.

(a) TYPE OF STITCHES

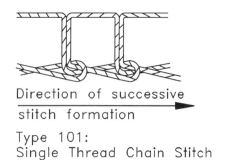

Direction of successive stitch formation

Type 101:
Single Thread Chain Stitch

Direction of successive stitch formation

Type 401:
Double Thread Chain
or "Lock" Stitch

(b) TYPE OF SEAMS

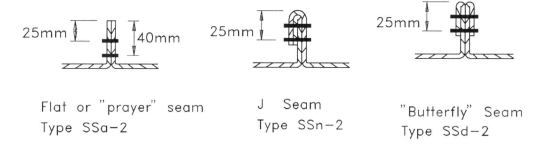

25mm ⇕ 40mm

Flat or "prayer" seam
Type SSa-2

25mm ⇕

J Seam
Type SSn-2

25mm ⇕

"Butterfly" Seam
Type SSd-2

(c) IMPROPER PLACEMENT

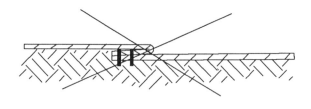

Cannot Inspect or Repair

Figure 1-2   Types of (a) stitches and (b) seams, according to Federal
Standard No. 751a (1965); and (c) improper seam placement.

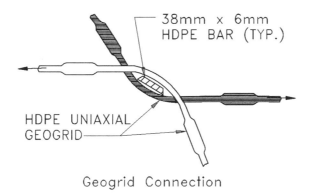

38mm x 6mm
HDPE BAR (TYP.)

HDPE UNIAXIAL
GEOGRID

Geogrid Connection

Figure 1-3   *Bodkin connection of HDPE uniaxial geogrid.*

1.9   REFERENCES

References quoted within this section are listed below. The Holtz and Paulson and the Cazzuffi and Anazani lists of geosynthetic literature are attached as Appendix A. The Koerner (1994) is a recent, comprehensive textbook on geosynthetics and is a key reference for design.   The bibliographies by Giroud (1993, 1994) comprehensively contain references of publications on geosynthetics before January 1, 1993.

AASHTO (1992), *Standard Specifications for Highway Bridges*, Fifteenth Edition, with 1993 and 1994 Interims, American Association of State Transportation and Highway Officials, Washington, D.C., 1992.

AASHTO (1990), Standard Specifications for Geotextiles - M 288, *Standard Specifications for Transportation Materials and Methods of Sampling and Testing*, American Association of State Transportation and Highway Officials, Washington, D.C., pp 689-692.

AASHTO (1990), Guide Specifications and Test Procedures for Geotextiles, *Task Force 25 Report*, Subcommittee on New Highway Materials, American Association of State Transportation and Highway Officials, Washington, D.C.

AASHTO (1990), Design Guidelines for Use of Extensible Reinforcements (Geosynthetic) for Mechanically Stabilized Earth Walls in Permanent Applications, *Task Force 27 Report -In Situ Soil Improvement Techniques*, American Association of State Transportation and Highway Officials, Washington, D.C.

ASTM (1994), *Annual Books of ASTM Standards,* American Society for Testing and Materials, Philadelphia, Pennsylvania:
Volume 4.08 (I), Soil and Rock
Volume 4.08 (II), Soil and Rock; Geosynthetics
Volume 7.01, Textiles

Volume 8.01, Plastics (I)
Volume 8.02, Plastics (II)
Volume 8.03, Plastics (III)

Berg, R.R. (1993), *Guidelines for Design, Specification, & Contracting of Geosynthetic Mechanically Stabilized Earth Slopes on Firm Foundations*, Report No. FHWA-SA-93-025, Federal Highway Administration, Washington, D.C., 87 p.

Cazzuffi, D. and Anazani, A. (1992), IGS Education Committee List of Reference Documents, *IGS News*, Vol. 8, No. 1, March, supplement.

Christopher, B.R., Gill, S.A., Giroud, J.P., Juran, I. Scholsser, F., Mitchell, J.K. and Dunnicliff, J. (1990), *Reinforced Soil Structures, Volume I. Design and Construction Guidelines*, Federal Highway Administration, Washington, D.C., Report No. FHWA-RD--89-043, 287 p.

Christopher, B.R. and Dahlstrand, T.K. (1989), *Geosynthetics in Dams - Design and Use*, Association of State Dam Safety Officials, Lexington, KY, April, 311 p.

Christopher, B.R. and Holtz, R.D. (1985), *Geotextile Engineering Manual*, Report No. FHWA-TS-86/203, Federal Highway Administration, Washington, D.C., March, 1044 p.

Diaz, V. and Myles, B. (1990), *Field Sewing of Geotextiles--A Guide to Seam Engineering*, Industrial Fabrics Association Internationals, St. Paul, MN, 1990, 29 p.

Federal Standards (1965), *Federal Standard Stitches, Seams, and Stitching*, No. 751a, January.

Fischer, G.R. (1994), *The Influence of Fabric Pore Structure on the Behavior of Geotextile Filters*, Ph.D. Dissertation, University of Washington, 498 p.

Giroud, J.P. (1994), with cooperation of Beech, J.F. and Khatami, A., *Geosynthetics Bibliography*, Volume II, IGS, Industrial Fabrics Association International, St. Paul, MN, 940 p.

Giroud, J.P. (1993), with cooperation of Beech, J.F. and Khatami, A., *Geosynthetics Bibliography*, Volume I, IGS, Industrial Fabrics Association International, St. Paul, MN, 781 p.

Giroud, J.P. and Carroll, R.G. (1983), Jr., Geotextile Products, *Geotechnical Fabrics Report*, Vol. 1, No. 1, Industrial Fabrics Association International, St. Paul, MN, Summer, pp. 12-15.

GRI Test Methods & Standards (1994), Geosynthetic Research Institute, Drexel University, Philadelphia, PA.

Holtz, R.D. and Paulson, J.N. (1988), Geosynthetic Literature, *Geotechnical News*, Vol. 6, No. 1, pp. 13-15.

Holtz, R.D., Tobin, W.R. and Burke, W.W. (1982), Creep Characteristics and Stress-Strain Behavior of a Geotextile -Reinforced Sand, *Proceedings of the Second International Conference on Geotextiles*, Las Vegas, NV, Vol. 3, pp. 805-809.

Hoover, T.P. (1982), Laboratory Testing of Geotextile Fabric Filters, *Proceedings of the Second International Conference on Geotextiles*, Las Vegas, Nevada, Vol. 3, August, pp. 839-844.

IGS (1993), *Recommended Mathematical and Graphical Symbols*, International Geosynthetic

Society, November, 19 p.

Ingold, T.S. and Miller, K.S. (1988), *Geotextiles Handbook*, Thomas Telford Ltd., London, 152 p.

John, N.W.M. (1987), *Geotextiles*, Blackie & Sons Ltd., Glasgow and London, 347 p.

Ko, F.K. (1987), Seaming and Joining Methods, *Geotextiles and Geomembranes*, Vol. 6, Nos. 1-3, pp 93-107.

Koerner, R.M. (1994), *Designing With Geosynthetics*, 3rd Edition, Prentice-Hall Inc., Englewood Cliffs, NJ, 783 p.

Koerner, R.M. and Ko, F.K. (1982), Laboratory Studies on Long-Term Drainage Capability of Geotextiles, *Proceedings of the Second International Conference on Geotextiles*, Las Vegas, NV, Vol. I, 1982, pp. 91-95.

Koerner, R.M. and Wayne M.H. (1989), *Geotextile Specifications for Highway Applications*, Report No. FHWA-TS-89-026, Federal Highway Administration, Washington, D.C., 90 p.

Koerner, R.M. and Welsh, J.P. (1980), *Construction and Geotechnical Engineering Using Synthetic Fabrics*, John Wiley & Sons, New York, 267 p.

Legge, K.R. (1990), *A New Approach to Geotextile Selection*, Proceedings of the Fourth International Conference on Geotextiles, Geomembranes and Related Products, The Hague, Netherlands, Vol. 1., May, pp. 269-272.

Maré, A.D. (1994), *The Influence of Gradient Ratio Testing Procedures on the Filtration Behavior of Geotextiles*, MSCE Thesis, University of Washington.

McGown, A., Andrawes, K.Z. and Kabir, M.H. (1982), Load-Extension Testing of Geotextiles Confined in Soil, *Proceedings of the Second International Conference on Geotextiles*, Las Vegas, NV, Vol. 3, pp. 793-798.

Rankilor, P.R. (1981), *Membranes in Ground Engineering*, John Wiley & Sons, Inc., Chichester, England, 377 p.

Richardson, G.R. and Koerner, R.M., Editors (1990), *A Design Primer: Geotextiles and Related Materials*, Industrial Fabrics Association International, St. Paul, MN, 166 p.

Sansone, L.J. and Koerner, R.M. (1992), *Fine Fraction Filtration Test to Assess Geotextile Filter Performance*, Geotextiles and Geomembranes, Vol. 11., Nos. 4-6, pp. 371-393.

Veldhuijzen van Zanten, R., Editor (1986), *Geotextiles and Geomembranes in Civil Engineering*, John Wiley, New York, 642 p.

U.S. Army Corps of Engineers (1977), *Civil Works Construction Guide Specification for Plastic Filter Fabric*, Corps of Engineer Specifications No. CW-02215, Office, Chief of Engineers, U.S. Army Corps of Engineers, Washington, D.C.

2.0 GEOSYNTHETICS in SUBSURFACE DRAINAGE SYSTEMS

2.1 BACKGROUND

One major area of geotextile use is as filters in drain applications such as trench and interception drains, blanket drains, pavement edge drains, structure drains, and beneath permeable roadway bases. The *filter* restricts movement of soil particles as water flows into the *drain* structure and is collected and/or transported downstream. Geocomposites consisting of a drainage core surrounded by a geotextile filter are often used as the drain itself in these applications. Geotextiles are also used as filters beneath hard armor erosion control systems, and this application will be discussed in Chapter 3.

Because of their comparable performance, improved economy, consistent properties, and ease of placement, geotextiles have been used successfully to replace graded granular filters in almost all drainage applications. Thus, they must perform the same functions as graded granular filters:

- to allow water to flow through the filter into the drain, and to continue doing this throughout the life of the project; and
- to retain the soil particles in place and prevent their migration (*piping*) through the filter (if some soil particles do move, they must be able to pass through the filter without blinding or clogging the downstream media during the life of the project).

Geotextiles, like graded granular filters, require proper engineering design or they may not perform as desired. Unless flow requirements, piping resistance, clogging resistance and constructability requirements (defined later) are properly specified, the geotextile/soil filtration system may not perform properly. In addition, construction must be monitored to ensure that materials are installed correctly.

In most drainage and filtration applications, geotextile use can be justified over conventional graded granular filter material use because of cost advantages from:

- the use of less-costly drainage aggregate;
- the possible use of smaller-sized drains;
- the possible elimination of collector pipes;
- expedient construction;
- lower risk of contamination and segregation of drainage aggregate during construction; and
- reduced excavation.

2.2   APPLICATIONS

Properly designed geotextiles can be used as a replacement for, or in conjunction with, conventional graded granular filters in almost any drainage application. Properly designed geocomposites can be used as a replacement for granular drains in many applications (*e.g.*, pavement edge drains). Below are a few examples of drainage applications.

- Filters around trench drains and edge drains — to prevent soil from migrating into the drainage aggregate or system, while allowing water to exit from the soil.

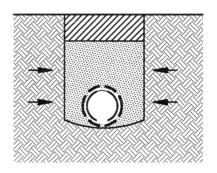

- Filters beneath pavement permeable bases, blanket drains and base courses. Prefabricated geo-composite drains and geotextile-wrapped trenches are used in pavement edge drain construction.

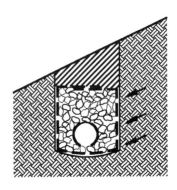

- Drains for structures such as retaining walls and bridge abutments. They separate the drainage aggregate or system from the backfill soil, while allowing free drainage of ground and infiltration water. Geocomposite drains are especially useful in this application.

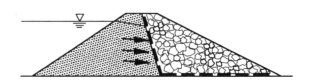

- Geotextile wraps for slotted or jointed drain and well pipes — to prevent filter aggregate from entering the pipe, while allowing the free flow of water into the pipe.

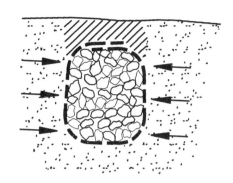

- Interceptor, toe drains, and surface drains — to aid in the stabilization of slopes by allowing excess pore pressures within the slope to dissipate, and by preventing surface erosion. Again, geocomposites have been successfully used in this application.

- Chimney and toe drains for earth dams and levees — to provide seepage control.

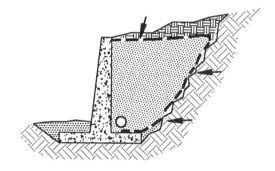

In each of these applications, flow is through the geotextile — that is, perpendicular to the plane of the fabric. In other applications, such as vertical drains in soft foundation soils, lateral drains below slabs and behind retaining walls, and gas transfer media, flow may occur both perpendicular to and transversely in the plane of the geotextile. In many of these applications, geocomposite drains may be appropriate. Design with geocomposite systems is covered in Section 2.11.

All geosynthetic designs should begin with a criticality and severity assessment of the project conditions (see Table 2-1) for a particular application. Although first developed by Carroll (1983) for drainage and filtration applications, the concept of critical-severe projects — and, thus, the level of engineering responsibility required — will be applied to other geosynthetic applications throughout this manual.

Table 2-1

Guidelines for Evaluating the Critical Nature or Severity of Drainage and Erosion Control Applications

(after Carroll, 1983)

A. CRITICAL NATURE OF THE PROJECT		
Item	Critical	Less Critical
1.Risk of loss of life and/or structural damage due to drain failure:	High	None
2.Repair costs versus installation costs of drain:	>>>	= or <
3.Evidence of drain clogging before potential catastrophic failure:	None	Yes
B. SEVERITY OF THE CONDITIONS		
Item	Severe	Less Severe
1.Soil to be drained:	Gap-graded, pipable, or dispersible	Well-graded or uniform
2.Hydraulic gradient:	High	Low
3.Flow conditions:	Dynamic, cyclic, or pulsating	Steady state

A few words about the condition of the soil to be drained (Table 2-1) are in order. First, gap-graded, well-graded and uniform soils are illustrated in Figure 2-1. Certain gap-graded and very well-graded soils may be *internally unstable*; that is, they can experience piping or internal erosion. On the other hand, a soil is *internally stable* if it is self-filtering and if its own fine particles do not move through the pores of its coarser fraction (LaFluer, et al., 1993). Criteria for deciding whether a soil is internally unstable will be given in the next section.

Dispersible soils are fine-grained natural soils which deflocculate in the presence of water and, therefore, are highly susceptible to erosion and piping (Sherard, et al., 1972). See also Sherard and Decker (1977) for more information on dispersible soils.

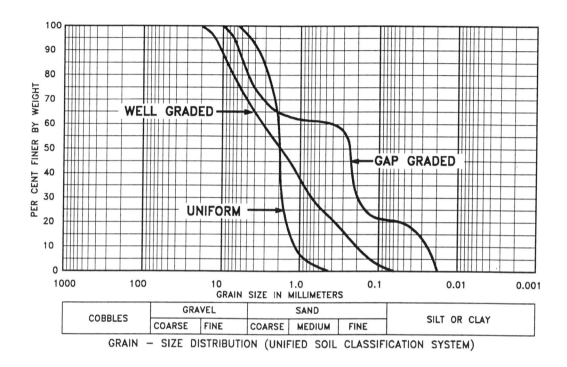

Figure 2-1   *Soil descriptions.*

2.3   GEOTEXTILE FILTER DESIGN

Designing with geotextiles for filtration is essentially the same as designing graded granular filters. A geotextile is similar to a soil in that it has voids (pores) and particles (filaments and fibers). However, because of the shape and arrangement of the filaments and the compressibility of the structure with geotextiles, the geometric relationships between filaments and voids is more complex than in soils. In geotextiles, pore size is measured directly, rather than using particle size as an estimate of pore size, as is done with soils. Since pore size can be directly measured, relatively simple relationships between the pore sizes and particle sizes of the soil to be retained can be developed. Three simple filtration concepts are used in the design process:

1. If the size of the largest pore in the geotextile filter is smaller than the larger particles of soil, the soil will be retained by the filter. As with graded granular filters, the larger particles of soil will form a filter bridge over the hole, which in turn, filters smaller particles of soil, which then retain the soil and prevent piping (Figure 2-2).

2. If the smaller openings in the geotextile are sufficiently large enough to allow smaller particles of soil to pass through the filter, then the geotextile will not *blind* or *clog* (see Figure 2-3).

3. A large number of openings should be present in the geotextile so proper flow can be maintained even if some of the openings later become plugged.

These simple concepts and analogies with soil filter design criteria are used to establish design criteria for geotextiles. Specifically, these criteria state:
- the geotextile must retain the soil (retention criterion), while
- allowing water to pass (permeability criterion), throughout
- the life of the structure (clogging resistance criterion).

To perform effectively, the geotextile must also survive the installation process (survivability criterion).

After a detailed study of research carried out both in North America and in Europe on conventional and geotextile filters, Christopher and Holtz (1985) developed the following design procedure for geotextile filters for drainage (this chapter) and permanent erosion control applications (Chapter 3). The level of design required depends on the critical nature of the project and the severity of the hydraulic and soil conditions (Table 2-1). Especially for critical projects, consideration of the risks and the consequences of geotextile filter failure require great care in selecting the appropriate geotextile. For such projects, and for severe hydraulic conditions, conservative designs are recommended. Geotextile selection should not be based on cost alone. The cost of the geotextile is usually minor in comparison to the other components and the construction costs of a drainage system. Also, do not try to save money by eliminating laboratory soil-geotextile performance testing when such testing is required by the design procedure.

A recent National Cooperative Highway Research Program (NCHRP) study (Koerner et al., 1994) of the performance of geotextile drainage systems indicated that the FHWA design criteria developed by Christopher and Holtz (1985) were an excellent prediction of filter performance, particularly for granular soils.

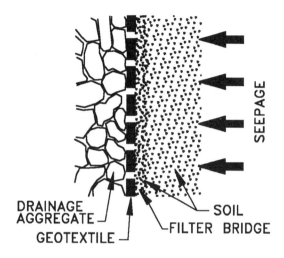

Figure 2-2   Filter bridge formation.

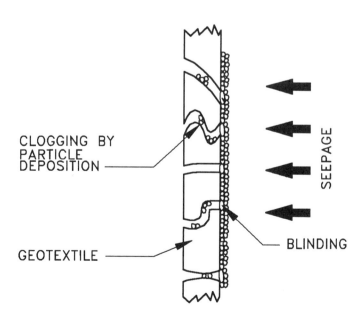

Figure 2-3   Definitions of clogging and blinding (Bell and Hicks, 1980).

2.3-1 Retention Criteria

2.3-1.a Steady State Flow Conditions

$$\text{AOS or } O_{95(geotextile)} \leq B \, D_{85 \, (soil)} \qquad\qquad [2-1]$$

where:

AOS	=	apparent opening size (see Table 1-3) (mm);
O_{95}	=	opening size in the geotextile for which 95% are smaller (mm);
AOS	$\approx$	O_{95};
B	=	a coefficient (dimensionless); and
D_{85}	=	soil particle size for which 85% are smaller (mm).

The coefficient B ranges from 0.5 to 2 and is a function of the type of soil to be filtered, its density, the uniformity coefficient C_u if the soil is granular, the type of geotextile (woven or nonwoven), and the flow conditions.

For sands, gravelly sands, silty sands, and clayey sands (with less than 50% passing the 0.075 mm sieve per the Unified Soil Classification System), B is a function of the uniformity coefficient, C_u. Therefore, for

$C_u \leq 2$ or ≥ 8:	B = 1	[2 - 2a]
$2 \leq C_u \leq 4$:	$B = 0.5 \, C_u$	[2 - 2b]
$4 < C_u < 8$:	$B = 8/C_u$	[2 - 2c]

where:

$$C_u = D_{60}/D_{10}$$

Sandy soils which are not uniform (Figure 2-1) tend to bridge across the openings; thus, the larger pores may actually be up to twice as large ($B \leq 2$) as the larger soil particles because, quite simply, two particles cannot pass through the same hole at the same time. Therefore, use of the criterion B = 1 would be quite conservative for retention, and such a criterion has been used by, for example, the Corps of Engineers.

If the protected soil contains any fines, use only the portion passing the 4.75 mm sieve for selecting the geotextile (i.e., scalp off the +4.75 mm material).

For silts and clays (with more than 50% passing the 0.075 mm sieve), B is a function of the type of geotextile:

$$\text{for } \textit{wovens}, \qquad B = 1; \; O_{95} \leq D_{85} \qquad\qquad [2-3]$$

for *nonwovens*, $\quad\quad$ B = 1.8; $O_{95} \leq 1.8\ D_{85}$ $\quad\quad$ [2 - 4]

and *for both*, $\quad\quad$ AOS or $O_{95} \leq 0.3$ mm $\quad\quad$ [2 - 5]

Due to their random pore characteristics and, in some types, their felt-like nature, nonwovens will generally retain finer particles than a woven geotextile of the same AOS. Therefore, the use of B = 1 will be even more conservative for nonwovens.

2.3-1.b Dynamic Flow Conditions

If the geotextile is not properly weighted down and in intimate contact with the soil to be protected, or if dynamic, cyclic, or pulsating loading conditions produce high localized hydraulic gradients, then soil particles can move behind the geotextile. Thus, the use of B = 1 is not conservative, because the bridging network will not develop and the geotextile will be required to retain even finer particles. When retention is the primary criteria, B should be reduced to 0.5; or:

$$O_{95} \leq 0.5\ D_{85} \quad\quad\quad\quad [2 - 6]$$

Dynamic flow conditions can occur in pavement drainage applications. For reversing inflow-outflow or high-gradient situations, it is best to maintain sufficient weight or load on the filter to prevent particle movement. Dynamic flow conditions with erosion control systems are discussed in Chapter 3.

2.3-1.c Stable versus Unstable Soils

The above retention criteria assumes that the soil to be filtered is internally stable — it will not pipe internally. If unstable soil conditions are encountered, performance tests should be conducted to select suitable geotextiles. According to Kenney and Lau (1985, 1986) and LaFluer, et al. (1989), broadly graded ($C_u > 20$) soils with concave upward grain size distributions tend to be internally unstable. The Kenney and Lau (1985, 1986) procedure utilizes a mass fraction analysis. Research by Skempton and Brogan (1994) verified the Kenney and Lau (1985, 1986) procedure.

2.3-2 Permeability/Permittivity Criteria

Permeability requirements:

— for *less critical applications* and *less severe conditions*:

$$k_{geotextile} \geq k_{soil} \quad\quad\quad\quad [2 - 7a]$$

— and, for *critical applications* and *severe conditions*:

$$k_{geotextile} \geq 10\ k_{soil} \quad\quad\quad\quad [2 - 7b]$$

Permittivity requirements:

$$\psi \geq 0.5\ sec^{-1}\ for < 15\%\ passing\ 0.075\ mm \quad\quad [2 - 8a]$$

$$\psi \geq 0.2\ sec^{-1}\ for\ 15\ to\ 50\%\ passing\ 0.075\ mm \quad\quad [2 - 8b]$$

$$\psi \geq 0.1\ sec^{-1}\ for > 50\%\ passing\ 0.075\ mm \quad\quad [2 - 8c]$$

In these equations:

$\quad\quad$ k $\quad =\quad$ Darcy coefficient of permeability (m/s); and

$$\psi \quad = \quad \text{geotextile permittivity, which is equal to } k_{geotextile}/t_{geotextile} \text{ (1/s)}$$
and is a function of the hydraulic head.

For actual flow capacity, the permeability criteria for noncritical applications is conservative, since an equal quantity of flow through a relatively thin geotextile takes significantly less time than through a thick granular filter. Even so, some pores in the geotextile may become blocked or plugged with time. Therefore, for critical or severe applications, Equation 2-7b is recommended to provide an additional level of conservatism. Equation 2-7a may be used where flow reduction is judged not to be a problem, such as in clean, medium to coarse sands and gravels.

The required flow rate, q, through the system should also be determined, and the geotextile and drainage aggregate selected to provide adequate capacity. As indicated above, flow capacities should not be a problem for most applications, provided the geotextile permeability is greater than the soil permeability. However, in certain situations, such as where geotextiles are used to span joints in rigid structures and where they are used as pipe wraps, portions of the geotextile may be blocked. For these applications, the following criteria should be used together with the permeability criteria:

$$q_{required} = q_{geotextile}(A_g/A_t) \qquad\qquad [2 - 9]$$

where:

$$A_g \quad = \quad \text{geotextile area available for flow; and}$$

$$A_t \quad = \quad \text{total geotextile area.}$$

2.3-3   Clogging Resistance

2.3-3.a Less Critical/Less Severe Conditions

For *less critical/less severe* conditions:

$$O_{95 \, (geotextile)} \geq 3 \, D_{15 \, (soil)} \qquad\qquad [2 - 10]$$

Equation 2-10 applies to soils with $C_u > 3$. For $C_u \leq 3$, select a geotextile with the maximum AOS value from Section 2.3-1.

In situations where clogging is a possibility (e.g., gap-graded or silty soils), the following optional qualifiers may be applied:

for *nonwovens* -

porosity of the geotextile, $n \geq 50\%$ $\qquad\qquad [2 - 11]$

for *woven monofilament* and *slit film wovens* -

percent open area, POA $\geq 4\%$ $\qquad\qquad [2 - 12]$

NOTE: See Section 1.5 for comments on porosity and POA.

Most common nonwovens have porosities much greater than 70%. Most woven monofilaments easily meet the criterion of Equation 2-12; tightly woven slit films do not, and are therefore not recommended for subsurface drainage applications.

Filtration tests provide another option for consideration, especially by inexperienced users.

2.3-3.b Critical/Severe Conditions

For critical/severe conditions, select geotextiles that meet the retention and permeability criteria in Sections 2.3-1 and 2.3-2. Then perform a filtration test using samples of on-site soils and hydraulic conditions. One type of filtration test is the gradient ratio test (ASTM D 5101).

Although several empirical methods have been proposed to evaluate geotextile filtration characteristics (*i.e.,* the clogging potential), the most realistic approach for all applications is to perform a laboratory test which simulates or models field conditions. We recommend the gradient ratio test, ASTM D 5101, Measuring the Soil-Geotextile System Clogging Potential by the Gradient Ratio. This test utilizes a rigid-wall soil permeameter with piezometer taps that allow for simultaneous measurement of the head losses in the soil and the head loss across the soil/geotextile interface (Figure 2-4). The

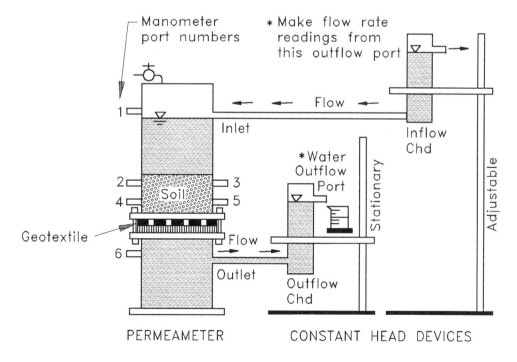

Figure 2-4   U.S. Army Corps of Engineers gradient ratio test device.

ratio of the head loss across this interface (nominally 25 mm) to the head loss across 50 mm of soil is termed the gradient ratio. As fine soil particles adjacent to the geotextile become trapped inside or blind the surface, the gradient ratio will increase. A gradient ratio less than 3 is recommended by the U.S. Army Corps of Engineers (1977), based upon limited testing with severely gap-graded soils. Because the test is conducted in a rigid-wall permeameter, it is most appropriate for sandy and silty soils with $k \geq 10^{-7}$ m/s.

For soils with permeabilities less than about 10^{-7} m/s, long-term filtration tests should be conducted in a flexible wall or triaxial type apparatus to insure that flow is through the soil rather than along the sides of the specimen. The soil flexible wall test is ASTM D 5084, while the Hydraulic Conductivity Ratio (HCR) test (ASTM D 5567) has been suggested for geotextiles (see Section 1.5). Unfortunately, neither test is able to measure the permeability near the soil-geotextile interface nor determine changes in permeability and hydraulic gradient within the soil sample itself - a serious disadvantage (Fischer, 1994). Fortunately, very fine-grained, low-permeability soils rarely present a filtration problem unless they are dispersive (Sherard and Decker, 1977) or subject to hydraulic fracturing, such as might occur in dams under high hydraulic gradients (Sherard, 1986).

Again, we emphasize that these filtration tests are performance tests. They must be conducted on samples of project site soil by the specifying agency or its representative. These tests are the responsibility of the engineer because manufacturers generally do not have soil laboratories or samples of on-site soils. Therefore, realistically, the manufacturers are unable to certify the clogging resistance of a geotextile.

For less critical/less severe conditions, a simple way to avoid clogging, especially with silty soils, is to allow fine particles already in suspension to pass through the geotextile. Then the bridge network (Figure 2-2) formed by the larger particles retains the smaller particles. The bridge network should develop rather quickly, and the quantity of fine particles actually passing through the geotextile is relatively small. This is why the less critical/less severe clogging resistance criteria requires an AOS (O_{95}) sufficiently larger than the finer soil particles (D_{15}). Those are the particles that will pass through the geotextile. Unfortunately, the AOS value only indicates the size and not the number of O_{95}-sized holes available. Thus, the finer soil particles will be retained by the smaller holes in the geotextile, and if there are sufficient fines, a significant reduction in flow rate can occur.

Consequently, to control the number of holes in the geotextile, it may be desirable to increase other qualifiers such as the porosity and open area requirements. There should always be a sufficient number of holes in the geotextile to maintain permeability and drainage, even if some of them clog.

It should be pointed out that some soil types and gradations, may result in calculated AOS values that cannot reasonably be met by any available product. Another problem is that the AOS given by Equation 2-10 is greater than the AOS permitted by Equation 2-1. In these cases, the design must be modified accordingly to accommodate available products and result in an economical project.

2.3-4 Survivability and Endurance Criteria

To be sure the geotextile will survive the construction process, certain geotextile strength and endurance properties are required for filtration and drainage applications. These minimum requirements are given in Table 2-2. Note that stated values are for less critical/less severe applications.

Table 2-2
Physical Requirements[1,2] for Drainage Geotextiles
(Modified from AASHTO, 1990 and 1996)

Property	Drainage[3]		Test Method
	High Survivability[4] (Class 2)[7]	Moderate Survivability[5] (Class 3)[7]	
Grab Strength (N)	700 *900 — 205* ✓ *= 150*	500	ASTM D 4632
Elongation (%)	n/a	n/a	ASTM D 4632
Seam Strength[6] (N)	630 *140*	450	ASTM D 4632
Puncture Strength (N)	250 *58*	180	ASTM D 4833
Trapezoidal Tear (N)	250 *58*	180	ASTM D 4533
Burst Strength (kN/m²)	1300	950	ASTM D 3786
Ultraviolet Degradation	50% strength retained at 500 hrs	50% strength retained at 500 hrs	ASTM D 4355

NOTES:
1. Acceptance of geotextile material shall be based on ASTM D 4759.
2. Contracting agency may require a letter from the supplier certifying that its geotextile meets specification requirements. Note: Woven slit film geotextiles should not be allowed.
3. Minimum; use value in weaker principal direction. All numerical values represent minimum average roll value (i.e., test results from any sampled roll in a lot shall meet or exceed the minimum values in the table). Stated values are for less critical/less severe applications. Lot samples according to ASTM D 4354.
4. High-survivability drainage applications for geotextiles are where installation stresses are more severe than moderate applications, i.e., very coarse, sharp, angular aggregate is used, a heavy degree of compaction (>95% AASHTO T99) is specified, or depth of trench is greater than 3 m.
5. Moderate-survivability drainage applications are those where geotextiles are used with smooth-graded surfaces having no sharp, angular projections, no sharp, angular aggregate is used, compaction requirements are light, (<95% AASHTO T99), and trenches are less than 3 m in depth.
6. Values apply to both field and manufactured seams.
7. AASHTO (1996) classification.

It is important to realize that these minimum survivability values are not based on any systematic research, but on the properties of existing geotextiles which are known to have performed satisfactorily in drainage applications. The values are meant to serve as guidelines for inexperienced users in selecting geotextiles for routine projects. They are not intended to replace site-specific evaluation, testing, and design.

Geotextile endurance relates to its longevity. Geotextiles have been shown to be basically inert materials for most environments and applications. However, certain applications may expose the geotextile to chemical or biological activity that could drastically influence its filtration properties or durability. For example, in drains, granular filters and geotextiles can become chemically clogged by iron or carbonate precipitates, and biologically clogged by algae, mosses, etc. Biological clogging is a potential problem when filters and drains are periodically inundated then exposed to air. Excessive chemical and biological clogging can significantly influence filter and drain performance. These conditions are present, for example, in landfills.

Biological clogging potential can be examined with ASTM D 1987, Standard Test Method for Biological Clogging of Geotextile or Soil/Geotextile Filters (1991). If biological clogging is a concern, a higher-porosity geotextile may be used, and/or the drain design and operation can include an inspection and maintenance program to flush the drainage system.

2.4   DRAINAGE SYSTEM DESIGN GUIDELINES

In this section, step-by-step design procedures are given. As with a chain, the integrity of the resulting design will depend on its weakest link; thus, no steps should be compromised or omitted.

STEP 1.   Evaluate the critical nature and site conditions (see Table 2.1) of the application.

Reasonable judgment should be used in categorizing a project, since there may be a significant cost difference for geotextiles required for critical/severe conditions. Final selection should *not* be based on the lowest material cost alone, nor should costs be reduced by eliminating laboratory soil-geotextile performance testing, if such testing is appropriate.

STEP 2.   Obtain soil samples from the site and:

A. Perform grain size analyses.
- Calculate $C_u = D_{60}/D_{10}$ (Eq. 2 - 3)
- Select the *worst case* soil for retention (*i.e.*, usually the soil with smallest B x D_{85})

NOTE: When the soil contains particles 25 mm and larger, use only the gradation of soil passing the 4.75 mm sieve in selecting the geotextile (*i.e.*, scalp off the +4.75 mm material).

B. Perform field or laboratory permeability tests.
- Select worst case soil (*i.e.*, soil with highest coefficient of

permeability, k).

- The permeability of clean sands with 0.1 mm < D_{10} < 3 mm and C_u < 5 can be estimated by the Hazen formula, k = $(D_{10})^2$ (k in cm/s; D_{10} in mm). This formula should **not** be used for soils with appreciable fines.

C. Select drainage aggregate.

- Use free-draining, open-graded material and determine its permeability (*e.g.*, Figure 2-6). If possible, sharp, angular aggregate should be avoided. If it must be used, then a geotextile meeting the property requirements for high survivability in Table 2-2 should be specified. For an accurate design cost comparison, compare cost of open-graded aggregate with select well-graded, free-draining filter aggregate.

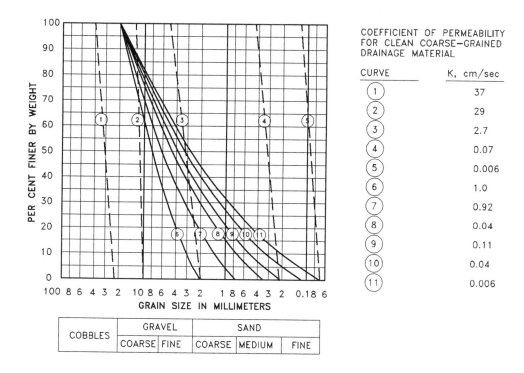

COEFFICIENT OF PERMEABILITY FOR CLEAN COARSE-GRAINED DRAINAGE MATERIAL	
CURVE	K, cm/sec
1	37
2	29
3	2.7
4	0.07
5	0.006
6	1.0
7	0.92
8	0.04
9	0.11
10	0.04
11	0.006

Figure 2-5   *Typical gradations and Darcy permeabilities of several aggregate and graded filter materials (U.S. Navy, 1982).*

STEP 3.   Calculate anticipated flow into and through drainage system and dimension the system. Use collector pipe to reduce size of drain.

A. General Case

Use Darcy's Law

$$q = k i A \qquad\qquad [2 - 13]$$

where:

q	$=$	infiltration rate (L^3/T)
k	$=$	effective permeability of soil (from Step 2B above) (L/T)
i	$=$	average hydraulic gradient in soil and in drain (L/L)
A	$=$	area of soil and drain material normal to the direction of flow (L^2)

Use conventional flow net analysis (Cedergren, 1989) and Darcy's Law for estimating infiltration rates into drain; then use Darcy's Law to design drain (*i.e.,* calculate cross-sectional area A for flow through open-graded aggregate).

B. Specific Drainage Systems

Estimates of surface infiltration, runoff infiltration rates, and drainage dimensions can be determined using accepted principles of hydraulic engineering (Moulton, 1980). Specific references are:
1. *Flow into trenches* - Mansur and Kaufman (1962)
2. *Horizontal blanket drains* - Cedergren (1989)
3. *Slope drains* - Cedergren (1989)

STEP 4.   Determine geotextile requirements.

A. Retention Criteria

From Step 2A, obtain D_{85} and C_u; then determine largest pore size allowed.

$$AOS \leq B D_{85} \qquad\qquad (Eq. 2 - 1)$$

where:

B = 1 for a conservative design. For a less-conservative design, and for $\leq 50\%$ passing 0.075 mm sieve:

$B = 1$	for $C_u \leq 2$ or ≥ 8	(Eq. 2 - 2a)
$B = 0.5 C_u$	for $2 \leq C_u \leq 4$	(Eq. 2 - 2b)

$$B = 8/C_u \qquad \text{for } 4 < C_u < 8 \qquad \text{(Eq. 2 - 2c)}$$

and, for $\geq 50\%$ passing 0.075 mm sieve:

$$B = 1 \qquad \text{for wovens,}$$

$$B = 1.8 \qquad \text{for nonwovens,}$$

and AOS (geotextile) ≤ 0.3 mm $\qquad$ (Eq. 2 - 5)

Soils with a C_u of greater 20 may be unstable (see section 2.3-1.c): if so, performance tests are recommended to select suitable geotextiles.

B. Permeability/Permittivity Criteria

1. Less Critical/Less Severe

$$k_{geotextile} \geq k_{soil} \qquad \text{(Eq. 2 - 7a)}$$

2. Critical/Severe

$$k_{geotextile} \geq 10 \; k_{soil} \qquad \text{(Eq. 2 - 7b)}$$

3. Permittivity Requirements

$\psi \geq 0.5$ sec^{-1} for $< 15\%$ passing 0.075 mm $\qquad$ (Eq. 2 - 8a)

$\psi \geq 0.2$ sec^{-1} for 15 to 50% passing 0.075 mm $\qquad$ (Eq. 2 - 8b)

$\psi \geq 0.1$ sec^{-1} for $> 50\%$ passing 0.075 mm $\qquad$ (Eq. 2 - 8c)

4. Flow Capacity Requirement

$$q_{required} = q_{geotextile}/(A_g/A_t), \text{ or} \qquad \text{(Eq. 2 - 9)}$$

$$(k_{geotextile}/t) \; h \; A_g \geq q_{required} \qquad [2 - 14]$$

where:

$q_{required}$ is obtained from STEP 3B (Eq. 2-14) above;

$k_{geotextile}/t \quad = \psi = $ permittivity;

$t \qquad\qquad = $ geotextile thickness;

$h \qquad\qquad = $ average head in field;

$A_g \qquad\quad = $ geotextile area available for flow (*i.e.*, if 80% of geotextile is covered by the wall of a pipe, $A_g = 0.2$ x total area); and

$A_t \qquad\quad = $ total area of geotextile.

C. Clogging Criteria

1. Less Critical/Less Severe
   a. From Step 2A obtain D_{15}; then determine minimum pore size requirement from

   $$O_{95} \geq 3 \ D_{15}, \text{ for } C_u > 3 \qquad \text{(Eq. 2 - 10)}$$

   b. Other qualifiers:

   Nonwovens:

   $$\text{Porosity (geotextile)} \geq 50\% \qquad \text{(Eq. 2 - 11)}$$

   Wovens:

   $$\text{Percent open area} \geq 4\% \qquad \text{(Eq. 2 - 12)}$$

   Alternative: Run filtration tests

2. Critical/Severe

Select geotextiles that meet retention, permeability, and survivability criteria, as well as the criteria in Step 4C.1 above, and perform a filtration test.

Suggested filtration test for sandy and silty soils is the gradient ratio test. The hydraulic conductivity ratio test is recommended by some people for fine-grained soils, but as noted in Section 2.3-3, the test has serious disadvantages. Alternative: Long-term filtration tests, F^3 tests, etc.

NOTE: Experience is required to obtain reproducible results from performance tests. See Fischer (1994) and Maré (1994).

D. Survivability

Select geotextile properties required for survivability from Table 2-2. Add durability requirements if appropriate.

STEP 5.   Estimate costs.

Calculate the pipe size (if required), the volume of aggregate, and the area of the geotextile. Apply appropriate unit cost values.

Pipe (if required) (/m)   _____

Aggregate (/m³)   _____

Geotextile (/m²)   _____

Geotextile placement (/m²)   _____

Construction (LS)   _____

Total Cost: _____

STEP 6. Prepare specifications.

Include for the geotextile:
A. General requirements
B. Specific geotextile properties
C. Seams and overlaps
D. Placement procedures
E. Repairs
F. Testing and placement observation requirements
See Sections 1.6 and 2.7 for specification details.

STEP 7. Collect samples of aggregate and geotextile before acceptance.

STEP 8. Monitor installation during and after construction.

STEP 9. Observe drainage system during and after storm events.

2.5   DESIGN EXAMPLE

DEFINITION OF DESIGN EXAMPLE

- Project Description: drains to intercept groundwater are to be placed adjacent to a two-lane highway

- Type of Structure: trench drain

- Type of Application: geotextile wrapping of aggregate drain stone

- Alternatives:
  - i) graded soil filter between aggregate and soil being drained; or
  - ii) geotextile wrapping of aggregate

GIVEN DATA

- site has a high groundwater table
- drain is to prevent seepage and shallow slope failures, which are currently a maintenance problem
- depth of trench drain is 1 m
- soil samples along the proposed drain alignment are nonplastic
- gradations of three *representative* soil samples along the proposed drain alignment

SIEVE SIZE (mm)	PERCENT PASSING, BY WEIGHT		
	Sample A	Sample B	Sample C
25	99	100	100
13	97	100	100
4.76	95	100	100
1.68	90	96	100
0.84	78	86	93
0.42	55	74	70
0.15	10	40	11
0.075	1	15	0

DEFINE

A. Geotextile function(s)

B. Geotextile properties required

C. Geotextile specification

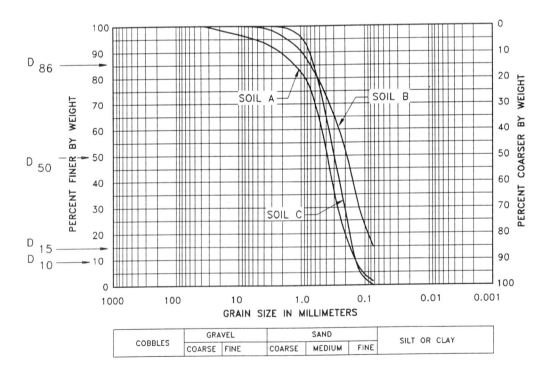

Grain Size Distribution Curve

SOLUTION

A. Geotextile function(s):

      Primary   -   filtration

      Secondary -   separation

B. Geotextile properties required:

      apparent opening size (AOS)

      permittivity

      survivability

DESIGN

STEP 1.   EVALUATE CRITICAL NATURE AND SITE CONDITIONS

From given data, assume that this is a noncritical application.

Soils are well-graded, hydraulic gradient is low for this type of application, and flow conditions are steady state for this type of application.

STEP 2.   OBTAIN SOIL SAMPLES

A. GRAIN SIZE ANALYSES

Plot gradations of representative soils. The D_{60}, D_{10}, and D_{85} sizes from the gradation plot are noted in the table below for Samples A, B, and C. Determine uniformity coefficient, C_u, coefficient B, and the maximum AOS.

Worst case soil for retention (*i.e.*, smallest B x D_{85}) is Soil C, from the table below.

Soil Sample	$D_{60} \div D_{10} = C_u$	B =	AOS $\leq$ B x D_{85} (mm)
A	$0.48 \div 0.15 = 3.2$	$0.5\, C_u = 0.5 \times 3.2 = 1.6$	$1.6 \times 1.0 = 1.6$
B	$0.25 \div 0.06 = 4.2$	$8 \div C_u = 8 \div 4.2 = 1.9$	$1.9 \times 0.75 = 1.4$
C	$0.36 \div 0.14 = 2.6$	$0.5\, C_u = 0.5 \times 2.6 = 1.3$	$1.3 \times 0.55 = 0.72$

B. PERMEABILITY TESTS

Noncritical application, drain will be conservatively designed with an estimated permeability.

The largest D_{10} controls permeability; therefore, Soil A with $D_{10} = 0.15$ mm controls. Therefore,

$$k \approx (D_{10})^2 = (0.15)^2 = 2 (10)^{-2} \text{ cm/s} = 2 (10)^{-4} \text{ m/s}$$

C. SELECT DRAIN AGGREGATE
Assume drain stone is a rounded aggregate.

STEP 3. DIMENSION DRAIN SYSTEM

Determine depth and width of drain trench and whether a pipe is required to carry flow — details of which are not included within this example.

STEP 4. DETERMINE GEOTEXTILE REQUIREMENTS

A. RETENTION CRITERIA
Sample C controls (see preceding table), therefore,   AOS ≤ 0.72 mm

B. PERMEABILITY CRITERIA
From given data, it has been judged that this application is a less critical/less severe application.
Therefore, $k_{geotextile} \geq k_{soil}$

Soil C controls, therefore $k_{geotextile} \geq 2 (10)^{-4}$ m/sec

Flow capacity requirements of the system — details of which are not included within this example.

C. PERMITTIVITY CRITERIA
All three soils have < 15% passing the 0.075 mm, therefore $\psi \geq 0.5$ sec^{-1}

D. CLOGGING CRITERIA
From given data, it has been judged that this application is a less critical/less severe application, and Soils A and B have a C_u greater than 3. Therefore, for soils A and B, $O_{95} \geq 3 D_{15}$

$O_{95} \geq$ 3 x 0.15 = 0.45 mm for Sample A
                   3 x 0.075 = 0.22 mm for Sample B

Soil A controls [Note that sand size particles typically don't create clogging problems, therefore, Soil B could have been used as the design control.], therefore,   AOS ≥ 0.45 mm

For Soil C, a geotextile with the maximum AOS value determined from the retention criteria should be used. Therefore

$$AOS \approx 0.72 \text{ mm}$$

Also,

nonwoven porosity $\geq 50\%$

and

woven percent open area $\geq 4\%$

For the primary function of filtration, the geotextile should have 0.45 mm $\leq$ AOS $\leq$ 0.72 mm; and $k_{\text{geotextile}} \geq 2 (10)^{-2}$ cm/sec and, $\psi \geq 0.5$ sec^{-1}.

E. SURVIVABILITY

Depth of trench is moderate. Assume drain stone is not sharp, angular aggregate. Therefore, a *Moderate* Survivability criterion is applicable. From Table 2-2, the following **minimum** values are recommended:

For Survivability, the geotextile shall have the following minimum values (values are MARV) -

Grab Strength	500 N
Puncture Strength	180 N
Trapezoidal Tear	180 N
Burst Strength	950 kN/m^2
Ultraviolet at 500 hours	>50% strength retained

Complete Steps 5 through 9 to finish design.

STEP 5.   ESTIMATE COSTS

STEP 6.   PREPARE SPECIFICATIONS

STEP 7.   COLLECT SAMPLES

STEP 8.   MONITOR INSTALLATION

STEP 9.   OBSERVE DRAIN SYSTEM DURING AND AFTER STORM EVENTS

2.6   COST CONSIDERATIONS

Determining the cost effectiveness of geotextiles versus conventional drainage systems is a straightforward process. Simply compare the cost of the geotextile with the cost of a conventional granular filter layer, while keeping in mind the following:

- Overall material costs including a geotextile versus a conventional system - For example, the geotextile system will allow the use of poorly graded (less-select) aggregates, which may reduce the need for a collector pipe, provided the amount of fines is small (Q decreases considerably if the percent passing the 0.075 mm sieve is greater than 5%, even in gravel).
- Construction requirements - There is, of course, a cost for placing the geotextile; but in most cases, it is less than the cost of constructing dual-layered, granular filters, for example, which are often necessary with conventional filters and fine-grained soils.
- Possible dimensional design improvements - If an open-graded aggregate is used (especially with a collector pipe), a considerable reduction in the physical dimensions of the drain can be made without a decrease in flow capacity. This size reduction also reduces the volume of the excavation, the volume of filter material required, and the construction time necessary per unit length of drain.

In general, the cost of the geotextile material in drainage applications will typically range from $1.00 to $1.50 per square meter, depending upon the type specified and quantity ordered. Installation costs will depend upon the project difficulty and contractor's experience; typically, they range from $0.50 to $1.50 per square meter of geotextile. Higher costs should be anticipated for below-water placement. Labor installation costs for the geotextile are easily repaid because construction can proceed at a faster pace, less care is needed to prevent segregation and contamination of granular filter materials, and multilayered granular filters are typically not necessary.

2.7   SPECIFICATIONS

The following guide specifications are provided as an example. They have been developed by the AASHTO-AGC-ARTBA Task Force 25 (1990) for routine drainage and filtration applications. The actual hydraulic and physical properties of the geotextile must be selected by considering of the nature of the project (critical/less critical), hydraulic conditions (severe/less severe), soil conditions at the site, and construction and installation procedures appropriate for the project.

AASHTO-AGC-ARTBA TASK FORCE 25
SPECIFICATION GUIDE FOR DRAINAGE GEOTEXTILES
(July, 1986)

1. *Description*

1.1 This work shall consist of furnishing and placing a geotextile for the following drainage applications: edge of pavement drains, interceptor drains, wall drains, recharge basins, and relief wells. The geotextile shall be designed to allow passage of water while retaining in situ soil without clogging. The quantities of drainage geotextiles as shown on the plans may be increased or decreased at the direction of the Engineer based on construction procedures and actual site conditions that occur during construction of the project. Such variations in quantity will not be considered as alterations in the details of construction or a change in the character of the work.

2. *Materials*

2.1 Fibers used in the manufacture of geotextiles, and the threads used in joining the geotextiles by sewing, shall consist of long-chain synthetic polymers composed of at least 85% by weight polyolefins, polyesters, or polyamides. They shall be formed into a network such that the filaments or yarns retain dimensional stability relative to each other, including selvedges. These materials shall conform to the performance requirements for soil retention, permeability, and clogging resistance (Section 2.3) and the physical requirements of Table 2-2 (Section 2.3-4) constructability, survivability, and durability.

2.2 Geotextile rolls shall be furnished with suitable wrapping for protection against moisture, and extended ultraviolet exposure prior to placement. Each roll shall be labeled or tagged to provide product identification sufficient for inventory and quality control purposes. Rolls shall be stored in a manner which protects them from the elements. If stored outdoors, they shall be elevated and protected with a waterproof cover.

3. *Construction Requirements*

3.1 *Geotextile Exposure Following Placement* - Exposure of geotextiles to the elements between lay down and cover shall be a maximum of 14 days to minimize damage potential.

3.2 *Geotextile Placement* - In trenches, after placing the backfill material, the geotextile shall be folded over the top of the filter material to produce a minimum overlap of 0.3 m for trenches greater than 0.3 m wide. In trenches less than 0.3 m in width, the overlap shall be equal to the width of the trench. The geotextile shall then be covered with the subsequent course.

Successive sheets of geotextiles shall be overlapped a minimum of 0.3 m in the direction of flow.

3.3 *Seams* - Where seams are required in the longitudinal trench direction, they shall be joined by either sewing or overlapping. All seams shall be subject to approval by the Engineer.

3.4 *Repair* - A geotextile patch shall be placed over the damaged area and extend 1 m beyond the perimeter of the tear or damage.

4.   *Method of Measurement*

4.1 The geotextile shall be measured by the number of square meters computed from the payment lines shown on the plans or from the payment lines established in writing by the Engineer. This excludes seam overlaps.

4.2 Excavation, backfill, bedding, and cover material are separate pay items.

5.   *Basis* of Payment

5.1 The accepted quantities of geotextiles shall be paid for at the contract unit price per square meter in place.

5.2 Payment will be made under:

Pay Item	Pay Unit
Drainage Geotextile	Square Meter

2.8   INSTALLATION PROCEDURES

For all drainage applications, the following construction steps should be followed:

1.   The surface on which the geotextile is to be placed should be excavated to design grade to provide a smooth, graded surface free of debris and large cavities.
2.   Between preparation of the subgrade and construction of the system itself, the geotextile should be well-protected to prevent any degradation due to exposure to the elements.
3.   After excavating to design grade, the geotextile should be cut (if required) to the desired width (including allowances for *non-tight* placement in trenches and overlaps of the ends of adjacent rolls) or cut at the top of the trench after placement of the drainage aggregate.
4.   Care should be taken during construction to avoid contamination of the geotextile. If it becomes contaminated, it must be removed and replaced with new material.

5. In drainage systems, the geotextile should be placed with the machine direction following the direction of water flow; for pavements, the geotextile should be parallel to the roadway. It should be placed loosely (not taut), but with no wrinkles or folds. Care should be taken to place the geotextile in intimate contact with the soil so that no void spaces occur behind it.

6. The ends for subsequent rolls and parallel rolls of geotextile should be overlapped a minimum of 0.3 in roadways and 0.3 to 0.6 m in drains, depending on the anticipated severity of hydraulic flow and the placement conditions. For high hydraulic flow conditions and heavy construction, such as with deep trenches or large stone, the overlaps should be increased. For large open sites using base drains, overlaps should be pinned or anchored to hold the geotextile in place until placement of the aggregate. Upstream geotextile should always overlap over downstream geotextile.

7. To limit exposure of the geotextile to sunlight, dirt, damage, etc., placement of drainage or roadway base aggregate should proceed immediately following placement of the geotextile. The geotextile should be covered with a minimum of 0.3 m of loosely placed aggregate prior to compaction. If thinner lifts are used, higher survivability fabrics may be required. For drainage trenches, at least 0.1 m of drainage stone should be placed as a bedding layer below the slotted collector pipe (if required), with additional aggregate placed to the minimum required construction depth. Compaction is necessary to seat the drainage system against the natural soil and to reduce settlement within the drain. The aggregate should be compacted with vibratory equipment to a minimum of 95% Standard AASHTO T99 density unless the trench is required for structural support. If higher compactive efforts are required, the geotextiles meeting the property values listed under the high survivability category in Table 2-2 should be utilized.

8. After compaction, for trench drains, the two protruding edges of the geotextile should be overlapped at the top of the compacted granular drainage material. A minimum overlap of 0.3 m is recommended to ensure complete coverage of the trench width. The overlap is important because it protects the drainage aggregate from surface contamination. After completing the overlap, backfill should be placed and compacted to the desired final grade.

A schematic of the construction procedures for a geotextile-lined underdrain trench is shown in Figure 2-6. Construction photographs of an underdrain trench are shown in Figure 2-7, and diagrams of geosynthetic placement beneath a permeable roadway base are shown in Figure 2-8.

1. EXCAVATE TRENCH

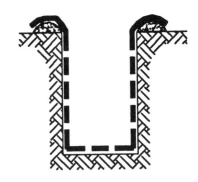

2. PLACE GEOTEXTILE

3. ADD BEDDING & PIPE

4. PLACE/COMPACT DRAINAGE AGGREGATE

5. WRAP GEOTEXTILE OVER TOP

6. COMPACT BACKFILL

Figure 2-6   Construction procedure for geotextile-lined underdrains.

a.

b.

Figure 2-7   Construction of geotextile drainage systems: a.) geotextile
placement in drainage ditch; b.) aggregate placement; c.)
compaction of aggregate; and d.) geotextile overlap prior to
final cover.

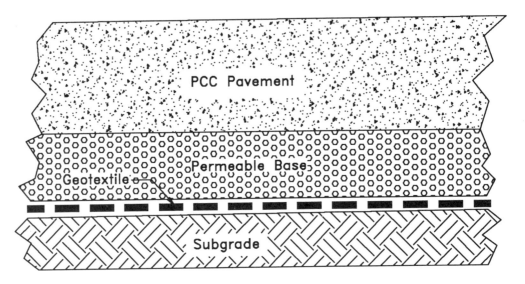

a.

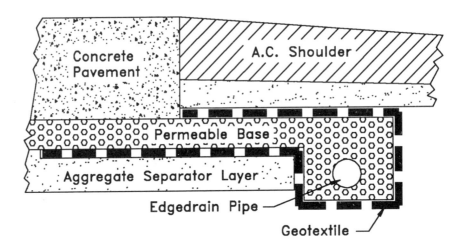

b.

Figure 2-8   *Construction geotextile filters and separators beneath permeable pavement base: a.) geotextile used as a separator; and b.) permeable base and edge drain combination. (Baumgardner, 1994)*

2.9 FIELD INSPECTION

The field inspector should review the field inspection guidelines in Section 1.7. Special attention should be given to aggregate placement and potential for geotextile damage. Also, maintaining the appropriate geotextile overlap at the top of the trench and at roll ends is especially important.

2.10 ADDITIONAL SELECTION CONSIDERATIONS

The late Dr. Allan Haliburton, a geotextile pioneer, noted that all geotextiles will work in some applications, but no one geotextile will work in all applications. Even though several types of geotextiles (monofilament wovens and an array of light- to heavy-weight nonwovens) may meet all of the desired design criteria, it may be preferable to use one type over another to enhance system performance. Selection will depend on the actual soil and hydraulic conditions, as well as the intended function of the design. Intuitively, the following considerations seem appropriate for the soil conditions given.

1. Graded gravels and coarse sands — Very open monofilament or even multifilament wovens may be required to permit high rates of flow and low-risk of blinding.
2. Sands and gravels with less than 20% fines — Open monofilament wovens and needlepunched nonwovens with large openings are preferable to reduce the risk of blinding. For thin, heat-bonded geotextiles and thick, needlepunched nonwoven geotextiles, filtration tests should be performed.
3. Soils with 20% to 60% fines — Filtration tests should be performed on all types of geotextiles.
4. Soils with greater than 60% fines — Heavy-weight, needlepunched geotextiles and heat-bonded geotextiles tend to work best as fines will not pass. If blinding does occur, the permeability of the blinding cake would equal that of the soil.
5. Gap-graded cohesionless soils — Consider using a uniform sand filter with a very open geotextile designed to allow fines to pass.
6. Silts with sand seams - Consider using a uniform sand filter over the soil with a very open geotextile, designed to allow the silt to pass but to prevent movement of the filter sand; alternatively, consider using a heavy-weight (thick) needlepunched nonwoven directly against soil so water can flow laterally through the geotextile should it become locally clogged.

These general observations are not meant to serve as recommendations, but are offered to provide insight for selecting optimum materials. They are **not** intended to exclude other possible geotextiles that you may want to consider.

2.11 IN-PLANE DRAINAGE; PREFABRICATED GEOCOMPOSITE DRAINS

Geotextiles with high in-plane drainage ability and prefabricated geocomposite drains are potentially quite effective in several applications.

The ability of geotextiles to transmit water in the plane of the geotextile itself may be an added benefit in certain drainage applications where lateral transmission of water is desirable or where reduction of pore water pressures in the soil can be accelerated. These applications include interceptor drains, transmission of seepage water below pavement base course layers, horizontal and vertical strip drains to accelerate consolidation of soft foundation soils, dissipation of seepage forces in earth and rock slopes, as part of chimney drains in earth dams, dissipaters of pore water pressures in embankments and fills, gas venting below containment liner systems, etc. However, it should be realized that the seepage quantities transmitted by in-plane flow of geotextiles (typically on the order of 2 x 10^{-5} m^3/s/linear meter of geotextile under a pressure equivalent to 0.6 m of soil) are relatively small when compared to the seepage capacity of 0.150 to 0.3 m of sand or other typical filter materials. Therefore, geotextiles should only replace sand or other filter layers where they can handle high seepage quantities. Remember, too, that seepage quantities are highly affected by compressive forces, incomplete saturation, and hydraulic gradients.

In recent years, special geocomposite materials have been developed which consist of cores of extruded and fluted plastics sheets, three-dimensional meshes and mats, plastic waffles, and nets and channels to convey water, which are covered by a geotextile on one or both sides to act as a filter. Geocomposite drains may be prefabricated or fabricated on site. They generally range in thickness from 5 mm to 25 mm or greater and have transmission capabilities of between 0.0002 and 0.01 m^3/s/linear width of drain. Some geocomposite systems are shown in Figure 2-9. Geocomposite drains have been used in six major areas:

1. Edge drains for pavements.
2. Interceptor trenches on slopes.
3. Drainage behind abutments and retaining structures.
4. Relief of water pressures on buried structures.
5. Substitute for conventional sand drains.
6. Waste containment systems for leachate collection and gas venting.

Prefabricated geocomposite drains are essentially used to replace or support conventional drainage systems. According to Hunt (1982), prefabricated drains offer a readily available material with known filtration and hydraulic flow properties; easy installation, and, therefore, construction economies; and protection of any waterproofing

Figure 2-9   Geocomposite drains.

applied to the structure's exterior. Cost of prefabricated drains typically ranges from $4.50 to $25.00 per square meter. The high material cost is usually offset by expedient construction and reduction in required quantities of select granular materials. For example, geocomposites used for pavement edge drains typically cost $1.75 to $5.00/linear meter installed.

2.11-1 Design Criteria

For the geotextile design and selection with in-plane drainage capabilities and geocomposite drainage systems, there are three basic design considerations:

1. Adequate filtration without clogging or piping.
2. Adequate inflow/outflow capacity under design loads to provide maximum anticipated seepage during design life.
3. System performance considerations.

As with conventional drainage systems, geotextile selection should be based on the grain size of the material to be protected, permeability requirements, clogging resistance, and physical property requirements, as described in Section 2.3. In pavement drainage systems, dynamic loading means severe hydraulic conditions (Table 2-1). If, for example, the geotextile supplied with the geocomposite drainage system is not appropriate for your design conditions, system safety will be compromised and you should specify alternate geotextiles. This is important especially when prefabricated drains are used in critical situations and where system failure could lead to structure failure.

The maximum seepage flow into the system must be estimated and the geotextile or geocomposite selected on the basis of seepage requirements. The flow capacity of the geocomposite or geotextile can be determined from the transmissivity of the material. The test for transmissivity is ASTM D 4716, Constant Head Hydraulic Transmissivity (In-Plane Flow) of Geotextiles and Geotextile Related Products. The flow capacity per unit width of the geotextile or geocomposite can then be calculated using Darcy's Law:

$$q = k_p i A = k_p i B t \qquad [2 - 15]$$

or,

$$q/B = \theta i \qquad [2 - 16]$$

where:

$$q \quad = \quad \text{flow rate } (L^3/T)$$

$$k_p \quad = \quad \text{in-plane geosynthetic coefficient of permeability } (L/T)$$

$$B \quad = \quad \text{width of geosynthetic } (L)$$

$$t \quad = \quad \text{thickness of geosynthetic } (L)$$

$$\theta \quad = \quad \text{transmissivity of geosynthetic } (= k_p t) \ (L^2/T)$$

$$i \quad = \quad \text{hydraulic gradient } (L/L)$$

The flow rate per unit width of the geosynthetic can then be compared with the flow rate per unit width required of the drainage system. It should be recognized that the in-plane flow capacity for geosynthetic drains reduces significantly under compression (Giroud, 1980). Additional decreases in transmissivity may occur with time due to creep. Therefore, the material should be evaluated by an appropriate laboratory model (performance) test, under the anticipated design loading conditions (with a safety factor) for the design life of the project.

Long-term compressive stress and eccentric loadings on the core of a geocomposite should be considered during design and selection. Though not yet addressed in standardized test methods or standards of practice, the following criteria (Berg, 1993) are suggested for addressing core compression. The design pressure on a geocomposite core should be limited to either:

i) the maximum pressure sustained on the core in a test of 10,000 hour minimum duration; or

ii) the crushing pressure of a core, as defined with a quick loading test, divided by a safety factor of 5.

Note that crushing pressure can only be defined for some core types. For cases where a crushing pressure cannot be defined, suitability should be based on the maximum load resulting in a residual thickness of the core adequate to provide the required flow after 10,000 hours, or the maximum load resulting in a residual thickness of the core adequate to provide the required flow as defined with the quick loading test divided by a safety factor of 5.

Intrusion of the geotextiles into the core and long-term outflow capacity should be measured with a sustained transmissivity test (Berg, 1993). The ASTM D 4716 test procedure, Constant Head Hydraulic Transmissivity of Geotextiles and Geotextile Related Products, should be followed. Test procedure should be modified for sustained testing and for use of sand substratum and super-stratum in lieu of closed cell foam rubber. Load should be maintained for 300 hours or until equilibrium is reached, whichever is greater.

Finally, special consideration must be given to drain location and pressures on the wall when using geosynthetics to drain earth retaining structures and abutments. It is important that the drain be located away from the back of the wall and be appropriately inclined so it can intercept seepage before it impinges on the back of the wall. Placement of a thin vertical drain directly against a retaining wall may actually increase seepage forces on the wall due to rainwater infiltration (Terzaghi and Peck, 1967; and Cedergren, 1989). For further discussion of this point, see Christopher and Holtz (1985).

2.11-2 Construction Considerations

The following are considerations specific to the installation of geocomposite drains:

1. As with all geotextile applications, care should be taken during storage and placement to avoid damage to the material.

2. Placement of the backfill directly against the geotextile must be closely observed, and compaction of soil directly against the material should be avoided. Otherwise, loading during placement of backfill could damage the filter or even crush the drain. Use of clean granular backfill reduces the compaction energy requirements.

3. At the joints, where the sheets or strips of geocomposite butt together, the geotextile must be carefully overlapped to prevent soil infiltration. Also, the geotextile should extend beyond the ends of the drain to prevent soil from entering at the edges.

4. Details must be provided on how the prefabricated drains tie into the collector drainage systems.

Construction of a typical edge drain installation is shown in Figures 2-10 and 2-11. Additional information and recommendations regarding proper edge drain installation can be found in Koerner, et al. (1994).

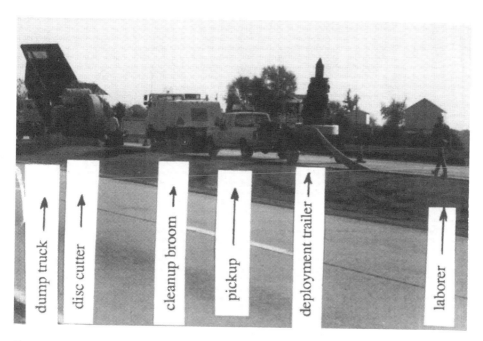

a. Equipment train used to install PGEDs according to Figure 2-11.

b. Sand installation and backfilling equipment at end of equipment train according to Figure 2-11.

Figure 2-10   *Prefabricated geocomposite edge drain construction using sand fill upstream of composite (as illustrated in Figure 2-11) (from Koerner, et al., 1994).*

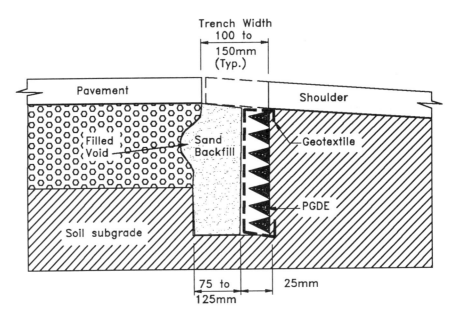

*Figure 2-11   Recommended   installation   method   for   prefabricated
geocomposite edge drains (from Koerner, et al., 1994).*

2.12   REFERENCES

References quoted within this section are listed below. A key reference for design is this manual (FHWA Geosynthetics Manual) and its predecessor Christopher and Holtz (1985).   The recent NCHRP report (Koerner et al., 1994) specifically addresses pavement edge drain systems and is based upon analysis of failed systems. It also is a key reference for design.

ASTM (1994), *Annual Books of ASTM Standards,* American Society for Testing and Materials, Philadelphia, Pennsylvania:
Volume 4.08 (I), Soil and Rock
Volume 4.08 (II), Soil and Rock; Geosynthetics

AASHTO (1996), Standard Specifications for Geotextiles, M288-96 (DRAFT), American Association of State Transportation and Highway Officials, Washington, D.C.

AASHTO (1990), Standard Specifications for Geotextiles - M 288, *Standard Specifications for Transportation Materials and Methods of Sampling and Testing*, American Association of State Transportation and Highway Officials, Washington, D.C., pp 689-692.

AASHTO (1990), Guide Specifications and Test Procedures for Geotextiles, *Task Force 25 Report*, Subcommittee on New Highway Materials, American Association of State Transportation and Highway Officials, Washington, D.C.

Bell, J.R. and Hicks, R.G. (1980), *Evaluation of Test Methods and Use Criteria for Geotechnical Fabrics in Highway Applications - Interim Report*, Report No. FHWA/RD-80/021, Federal Highway Administration, Washington, D.C., June, 190 p.

Berg, R.R. (1993), *Guidelines for Design, Specification, & Contracting of Geosynthetic Mechanically Stabilized Earth Slopes on Firm Foundations*, Report No. FHWA-SA-93-025, Federal Highway Administration, Washington, D.C., 87 p.

Carroll, R.G., Jr. (1983), Geotextile Filter Criteria, *Engineering Fabrics in Transportation Construction, Transportation Research Record 916*, Transportation Research Board, Washington, D.C., January, pp. 46-53.

Cedergren, H.R. (1989), *Seepage, Drainage, and Flow Nets*, Third Edition, John Wiley and Sons, New York, 465 p.

Christopher, B.R. and Holtz, R.D. (1985), *Geotextile Engineering Manual*, Report No. FHWA-TS-86/203, Federal Highway Administration, Washington, D.C., March, 1044 p.

Fischer, G.R. (1994), *The Influence of Fabric Pore Structure on the Behavior of Geotextile Filters*, Ph.D. Dissertation, University of Washington, 498 p.

Giroud, J.P. (1980), Introduction to Geotextiles and Their Applications, *Proceedings of the First Canadian Symposium on Geotextiles*, Calgary, Alberta, September, pp. 3-31.

GRI Test Method GT1 (1993), Geotextile Filter Performance via Long Term Flow (LTF) Tests, Standard Test Method, Geosynthetic Research Institute, Drexel University, Philadelphia, PA.

Hunt, J.R. (1982), The Development of Fin Drains for Structure Drainage, *Proceedings of the Second International Conference on Geotextiles*, Las Vegas, Nevada, Vol. 1, August, pp. 25-36.

Kenney, T.C. and Lau, D. (1985), Internal Stability of Granular Filters, *Canadian Geotechnical Journal*, Vol. 22, No. 2, pp. 215-225.

Kenney, T.C. and Lau, D. (1986), Reply (to discussions) - Internal Stability of Granular Filters, *Canadian Geotechnical Journal*, Vol. 23, No. 3, pp. 420-423.

Koerner, R.M., Koerner, G.R., Fahim, A.K. and Wilson-Fahmy, R.F. (1994), *Long Term Performance of Geosynthetics in Drainage Applications*, National Cooperative Highway Research Program, Report No. 367, 54 p.

LeFluer, J., Mlynarek, J. and Rollin, A.L. (1993), Filter Criteria for Well Graded Cohesionless Soils, Filters in Geotechnical and Hydraulic Engineering, *Proceedings of the First International Conference - Geo-Filters*, Karlsruhe, Brauns, Schuler, and Heibaum Eds., Balkema, pp. 97-106.

LeFluer, J., Mlynarek, J. and Rollin, A.L. (1989), Filtration of Broadly Graded Cohesionless Soils, *Journal of Geotechnical Engineering, American Society of Civil Engineers*, Vol. 115, No. 12, pp. 1747-1768.

Mansur, C.I. and Kaufman, R.I. (1962), Dewatering, Chapter 3 in *Foundation Engineering*, G.A. Leonards, Editor, McGraw-Hill, pp. 241-350.

Maré, A.D. (1994), *The Influence of Gradient Ratio Testing Procedures on the Filtration Behavior of Geotextiles*, MSCE Thesis, University of Washington.

Moulton, L.K. (1980), *Highway Subdrainage Design*, Report No. FHWA-TS-80-224, Federal Highway Administration, Washington, D.C.

Sherard, J.L. (1986), Hydraulic Fracturing in Embankment Dams, *Journal of Geotechnical Engineering*, American Society of Civil Engineers, Vol. 112, No. 10, pp. 905-927.

Sherard, J.L. and Decker, R.S., Editors (1977), Dispersive Clays, Related Piping, and Erosion in Geotechnical Projects, *ASTM Special Technical Publication 623*, American Society for Testing and Materials, Philadelphia, PA, 486 p.

Sherard, J.L., Decker, R.S. and Ryker, N.L. (1972), Piping in Earth Dams of Dispersive Clay, *Proceedings of the ASCE Specialty Conference on Performance of Earth and Earth -Supported Structures*, American Society of Civil Engineers, New York, Vol. I, Part 1, pp. 589-626.

Skempton, A.W. and Brogan, J.M. (1994), Experiments on Piping in Sandy Gravels, *Geotechnique*, Vol. XLIV, No. 3, pp. 461-478.

Terzaghi, K. and Peck, R.B. (1967), *Soil Mechanics in Engineering Practice*, John Wiley & Sons, New York, 729 p.

U.S. Army Corps of Engineers (1977), *Civil Works Construction Guide Specification for Plastic Filter Fabric*, Corps of Engineer Specifications No. CW-02215, Office, Chief of Engineers, U.S. Army Corps of Engineers, Washington, D.C.

U.S. Department of the Navy (1982), *Soil Mechanics, Design Manual 7.1,* Naval Facilities Engineering Command, Alexandria, VA.

U.S. Department of the Navy (1982), *Foundations and Earth Structures, Design Manual 7.2,* Naval Facilities Engineering Command, Alexandria, VA.

3.0 GEOTEXTILES in RIPRAP REVETMENTS and OTHER PERMANENT EROSION CONTROL SYSTEMS

3.1 BACKGROUND

As in drainage systems, geotextiles can effectively replace graded granular filters typically used beneath riprap or other hard armor materials in revetments and other erosion control systems designed to keep soil in place. This was one of the first applications of woven monofilament geotextiles in the United States; rather extensive use started in the early 1960s. Numerous case histories have shown geotextiles to be very effective compared to riprap-only systems and equally effective as conventional graded granular filters in preventing fines from migrating through the armor system, while providing a cost savings.

Since the early developments in coastal and lake shoreline erosion control, the same design concepts and construction procedures have subsequently been applied to stream bank protection (see HEC 11, FHWA, 1989), cut and fill slope protection, protection of various small drainage structures (see HEC 14, FHWA, 1975) and ditches (see HEC 15, FHWA, 1988), wave protection for causeway and shoreline roadway embankments, and scour protection for structures such as bridge piers and abutments (see HEC 18, FHWA, 1992). Design guidelines and construction procedures for these and other similar permanent erosion control applications are presented in sections 3.3 through 3.10. Hydraulic design considerations can be found in the AASHTO *Model Drainage Manual* (1991) and the FHWA Hydraulic Engineering Circulars (1992, 1989, 1988, 1975).

Erosion control *mats* are another type of geosynthetic used in permanent erosion control systems. These three-dimensional mats retain soil and moisture, thus promoting vegetation growth. Vegetation roots grow through and are reinforced by the mat. The *reinforced grass* system is capable of withstanding short-term (*e.g.*, 2 hours), high velocity (*e.g.*, 6 m/s) flows without erosion. Erosion control mats are addressed in section 3.11. Sediment control and temporary erosion control designed to keep soil within a prescribed boundary, including the use of geotextiles as silt fences, erosion control *blankets*, and other geosynthetics, are covered in Chapter 4.

3.2   APPLICATIONS

- Riprap-geotextile systems have found successful application in protecting precipitation runoff collection and high-velocity diversion ditches.

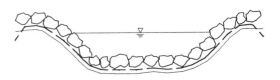

- Geotextiles may be used in slope protection to prevent or reduce erosion from precipitation, surface runoff, and internal seepage or piping. In this instance, the geotextile may replace one or more layers of granular filter materials which would be placed on the slope in conventional applications.

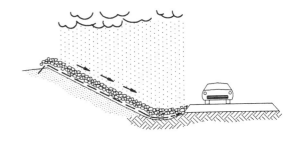

- Erosion control systems with geotextiles may also be required along streambanks to prevent encroachment of roadways or appurtenant facilities.

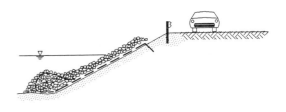

- Similarly, they may be used for scour protection around structures.

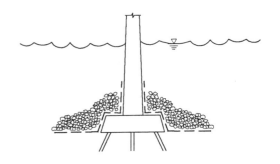

- A riprap-geotextile system can also be effective in reducing erosion caused by wave attack or tidal variations when facilities are constructed across or adjacent to large bodies of water.

- Finally, hydraulic structures such as culverts, drop inlets, and artificial stream channels may require protection from erosion. In such applications, if vegetation cannot be established or the natural soil is highly erodible, a geotextile can be used beneath armor materials to increase erosion resistance.

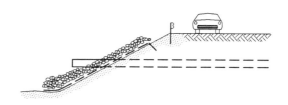

In several of the above applications, placement of the filter layer may be required below water. In these cases, in comparison with conventional granular filter layers, geotextiles provide easier placement and continuity of the filter medium is assured.

- Geosynthetic erosion control *mats* are made of synthetic meshes and webbings and reinforce the vegetation root mass to provide tractive resistance to high water velocity on slopes and in ditches. These three-dimensional mats retain soil, moisture, and seed, and thus promote vegetative growth.

3.3   DESIGN OF GEOTEXTILES BENEATH HARD ARMOR

Geotextile design for hard armor erosion control systems is essentially the same as geotextile design for filters in subsurface drainage systems discussed in Section 2.3. Table 3-1 reiterates the design criteria and highlights special considerations for geotextiles beneath hard armor erosion control systems. The following is a discussion of these special considerations.

3.3-1   Retention Criteria for Cyclic or Dynamic Flow

In cyclic or dynamic flow conditions, soil particles may be able to move behind the geotextile if it is not properly weighted down. Thus, the coefficient B = 1 may not be conservative, as the bridging network (Figure 2-2) may not develop and the geotextile may be required to retain even the finer particles of soil. If there is a risk that uplift of the armor system can occur, it is recommended that the B value be reduced to 0.5 or less; that is, the largest hole in the geotextile should be small enough to retain the smaller particles of soil.

In many erosion control applications it is common to have high hydraulic stresses induced by wave or tidal action. The geotextile may be loose when it spans between large armor stone or large joints in block-type armor systems. For these conditions, it is recommended that an intermediate layer of finer stone or gravel be placed over the geotextile and that riprap of sufficient weight be placed to prevent wave action from moving either stone or geotextile. For all applications where the geotextile can move, and when it is used as sandbags, it is recommended that samples of the site soils be washed through the geotextile to determine its particle-retention capabilities.

3.3-2   Permeability and Effective Flow Capacity Requirements for Erosion Control

In certain erosion control systems, portions of the geotextile may be covered by the armor stone or concrete block revetment systems, or the geotextile may be used to span joints in sheet pile bulkheads. For such systems, it is especially important to evaluate the flow rate required through the open portion of the system and select a geotextile that meets those flow requirements. Again, since flow is restricted through the geotextile, the required flow capacity is based on the flow capacity of the area available for flow; or

$$q_{required} = q_{geotextile}(A_g/A_t)$$ (Eq. 2 - 9)

where:

A_g = geotextile area available for flow, and

A_t = total geotextile area.

3.3-3 Clogging Resistance for Cyclic or Dynamic Flow

Since erosion control systems are often used on highly erodible soils with reversing and cyclic flow conditions, severe hydraulic conditions often exist. Accordingly, designs should reflect these conditions, and soil-geotextile filtration tests should always be conducted. Since these tests are performance-type tests and require project site soil samples, they must be conducted by the owner or an owner representative and not by the geotextile manufacturers or suppliers. For sandy and silty soils ($k \geq 10^{-7}$ m/s) the long-term, gradient ratio test (ASTM D 5101) is recommended as described in Chapter 1. For fine-grained soils, the hydraulic conductivity ratio (HCR) test (ASTM D 5567) should be considered with the modifications and caveats recommended in Chapter 1. Other filtration tests, some of which are appropriate for finer soils, are described by Christopher and Holtz (1985) and Koerner (1994), among others.

3.3-4 Survivability Criteria for Erosion Control

Because the construction procedures for erosion control systems are different than those for drainage systems, the geotextile property requirements for survivability in Table 3-1 differ somewhat from those discussed in Section 2.3-4. As placement of armor stone is generally more severe than placement of drainage aggregate, required property values are higher for each category of geotextile.

Riprap or armor stone should be large enough to withstand wave action and thus not abrade the geotextile. The specific site conditions should be reviewed, and if such movement cannot be avoided, then an abrasion requirement based on ASTM D 4886 (modified flex stoll) should be included in the specifications. Allowable physical property reduction due to abrasion should be specified. No reduction in piping resistance, permeability, or clogging resistance should be allowed after exposure to abrasion.

It is important to realize that these **minimum** survivability values are not based on any systematic research but on the properties of existing geotextiles which are known to have performed satisfactorily in hard armor erosion control applications. The values are meant to serve as guidelines for inexperienced users in selecting geotextiles for routine projects. They are not intended to replace site-specific evaluation, testing, and design.

Table 3-1

Summary of Geotextile Design and Selection Criteria for Hard Armor Erosion Control Applications

I.	SOIL RETENTION (PIPING RESISTANCE CRITERIA)[1]		
	Soils	Steady State Flow	Dynamic, Pulsating and Cyclic Flow (if geotextile can move)
	<50% Passing[2] 0.075 mm	AOS or $O_{95} \leq B\,D_{85}$ $C_u \leq 2$ or ≥ 8: B=1 $2 \leq C_u \leq 4$: B=0.5 C_u $4 \leq C_u \leq 8$: B=8/C_u	$O_{95} \leq 0.5\,D_{85}$
	$\geq$50% Passing 0.075 mm	Woven: $O_{95} \leq D_{85}$ Nonwoven: $O_{95} \leq 1.8\,D_{85}$	$O_{95} \leq 0.5\,D_{85}$
	For cohesive soils (PI > 7)	O_{95} (geotextile) $\leq$ 0.3 mm	

II.	PERMEABILITY/PERMITTIVITY CRITERIA[3]		
	A. Critical/Severe Applications $k_{geotextile} \geq 10\,k_{soil}$		
	B. Less Critical/Less Severe Applications (with Clean Medium to Coarse Sands and Gravels) $k_{geotextile} \geq k_{soil}$		
	C. Permittivity Requirement	$\psi \geq 0.7\ sec^{-1}$	for < 15% passing 0.075 mm
		$\psi \geq 0.2\ sec^{-1}$	for 15 to 50% passing 0.075 mm
		$\psi \geq 0.1\ sec^{-1}$	for > 50 % passing 0.075 mm

III	CLOGGING CRITERIA
	A. Critical/Severe Applications[4] Select geotextile meeting I, II, IIIB, and perform soil/geotextile filtration tests before specification, prequalifying the geotextile, or after selection before bid closing. Alternative: use approved list specification for filtration applications. Suggested performance test method: Gradient Ratio, ASTM D 5101 for cohesionless soils or Hydraulic Conductivity Ratio, ASTM D 5567 for cohesive soils.

B. Less Critical/Less Severe Applications

1.	Perform soil-geotextile filtration tests.		
2.	Alternative: $O_{95} \geq 3\,D_{15}$ for $C_u > 3$		
3.	For $C_u \leq 3$, specify geotextile with maximum opening size possible from retention criteria		
4.	Apparent Open Area Qualifiers		
	For soils with % passing 0.075 mm	$\geq$ 5%	$\leq$ 5%
	- Woven monofilament geotextiles, POA	$\geq$ 4%	10%
	- Nonwoven geotextiles, Porosity[5] $\geq$	50%	70%

Table 3-1 (continued)

Summary of Geotextile Design and Selection Criteria for Hard Armor Erosion Control Applications

IV.   SURVIVABILITY REQUIREMENTS

PHYSICAL REQUIREMENTS[6,7,8,9] FOR EROSION CONTROL GEOTEXTILES
(after AASHTO, 1990 and 1996)

Property	High Survivability[10] (Class 1)[14]	Moderate Survivability[11] (Class 2)[14]	Test Method
Grab Strength (N)	900 *200 #*	700 *−150 #*	ASTM D 4632
Elongation (%) (min)	15%	15%	ASTM D 4632
Seam Strength (N)[12]	810 *190 #*	630	ASTM D 4632
Puncture Strength (N)	350 *80 #*	250	ASTM D 4833
Burst Strength (kPa)	1700	1300	ASTM D 3787
Trapezoid Tear (N)	350	250	ASTM D 4533
Ultraviolet Degradation at 500 hr[13]	50% strength retained for both classes		ASTM D 4355

NOTES:
1.   When the protected soil contains particles passing the 0.075 mm sieve, use only the gradation of soil passing the 4.75 mm sieve in selecting the geotextile (i.e., scalp off the +4.75 mm material).
2.   Select geotextile on the basis of largest opening value required.
3.   Permeability should be based on the actual geotextile open area available for flow. For example, if 50% of geotextile area is to be covered by flat concrete blocks, the effective flow area is reduced by 50%.
4.   Filtration tests are performance tests and, as they depend on specific soil and design conditions, they cannot be performed by the manufacturer. Tests are to be performed by specifying agency or their representative. It should also be recognized that experience is required to obtain reproducible results from performance tests.
5.   Porosity requirements are based on graded granular filter porosity.
6.   Acceptance of geotextiles should be based on ASTM D 4759.
7.   Contracting agency may require a letter from the supplier certifying that its geotextile meets specification requirements. Note: Woven slit film geotextiles should not be allowed.
8.   Minimum — use value in weaker principal direction. All numerical values represent minimum average roll values (i.e., test results from any sampled roll in a lot shall meet or exceed the minimum values in the table). Stated values are for less critical/less severe conditions. Lot should be sampled according to ASTM D 4354.
9.   If the armor stone can move after installation (e.g., due to high wave action), then larger stone should be used or abrasion resistance requirements for the geotextile should be considered, using the results of ASTM D 4886.
10.   High Survivability erosion control applications are used when geotextile installation stresses are more severe than Moderate Survivability (i.e., stone placement height should be less than 1 m and stone weights should not exceed 100 kg).
11.   Moderate Survivability erosion control applications are those in which geotextiles are used in structures or under conditions where the geotextile is protected by a sand cushion or by *zero drop height* placement of stone.
12.   Values apply to both field and manufactured seams.
13.   500 hours is the recommendation of the authors; Task Force 25 recommended 150 hours.
14.   AASHTO (1996) classification.

3.4   GEOTEXTILE DESIGN GUIDELINES

STEP 1.   Application evaluation.

    A. Critical/less critical

    1. If the erosion control system fails, will there be a risk of loss of life?

    2. Does the erosion control system protect a significant structure, and will failure lead to significant structural damage?

    3. If the geotextile clogs, will failure occur with no warning? Will failure be catastrophic?

    4. If the erosion control system fails, will the repair costs greatly exceed installation costs?

    B. Severe/less severe

    1. Are soils to be protected gap-graded, pipable, or dispersive?

    2. Are soils present which consist primarily of silts and uniform sands with 85% passing the 0.15 mm sieve?

    3. Will the erosion control system be subjected to reversing or cyclic flow conditions such as wave action or tidal variations?

    4. Will high hydraulic gradients exist in the soils to be protected? Will rapid drawdown conditions or seeps or weeps in the soil exist? Will blockage of seeps and weeps produce high hydraulic pressures?

    5. Will high-velocity conditions exist, such as in stream channels?

NOTE: If the answer is yes to any of the above questions, the design should proceed under the *critical/severe* requirements; otherwise use the *less critical/less severe* design approach.

STEP 2.   Obtain soil samples from the site.

    A. Perform grain size analyses

    1. Determine percent passing the 0.075 mm sieve.

    2. Determine the plastic index (PI).

    3. Calculate $C_u = D_{60}/D_{10}$.

NOTE: When the protected soil contains particles passing the 0.075 mm sieve, use only the gradation of soil passing the 4.75 mm sieve in selecting the geotextile (*i.e.*, scalp off the +4.75 mm material).

    4. Obtain D_{85} for each soil and select the worst case soil (*i.e.*, soil with smallest B x D_{85}) for retention.

    B. Perform field or laboratory permeability tests

1. Select worse case soil (*i.e.*, soil with highest coefficient of permeability k).

NOTE: The permeability of clean sands (<5% passing 0.075 mm sieve) with 0.1 mm $< D_{10} < 3$ mm and $C_u < 5$ can be estimated by Hazen's formula, $k = (D_{10})^2$ (k in cm/s; D_{10} in mm). This formula should not be used for finer-grained soils.

STEP 3. Evaluate armor material and placement.

Design reference: FHWA Hydraulic Engineering Circular No. 15 (FHWA, 1988).

A. Size armor stone or riprap

Where minimum size of stone exceeds 100 mm, or greater than a 100 mm gap exists between blocks, an intermediate gravel layer 150 mm thick should be used between the armor stone and geotextile. Gravel should be sized such that it will not wash through the armor stone (*i.e.*, D_{85} gravel $\geq D_{15}$ riprap/5).

B. Determine armor stone placement technique (*i.e.*, maximum height of drop).

STEP 4. Calculate anticipated reverse flow through erosion control system.

Here we need to estimate the maximum flow from seeps and weeps, maximum flow from wave runout, or maximum flow from rapid drawdown.

A. General case -- use Darcy's law

$$q = kiA \qquad\qquad\qquad \text{(Eq. 2 - 15)}$$

where:

q = outflow rate (L^3/T)

k = effective permeability of soil (from Step 2B above) (L/T)

i = average hydraulic gradient in soil (*e.g.*, tangent of slope angle for wave runoff)(dimensionless)

A = area of soil and drain material normal to the direction of flow (L^2). Can be evaluated using a unit area.

Use a conventional flow net analysis (Cedergren, 1989) for seepage through dikes and dams or from a rapid drawdown analysis.

B. Specific erosion control systems — Hydraulic characteristics depend on expected precipitation, runoff volumes and flow rates, stream flow volumes and water level fluctuations, normal and maximum wave heights anticipated, direction of waves and tidal variations. Detailed information on determination of these parameters is available in the FHWA (1989) Hydraulic Engineering Circular No. 11.

STEP 5.   Determine geotextile requirements.

A. Retention Criteria

From Step 2A, obtain D_{85} and C_u; then determine largest pore size allowed.

$$AOS \text{ or } O_{95(geotextile)} < B\,D_{85(soil)} \qquad (Eq.\ 2 - 1)$$

where: B = 1 for a conservative design.

For a less-conservative design and for ≤ 50% passing 0.075 mm sieve:

B = 1   for $C_u \leq 2$ or ≥ 8

$\qquad\qquad\qquad\qquad\qquad\qquad\qquad$ (Eq. 2 - 2a)

B = 0.5 C_u for $2 \leq C_u \leq 4$ $\qquad$ (Eq. 2 - 2b)

B = 8/C_u for $4 < C_u < 8$ $\qquad$ (Eq. 2 - 2c)

For ≥ 50% passing 0.075 mm sieve:

B = 1   for wovens

B = 1.8   for nonwovens

and AOS or O_{95} (geotextile) ≤ 0.3 mm $\qquad$ (Eq. 2 - 5)

For nondispersive cohesive soils (PI > 7) use:

AOS or O_{95} ≤ 0.3 mm

If geotextile and soil retained by it can move:

B = 0.5

B. Permeability/Permittivity Criteria

1. Less Critical/Less Severe

$$k_{geotextile} \geq k_{soil} \qquad (Eq.\ 2 - 7a)$$

2. Critical/Severe

$$k_{geotextile} \geq 10\,k_{soil} \qquad (Eq.\ 2 - 8a)$$

3. Permittivity ψ Requirement

$\psi \geq 0.7\ sec^{-1}$ $\qquad\qquad$ for < 15% passing 0.075 mm $\qquad$ [3 - 1a]

$$\psi \geq 0.2 \ \text{sec}^{-1} \qquad \text{for 15 to 50\% passing 0.075 mm} \qquad [3 - 1b]$$

$$\psi \geq 0.1 \ \text{sec}^{-1} \qquad \text{for} > 50\% \ \text{passing 0.075 mm} \qquad [3 - 1c]$$

4. Flow Capacity Requirement

$$q_{\text{geotextile}} \geq (A_t/A_g) \ q_{\text{required}} \qquad \text{(from Eq. 2 - 9)}$$

or

$$(k_{\text{geotextile}}/t) \ h \ A_g \geq q_{\text{required}}$$

where:

q_{required} is obtained from Step 4 (Eq. 15) above.

$k_{\text{geotextile}}/t = \psi = $ permittivity

h = average head in field

A_g = area of fabric available for flow (*e.g.*, if 50% of geotextile covered by flat rocks or riprap, $A_g = 0.5$ total area)

A_t = total area of geotextile

C. Clogging Criteria

1. Less critical/less severe

   a. Perform soil-geotextile filtration tests.

   b. Alternative: From Step 2A obtain D_{15}; then determine minimum pore size requirement, for soils with $C_u > 3$, from

$$O_{95} \geq 3 \ D_{15} \qquad \text{(Eq. 2 - 10)}$$

   c. Other qualifiers

For soils with % passing 0.075 mm	$\geq 5\%$	$\leq 5\%$
Woven monofilament geotextiles, POA $\geq$	4%	10%
Nonwoven geotextiles, Porosity $\geq$	50%	70%

2. Critical/severe

Select geotextiles that meet retention, permeability, and survivability criteria; as well as the criteria in Step 5C.1 above; perform a filtration test.

Suggested filtration test for sandy and silty soils (*i.e.,* $k > 10^{-7}$ m/s) is the gradient ratio test as described in Chapter 1. The hydraulic conductivity ratio test (see Chapter 1) is recommended for fine-grained soils (*i.e.,* $k < 10^{-7}$ m/s), if appropriately modified.

D. Survivability

Select geotextile properties required for survivability from Table 3-1. Add durability requirements if applicable. Don't forget to check for abrasion and check drop height. Evaluate worst case scenario for drop height.

STEP 6. Estimate costs.

Calculate the volume of armor stone, the volume of aggregate and the area of the geotextile. Apply appropriate unit cost values.

Grading and site preparation (LS) _____

Geotextile ($/m^2$) _____

Geotextile placement ($/m^2$) _____

In-place aggregate bedding layer ($/m^2$) _____

Armor stone ($/kg$) _____

Armor stone placement ($/kg$) _____

Total cost _____

STEP 7. Prepare specifications.

Include for the geotextile:

A. General requirements

B. Specific geotextile properties

C. Seams and overlaps

D. Placement procedures

E. Repairs

F. Testing and placement observation requirements

See Sections 1.6 and 3.7 for specification details.

STEP 8. Obtain samples of the geotextile before acceptance.

STEP 9. Monitor installation during construction, and control drop height. Observe erosion control systems during and after significant storm events.

3.5   GEOTEXTILE DESIGN EXAMPLE

DEFINITION OF DESIGN EXAMPLE

- Project Description: Riprap on slope is required to permit groundwater seepage out of slope face, without erosion of slope. See figure for project cross section.

- Type of Structure: small stone riprap slope protection

- Type of Application: geotextile filter beneath riprap

- Alternatives:
  - i)   graded soil filter; or
  - ii)   geotextile filter between embankment and riprap

Project Cross Section

GIVEN DATA

- see cross section
- riprap is to allow unimpeded seepage out of slope
- riprap will consist of small stone (50 to 300 mm)
- stone will be placed by dropping from a backhoe
- seeps have been observed in the existing slope
- soil beneath the proposed riprap is a fine silty sand
- gradations of two *representative* soil samples

SIEVE SIZE (mm)	PERCENT PASSING, BY WEIGHT	
	Sample A	Sample B
4.75	100	100
1.68	96	100
0.84	92	98
0.42	85	76
0.15	43	32
0.075	25	15
0.037	3	0

DEFINE

A. Geotextile function(s)

B. Geotextile properties required

C. Geotextile specification

SOLUTION

A. Geotextile function(s):
   Primary   -   filtration
   Secondary   -   separation

B. Geotextile properties required:
   apparent opening size (AOS)
   permittivity
   survivability

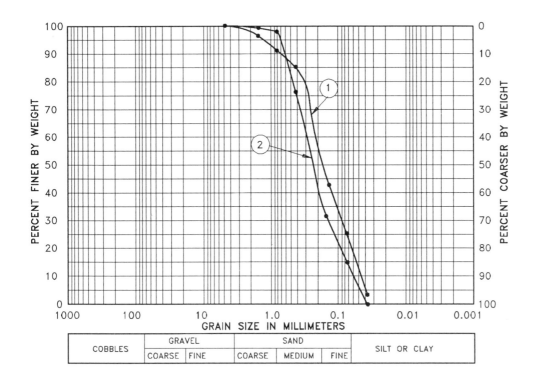

Grain Size Distribution Curve

DESIGN

STEP 1. EVALUATE CRITICAL NATURE AND SITE CONDITIONS.

From given data, this is a critical application due to potential for loss of life and potential for significant structural damage.

Soils are well-graded, hydraulic gradient is low for this type of application, and flow conditions are steady state.

STEP 2. OBTAIN SOIL SAMPLES.

A. GRAIN SIZE ANALYSES

Plot gradations of representative soils. The D_{60}, D_{10}, and D_{85} sizes from the gradation plot are noted in the table below for Samples A and B.

Soil Sample	$D_{60} \div D_{10} = C_u$	$B =$	$B \times D_{85} \geq$ AOS (mm)
A	$0.20 \div 0.045 = 4.4$	$8 \div C_u = 8 \div 4.4 = 1.82$	$1.82 \times 0.44 = 0.8$
B	$0.30 \div 0.06 = 5$	$8 \div C_u = 8 \div 5 = 1.6$	$1.6 \times 0.54 = 0.86$

Worst case soil for retention is Soil A, with D_{85} equal to 0.44 mm.

B. PERMEABILITY TESTS

This is a critical application and soil permeability tests should be conducted. An estimated permeability will be used for preliminary design purposes.

STEP 3. EVALUATE ARMOR MATERIAL AND PLACEMENT.

A. Small stone (50 to 300 mm) riprap will be used.

B. A placement drop of less than 1 m will be specified.

STEP 4. CALCULATE ANTICIPATED FLOW THROUGH SYSTEM.
Flow computations are not included within this example. The entire height of the slope face will be protected, to add to conservatism of design.

STEP 5. DETERMINE GEOTEXTILE REQUIREMENTS.

A. RETENTION

$$AOS < B \, D_{85} \hspace{3cm} \text{(Eq. 2 - 1)}$$

Determine uniformity coefficient, C_u, coefficient B, and the maximum AOS. Sample A controls (see table above), therefore, **AOS ≤ 0.8 mm**

B. PERMEABILITY/PERMITTIVITY

This is a critical application, therefore,

$$k_{geotextile} \geq 10 \, x \, k_{soil}$$

Estimate permeability (after Hazen's formula, which is for clean sands), for preliminary design,

$$k \approx (D_{10})^2$$

where:

k = approximate soil permeability (cm/sec); and

D_{10} is in mm.

$$k_{soil} = 2.0 \, (10)^{-3} \text{ cm/sec for Sample A}$$
$$3.6 \, (10)^{-3} \text{ cm/sec for Sample B}$$

Therefore (with rounding the number), **$k_{geotextile} \geq 4 \, (10)^{-2}$ cm/sec**

Since 15% to 25% of the soil to be protected is finer than 0.075 mm, from Table 3-1: $\psi_{geotextile} \geq 0.2 \text{ sec}^{-1}$

C. CLOGGING

As the project is critical, a filtration test is recommended to evaluate clogging potential. Select geotextile(s) meeting retention, permeability, survivability criteria, and the following qualifiers. Run filtration test (*e.g.,* gradient ratio) and prequalify materials or test representative materials to confirm compatibility.

Minimum Opening Size Qualifier (for $C_u > 3$): $O_{95} \geq 3\,D_{15}$

$O_{95} \geq$ $3 \times 0.057 = 0.17$ mm for Sample A
           $3 \times 0.079 = 0.24$ mm for Sample B

Sample A controls, therefore, $O_{95} \geq 0.17$ mm

Other Qualifiers, since greater than 5% of the soil to be protected is finer than 0.075 mm, from Table 3-1:

for Nonwovens -   Porosity > 50 %
for Wovens -   POA (Percent Open Area) > 4 %

D. SURVIVABILITY

A *Moderate Survivability*, Class 2 geotextile may suffice, but a *High Survivability*, Class 1 will be specified because this a critical application. Effect on project cost is minor. Therefore, from Table 3-1, the following **minimum** values will be specified:

For Survivability, the geotextile shall have the following minimum values (MARV) -

Grab Strength	900 N
Elongation	15 %
Seam Strength	810 N
Puncture Strength	350 N
Burst Strength	1700 N
Trapezoid Tear	350 N
Ultraviolet Degradation	70 % strength retained at 500 hours

Complete Steps 6 through 9 to finish design.

STEP 6. ESTIMATE COSTS.

STEP 7. PREPARE SPECIFICATIONS.

STEP 8. COLLECT SAMPLES.

STEP 9. MONITOR INSTALLATION, AND DURING & AFTER STORM EVENTS.

3.6 GEOTEXTILE COST CONSIDERATIONS

The total cost of a riprap-geotextile revetment system will depend on the actual application and type of revetment selected. The following items should be considered:
1. grading and site preparation;
2. cost of geotextile, including cost of overlapping and pins versus cost of sewn seams;
3. cost of placing geotextile, including special considerations for below-water placement;
4. bedding materials, if required, including placement;
5. armor stone, concrete blocks, sand bags, etc.; and
6. placement of armor stone (dropped versus hand- or machine-placed).

For Item No. 2, cost of overlapping includes the extra material required for the overlap, cost of pins, and labor considerations versus the cost of field and/or factory seaming, plus the additional cost of laboratory seam testing. These costs can be obtained from manufacturers, but typical costs of a sewn seam are equivalent to 1 to 1.5 m^2 of geotextile. Alternatively, the contractor can be required to supply the cost on an area covered or *in-place* basis. For example, current U.S. Army Corps of Engineers Specifications CW-02215 (1977) require measurement for payment for geotextiles in streambank and slope protection to be on an *in-place* basis without allowance for any material in laps and seams. Further, the unit price includes furnishing all plant, labor, material, equipment, securing pins, etc., and performing all operations in connection with placement of the geotextile, including prior preparation of banks and slopes. Of course, field performance should also be considered, and sewn seams are generally preferred to overlaps.

Items 2, 4, and 6 can be compared with respect to using Moderate Survivability versus High Survivability (Table 3-1, Section IV) geotextiles based on the cost of bedding materials and placement of armor stone.

To determine cost effectiveness, benefit-cost ratios should be compared for the riprap-geotextile system versus conventional riprap-granular filter systems or other available alternatives of equal technical feasibility and operational practicality. Average cost of geotextile protection systems placed above the water level, including slope preparation, geotextile cost of seaming or securing pins, and placement is approximately $3.00-6.00 per square meter, excluding the armor stone. Cost of placement below water level can vary considerably depending on the site conditions and the contractor's experience. For below-water placement, it is recommended that prebid meetings be held with potential contractors to explore ideas for placement and discuss anticipated costs.

3.7 GEOTEXTILE SPECIFICATIONS

In addition to the general recommendations concerning specifications in Chapter 1, erosion control specifications must include construction details (see Section 3.8), as the appropriate geotextile will depend on the placement technique. In addition, the specifications should require the contractor to demonstrate through trial sections that the proposed riprap placement technique will not damage the geotextile.

Many erosion control projects may be better-served by performance-type filtration tests that provide an indication of long-term performance. Thus, in many cases, *approved list*-type specifications, as discussed in Section 1.6, may be appropriate. To develop the list of approved geotextiles, filtration studies (as suggested in Section 3.4, Step 4) should be performed using problem soils and conditions that exist in the localities where geotextiles will be used. An approved list for each condition should be established. In addition, geotextiles should be classified as High or Moderate Survivability geotextiles, in accordance with the index properties listed in Table 3-1 and construction conditions.

The following is a sample specification developed by the AASHTO-AGC-ARTBA Task Force 25. It includes the requirements discussed in Section 1.6 for a good specification. As with the guide specification presented in Chapter 2, site-specific hydraulic and physical properties must be appropriately selected and included.

AASHTO-AGC-ARTBA TASK FORCE 25
SPECIFICATION GUIDE FOR EROSION CONTROL GEOTEXTILES
(July, 1986)

1. *Description*

    1.1 This work shall consist of furnishing and placing a geotextile for the following erosion control applications: cut and fill slope protection, protection for various small drainage structures and ditches, wave protection for causeways and shore line roadway embankments, and scour protection for structures such as bridge piers and abutments.

The geotextile shall be designed to allow passage of water while retaining in situ soil without clogging. The quantities of erosion control geotextiles as shown on the plans may be increased or decreased at the direction of the Engineer based on construction procedures and actual site conditions that occur during construction of the project. Such variations in quantity will not be considered as alterations in the details of construction or a change in the character of the work.

2. *Materials*

2.1 Fibers used in the manufacture of geotextile, and the threads used in polymers, composed of at least 85% by weight polyolefins, polyesters, or polyamides. They shall be formed into a network such that the filaments or yarns retain dimensional stability relative to each other, including selvedges. These materials shall conform to the performance requirements for soil retention, permeability, and clogging resistance and the physical requirements for constructability, survivability, and durability of Table 3-1 (Section 3.3).

2.2 Geotextile rolls shall be furnished with suitable wrapping for protection against moisture, and extended ultra-violet exposure prior to placement. Each roll shall be labeled or tagged to provide product identification sufficient for inventory and quality control purposes. Rolls shall be stored in a manner which protects them from the elements. If stored outdoors, they shall be elevated and protected with a waterproof cover.

3. *Construction Requirements*

3.1 <u>Geotextile Exposure Following Placement</u> - Exposure of geotextiles to the elements between lay down and cover shall be a maximum of 14 days to minimize damage potential.

3.2 <u>Erosion Control Placement</u> - The geotextile shall be placed and anchored on a smooth graded surface approved by the Engineer. The geotextile shall be placed in such a manner that placement of the overlying materials will not excessively stretch or tear the fabric. Anchoring of the terminal ends of the geotextile shall be accomplished through the use of key trenches or aprons at the crest and toe of slope.

NOTE: In certain applications to expedite construction, 0.5 m long anchoring pins placed on 0.5 to 2 m centers depending on the slope of the covered area have been used successfully.

3.2.1 Slope Protection Placement
Successive geotextile sheets shall be overlapped in such a manner that the upstream sheet is placed over the downstream sheet and/or upslope over downslope. In underwater applications, the geotextile and required thickness of backfill material shall be placed the same day. The backfill placement shall begin at the toe and proceed up the slope.

Riprap and heavy stone filling shall not be dropped onto the geotextile from the height of more than 0.3 m. Slope protection and smaller sizes of stone filling shall not be dropped onto the geotextile from a height exceeding 1 m. Any geotextile damaged during placement shall be replaced as directed by the Engineer at the Contractor's expense.

3.3 <u>Seams</u> - The geotextile shall be joined by either sewing or overlapping. All seams shall be subject to the approval of the Engineer.

Overlapped seams shall have a minimum overlap of 0.3 m except where placed underwater where the overlap shall be a minimum of 1 m.

3.4 <u>Repair</u> - A geotextile patch shall be placed over the damaged area and extend 1 m beyond the perimeter of the tear or damage.

4. *Method of Measurement*

4.1 The geotextile shall be measured by the number of square meters computed from the payment lines shown on the plans or from payment lines established in writing by the Engineer. This excludes seam overlaps, but shall include geotextiles used in crest and toe of slope treatments.

4.2 Slope preparation, excavation and backfill, bedding, and cover material are separate pay items.

5. *Basis of Payment*

5.1 The accepted quantities of geotextile shall be paid for per square meter in place.

5.2 Payment will be made under:

Pay Item	Pay Unit
Erosion Control Geotextile	Square Meter

3.8   GEOTEXTILE INSTALLATION PROCEDURES

Construction requirements will depend on specific application and site conditions. Photographs of several installations are shown in Figure 3-1. The following general construction considerations apply for most riprap-geotextile erosion protection systems. Special considerations related to specific applications and alternate riprap designs will follow.

3.8-1 General Construction Considerations

1. Grade area and remove debris to provide smooth, fairly even surface.
   a. Depressions or holes in the slope should be filled to avoid geotextile bridging and possible tearing when cover materials are placed.
   b. Large stones, limbs, and other debris should be removed prior to placement to prevent fabric damage from tearing or puncturing during stone placement.

2. Place geotextile loosely, laid with machine direction in the direction of anticipated water flow or movement.

3. Seam or overlap the geotextile as required.
   a. For overlaps, adjacent rolls of geotextile should be overlapped a minimum of 0.3 m. Overlaps should be in the direction of water flow and stapled or pinned to hold the overlap in place during placement of stone. Steel pins are normally 5 mm diameter, 0.5 m long, pointed at one end, and fitted with 40 mm diameter washers at the other end. Pins should be spaced along all overlap alignments at a distance of approximately 1 m center to center.
   b. The geotextile should be pinned loosely so it can easily conform to the ground surface and *give* when stone is placed.
   c. If seamed, seam strength should equal or exceed the minimum seam requirements indicated in the specification section of Chapter 1.

4. The maximum allowable slope on which a riprap-geotextile system can be placed is equal to the lowest soil-geotextile friction angle for the natural ground or stone-geotextile friction angle for cover (armor) materials. Additional reductions in slope may be necessary due to hydraulic considerations and possible long-term stability conditions. For slopes greater than 2.5 to 1, special construction procedures will be required, including toe berms to provide a buttress against slippage, loose placement of geotextile sufficient to allow for downslope movement, elimination of pins at overlaps, increase in overlap requirements, and possible benching of the slope. Care should be taken not to put irregular wrinkles in the geotextile because erosion channels can form beneath the geotextile.

a.

b.

c.

Figure 3-1   Erosion control installations: a) installation in wave protection revetment;   b) shoreline application;   and c) drainage ditch application.

5. For streambank and wave action applications, the geotextile must be keyed in at the bottom of the slope. If the riprap-geotextile system cannot be extended a few meters above the anticipated maximum high water level, the geotextile should also be keyed in at the crest of the slope. Alternative key details are shown in Figure 3-2.

6. Place revetment (cushion layer and/or riprap) over the geotextile width, while avoiding puncturing or tearing it.
   a. Revetment should be placed on the geotextile within 14 days.
   b. Placement of armor cover will depend on the type of riprap, whether quarry stone, sandbags (which may be constructed of geotextiles), interlocked or articulating concrete blocks, soil-cement filled bags, or other suitable slope protection is used.
   c. For sloped surfaces, placement should always start from the base of the slope, moving up slope and, preferably, from the center outward.
   d. In no case should stone weighing more than 400 N be allowed to roll downslope on the geotextile.
   e. Field trials should be performed to determine if placement techniques will damage the geotextile and to determine the maximum height of safe drop. As a general guideline, for Moderate Survivability geotextiles (Table 3-1) with no cushion layer, height of drop for stones less than 100 kg should be less than 0.3 m. For High Survivability geotextiles (Table 3-1) or Moderate Survivability geotextiles with a cushion layer, height of drop for stones less than 100 kg should be less than 1 m. Stones greater than 100 kg should be placed with no free fall unless field trials demonstrate they can be dropped without damaging the geotextile.
   f. Grading of slopes should be performed during placement of riprap. Grading should not be allowed after placement if it results in stone movement directly on the geotextile.

As previously indicated, construction requirements will depend on specific application and site conditions. In some cases, geotextile selection is affected by construction procedures. For example, if the system will be placed below water, a geotextile that facilitates such placement must be chosen. The geotextile may also affect the construction procedures. For example, the geotextile must be completely covered with riprap for protection from long-term exposure to ultraviolet radiation. Sufficient anchorage must also be provided by the riprap for *weighting* the geotextile in below-water applications. Other requirements related to specific applications are depicted in Figure 3-3 and are reviewed in the following subsections (from Christopher and Holtz, 1985).

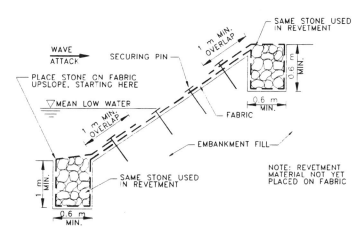

(a) CROSS-SECTION OF REVETMENT AND KEY TRENCHES

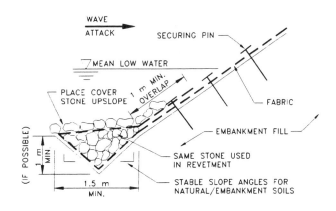

(b) CROSS-SECTION USING KEY TRENCH WHEN SOIL CONDITIONS
DO NOT PERMIT VERTICAL WALL CONSTRUCTION

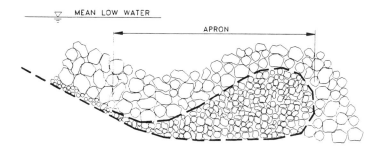

(c) DUTCH METHOD OF TOE DESIGN

Figure 3-2   Construction of hard armor erosion control systems (a., b. after
Keown and Dardeau, 1980; c. after Dunham and Barrett, 1974)

3.8-2 Cut and Fill Slope Protection

Cut and fill slopes are generally protected using an armor stone over a geotextile-type system. Special consideration must be given to the steepness of the slope. After grading, clearing, and leveling a slope, the geotextile should be placed directly on the slope. When possible, geotextile placement should be placed parallel to the slope direction. A minimum overlap of 0.3 m between adjacent roll ends and a minimum 0.3 m overlap of adjacent strips is recommended. It is also important to place the up-slope geotextile over the down-slope geotextile to prevent overlap separation during aggregate placement. When placing the aggregate, do not push the aggregate up the slope against the overlap. Generally, cut and fill slopes are protected with armor stone, and the recommended placement procedures in Section 3.8-1 should be followed.

3.8-3 Streambank Protection

For streambank protection, selecting a geotextile with appropriate clogging resistance to protect the natural soil and meet the expected hydraulic conditions is extremely important. Should clogging occur, excess hydrostatic pressures in the streambank could result in slope stability problems. Do not solve a surface erosion problem by causing a slope stability problem!

Detailed data on geotextile installation procedures and relevant case histories for streambank protection applications are given by Keown and Dardeau (1980). Construction procedures essentially follow the procedures listed in Section 3.8-1. The geotextile should be placed on the prepared streambank with the machine direction placed parallel to the bank (and parallel to the direction of stream flow). Adjacent rolls of geotextile should be seamed, sewed, or overlapped; if overlapped, secure the overlap with pins or staples. A 0.3 m overlap is recommended for adjacent roll edges, with the upstream roll edge placed over the downstream roll edge. Roll ends should be overlapped 1 m and offset as shown in Figure 3-3a. The upslope roll should overlap the downslope roll.

The geotextile should be placed along the bank to an elevation determined to be below mean low water level based on anticipated flow velocities in the stream. Existing agency design criteria for conventional nongeotextile streambank protection could be utilized to locate the toe of the erosion protection system. In the absence of other specifications, placement to a vertical distance of 1 m below mean water level, or to the bottom of the streambed for streams shallower than 1 m, is recommended. Geotextiles should either be placed to the top of the bank or at a given distance up the slope above expected high water level from the appropriate design storm event, including whatever requirements are normally used for conventional (nongeotextile) streambank protection systems. In the absence of other specifications, the geotextile should extend vertically a minimum of 0.5 m above the expected maximum water stage, or at least 1 m beyond the top of the embankment if less than 0.5 m above expected water level.

If strong water movements are expected, the geotextile must be toed in at the top and bottom of the embankment, or the riprap extended beyond the geotextile 0.5 m or more at the toe and the crest of the slope. If scour occurs at the toe and the rocks beyond the geotextile are undermined, they will in effect *toe into* the geotextile. The whole unit thus drops, until the toed-in section is stabilized. However, if the geotextile extends beyond the stone and scour occurs, the geotextile will flap in the water action, causing accelerated formation of a scour pit at the toe. Alternative toe treatments are shown in Figure 3-2. The trench methods in Figures 3-2a and 3-2b require excavating a trench at the toe of the slope. This may be a good alternative for new construction; however, it should be evaluated with respect to slope stability when a trench will be excavated at the toe of a potentially saturated slope below the water level. Keying in at the top can consist of burying the top bank edge of the geotextile in a shallow trench. This will provide resistance to undermining from infiltration of over-the-bank precipitation runoff, and also provide stability should a storm greater than anticipated occur. However, unless excessive quantities of runoff are expected and stream flows are relatively small, this step is usually omitted.

The armoring material (*e.g.*, riprap, sandbags, blocks) must be placed to avoid tearing or puncturing the geotextile, as indicated in Section 3.8-1.

3.8-4   Precipitation Runoff Collection and Diversion Ditches

Runoff drainage from cut slopes along the sides of roads and in the median of divided highways is normally controlled with one or more gravity flow ditches. Runoff from the pavement surface and shoulder slopes are collected and conveyed to drop inlets, stream channels, or other highway drainage structures. If a rock protection-geotextile system is used to control localized ditch erosion problems, select and specify the geotextile using the properties indicated in Table 3-1. Geotextile requirements for ditch linings are less critical than for other types of erosion protection, and minimum requirements for noncritical, nonsevere applications can generally be followed. If care is taken during construction, the protected strength requirements appear reasonable. The geotextile should be sized with AOS to prevent scour and piping erosion of the underlying natural soil and to be strong enough to survive stone placement.

The ditch alignment should be graded fairly smooth, with depressions and gullies filled and large stones and other debris moved from the ditch alignment. The geotextile should be placed with the machine direction parallel to the ditch alignment. Most geotextiles are available in widths of 2 m or more, and, thus, a single roll width of geotextile may provide satisfactory coverage on the entire ditch. If more than one roll width of geotextile is required, sew adjacent rolls together. This can be done by the manufacturer or on site. Again, for seams, the required strength of the seam should meet the minimum seam requirements in Table 3-1. The longitudinal seam produced by roll joining will run parallel with the ditch alignment. Geotextile widths should be ordered to avoid overlaps at the bottom of the ditch, since this is where maximum water velocity occurs. Roll ends should also be sewn or overlapped and pinned or stapled. If overlap is used, then an overlap of at least 1 m is recommended. The upslope roll end should be lapped over the downslope roll end, to retard in-service undermining. Pins or staples should be spaced so slippage will not occur during stone placement or after the ditch is placed in service.

Cover stone, sandbags, or other material intended to dissipate precipitation runoff energy should be placed directly on the geotextile, from downslope to upslope. Cover stone should have sufficient depth and gradation to protect the geotextile from ultraviolet radiation exposure. Again, the stone should be placed with care, especially if the geotextile strength criteria have been reduced to a *less critical* in-service application. A cross section of the proper placement is shown in Figure 3-3c. Vegetative cover can be established through the geotextile and stone cover if openings in the geotextile are sufficient to support growth. If a vegetative cover is desirable, geotextiles should be selected on the basis of the largest opening possible.

3.8-5 Wave Protection Revetments

Because of cyclic flow conditions, geotextiles used for wave protection systems should be selected on the basis of severe criteria, in most cases. Geotextile should be placed in accordance with the procedures listed in Section 3.8-1.

If a geotextile will be placed where existing riprap, rubble, or other materials placed on natural soil have been unsuccessful in retarding wave erosion, site preparation could consist of covering the existing riprap with a filter sand. The geotextile could then be designed with less rigorous requirements as a filter for the sand than if the geotextile is required to filter finer soils.

The geotextile is unrolled and loosely laid on the smooth graded slope. The machine direction of the geotextile should be placed parallel to the slope direction, rather than perpendicular to the slope, as was recommended in streambank protection. Thus, the long axis of the geotextile strips will be parallel to anticipated wave action. Sewing of adjacent rolls or overlapping rolls and roll ends should follow the steps described in Section 3.8-1, except that a 1 m overlap distance is recommended by the Corps of Engineers for underwater placement (Figure 3-2). Again, securing pins (requirements per Section 3.8-1) should be used to hold the geotextile in place.

If a large percentage of geotextile is to be placed below the existing tidal level, special fabrication and placement techniques may be required. It may be advantageous to pre-sew the geotextile into relatively large panels and pull the prefabricated panels downslope, anchoring them below the waterline. Depending upon the placement scheme used, selection of a floating or nonfloating geotextile may be advantageous.

Because of potential wave action undermining, the geotextile must be securely toed-in using one of the schemes shown in Figure 3-2. Also, a key trench should be placed at the top of the bank, as shown in Figure 3-2a, to prevent revetment stripping should the embankment be overtopped by wave action during high-level storm events.

Riprap or cover stone should be placed on the geotextile from downslope to upslope, and stone placement techniques should be designed to prevent puncturing or tearing of the geotextile. Drop heights should follow the recommendations stated in the general construction criteria section (Section 3.8-1).

3.8-6 Scour Protection

Scour, because of high stream flow around or adjacent to structures, generally requires scour protection for structures. Scour protection systems generally fall under the critical and/or severe design criteria for geotextile selection.

An extremely wide variety of transportation-associated structures are possible and, thus, numerous ways exist to protect such structures with riprap geotextile systems. A typical application is shown in Figure 3-3d. In all instances, the geotextile is placed on a smoothly graded surface as stated in the general construction requirements. Such site preparation may be difficult if the geotextile will be placed underwater, but normal stream action may provide a fairly smooth stream bed. In bridge pier protection or culvert approach and discharge channel protection applications, previous high-velocity stream flow may have scoured a depression around the structure. Depressions should be filled with granular cohesionless material. It is usually desirable to place the geotextile and riprap in a shallow depression around bridge piers to prevent unnecessary constriction of the stream channel.

The geotextile should normally be placed with the machine direction parallel to the anticipated water flow direction. Seaming and/or overlapping of adjacent rolls should be performed as recommended in general construction requirements (Section 3.8-1). When roll ends are overlapped, the upstream ends should be placed over the downstream end. As necessary and appropriate, the geotextile may be secured in place with steel pins, as previously described. Securing the geotextile in the proper position may be of extreme importance in bridge pier scour protection. However, under high-flow velocities or under deep water, it will be difficult, if not impossible, to secure the geotextile with steel pins alone. Underwater securing methods must then be developed, and they will be unique for each project. Alternative methods include floating the geotextile into place, then filling from the center outward with stones, building a frame to which the geotextile can be sewn; using a heavy frame to submerge and anchor the geotextile; or constructing a light frame, then floating the geotextile and sinking it with riprap. In any case, it may be desirable to specify a geotextile which will either float or sink, depending upon the construction methods chosen. This can be based on a bulk density criteria for the geotextiles (*i.e.*, bulk density greater than 1 g/cm^3 will sink and less than 1 g/cm^3 will float).

Riprap and/or bedding material, precast concrete blocks, or other elements to be placed on the geotextile should be placed without puncturing or tearing the geotextile. Drop heights should be selected on the basis of geotextile strength criteria, as discussed in the general construction requirements (Section 3.8-1).

3.9   GEOTEXTILE FIELD INSPECTION

In addition to the general field inspection checklist presented in Table 1-4, the field inspector should pay close attention to construction procedures. If significant movement (greater than 0.15 m) of stone riprap occurs during or after placement, stone should be removed to inspect overlaps and ensure they are still intact. As indicated in Section 3.8, field trials should be performed to demonstrate that placement procedures will not damage the geotextile. If damage is observed, the engineer should be contacted, and the contractor should be required to change the placement procedure.

For below-water placement or placement adjacent to structures requiring special installation procedures, the inspector should discuss placement details with the engineer, and inspection requirements and procedures should be worked out in advance of construction.

3.10   GEOTEXTILE SELECTION CONSIDERATIONS

To enhance system performance, special consideration should be given to the type of geotextile chosen for certain soil and hydraulic conditions. The considerations listed in Section 2.10 also apply to erosion control systems. Special attention should be given to gap-graded soils, silts with sand seams, and dispersive clays. In certain situations, multiple filter layers may be appropriate. These consist of a sand layer over the soil, with the geotextile designed as a sand filter only and with sufficient size and number of

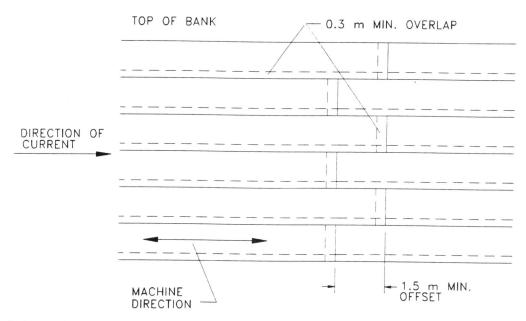

TOP OF BANK

0.3 m MIN. OVERLAP

DIRECTION OF
CURRENT

MACHINE
DIRECTION

1.5 m MIN.
OFFSET

(a) ELEVATION OF STREAMBANK REVETMENT

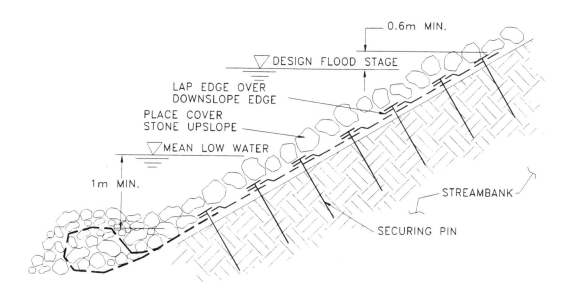

0.6m MIN.

DESIGN FLOOD STAGE

LAP EDGE OVER
DOWNSLOPE EDGE

PLACE COVER
STONE UPSLOPE

MEAN LOW WATER

1m MIN.

STREAMBANK

SECURING PIN

(b) CROSS-SECTION OF STREAMBANK REVETMENT

Figure 3-3   Special construction requirements related to specific hard armor
erosion control applications.

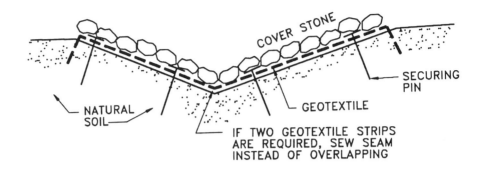

(c) CROSS-SECTION OF GEOTEXTILE LINED DITCH

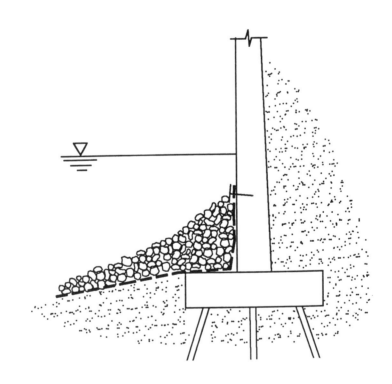

(d) SCOUR PROTECTION FOR ABUTMENTS

Figure 3-3   Special construction requirements related to specific hard armor erosion control applications (cont.).

openings to allow any fines that reach the geotextile to pass through it.

Another special consideration for erosion control applications relates to preference toward felted versus slick geotextiles on steep slope sections. In any case, for steep slopes, the potential for riprap to slide on the geotextile must be assessed either through field trials or laboratory tests.

3.11   EROSION CONTROL MATS

In unlined areas where water can flow, the earth surface is susceptible to erosion by high-velocity flow. Where flow is intermittent, a grass cover will provide protection against erosion. By reinforcing the grass cover, the resulting composite armor layer will enhance the erosion resistance. Geosynthetic erosion control *mats* are made of synthetic meshes and webbings that reinforce the vegetation root mass to provide tractive resistance to high water velocities (*e.g.*, 6 m/s). *Mats* are used within this manual to describe geosynthetics for permanent erosion control applications, and *blankets* (see Chapter 4) are used to describe geosynthetics used in temporary applications (*i.e.*, until vegetation is established).

The three-dimensional erosion control mats retain soil, moisture, and seed, and thus promote vegetative growth. The principal applications of reinforced grass are in highway stormwater runoff ditches, steep waterways such as auxiliary spillways on dams, and protection of embankments against erosion by heavy precipitation or flooding events. Reinforced grass is used for temporary (*e.g.*, 2 hours), high-velocity flow areas, and not for permanent or long-term flow applications suited for hard armor systems. These systems have been found very effective in preventing erosion of the steep face of reinforced slopes (Chapter 8).

This section provides the general design and construction procedures and principles for grass systems reinforced with erosion control mats. The information contained in this section along with additional details pertaining to planning, design, specifications, construction, on-going management, and support research, are contained in Hewlett et al. (1987).

The performance of reinforced grass is determined by a complex interaction of the constituent elements. At present, these physical processes, and the engineering properties of geotextiles and grass, cannot be fully described in quantitative terms. Thus, the design approach is largely empirical and involves a systematic consideration of each constituent element's behavior under service conditions, and how engineering properties can be effectively, yet safely, utilized. Specific products have been tested in laboratory flume tests to empirically quantify the tractive shear forces and velocities they can withstand as a function of flow time.

3.11-1 Planning

The planning stage involves assessing the feasibility of constructing a reinforced grass system in a particular situation and establishing the basic design parameters. The following points should be considered at this stage:

- overall concept of the waterway, and frequency and duration of flow;
- risk (acceptability of failure);
- design discharge and hydraulic loading;
- properties of subsoil;
- *dry* usage in normal no-flow conditions (*e.g.*, agricultural or amenity use, risk of vandalism);
- maintenance ability and requirements of the owner;
- appearance;
- capital and maintenance costs;
- access to site and method of construction;
- climate; and
- strategy for design, specification, construction, and future maintenance.

Any reinforced grass waterway will require an inspection and maintenance strategy different from that for conventionally lined waterways. Grass requires management, and some of the materials involved are more readily susceptible to damage, particularly by vandalism. If it is apparent at this stage that these considerations cannot be accommodated, then reinforced grass should not be used. However, the aesthetic advantages of a *soft* armor lining of reinforced grass usually outweighs potential disadvantages.

3.11-2 Design Procedure

Once the feasibility of constructing a reinforced grass waterway has been established, the detailed design can proceed. This will involve consideration of the hydraulic, geotechnical, and botanical aspects of the project. See by Hewlett, et. al. 1987, for other details.

Hydraulic Design: The main hydraulic design parameters are the velocity and duration of flow, as well as the erosion resistance of various armor layers.

The recommended hydraulic design procedure is as follows:

1. Choose the design hydrograph or overtopping condition. The consequences of waterway failure should be considered. Generally, grassed slopes can be

considered where the overtopping discharge intensity is less than 0.005 m³/s/m.

2. Consider various engineering options for the proposed waterway, with particular reference to topography of the site. A site survey may be required if sufficient topographical information is not available. These options may relate to either general overtopping or construction of a purpose-made channel. Channel widths, slopes downstream of the crest, and, where appropriate, alternative weir lengths and crest levels may be considered.

3. If a reservoir is involved, carry out a flood routing calculation for each option. If a spillway is involved, check that the freeboard is adequate (including any allowance for waves). The operation frequency of the waterway should then be apparent. Modify the layout accordingly if occurrence of flow is more or less frequent than desired. The effect of waves and spray on areas adjacent to the waterway, along with the potential effect of the works on the area downstream, should be considered.

4. A variety of engineering options may be suitable at the site. The detailed hydraulics of each option should be investigated using the following procedure:

   (i) Select an armor layer and a hydraulic roughness "n" value from Figure 3-4.

   (ii) Solve Manning's equation by trial and error for design flow or discharge intensity, using different depths of flow to determine the velocity. (Manning's equation is commonly used in civil engineering applications to estimate the velocity and depth of flow in open channels.)

$$V = \frac{R^{2/3} \ S^{1/2}}{n}$$

where:

V = mean velocity of flow (m/s)

R = hydraulic radius (m) which equals cross-sectional area of flow divided by wetted perimeter

S = slope of the energy line

n = Manning's roughness coefficient (Figure 3-4)

Alternative forms of the equation for discharge and discharge intensity in a wide channel, respectively, are:

$$Q = \frac{A \ R^{2/3} \ S^{1/2}}{n}$$

$$q = \frac{d^{5/3} \ S^{1/2}}{n}$$

where:

Q   =   discharge (m³/s)

A   =   area of flow (m²)

q   =   discharge per unit width of channel (m³/s/m)

d   =   depth of flow (m)

A channel may be considered to be *hydraulically* wide when velocity in the center of the channel is not affected by friction at the sides. In supercritical flow, this may require a channel width of up to 10 times the depth of flow.

When uniform flow conditions have developed (*i.e.*, terminal velocity is reached), the energy slope, S, equals the slope of the channel bed. Depth of uniform flow conditions is referred to as *normal depth*.

On steep slopes, the terminal velocity and normal *blackwater* depth calculated using Manning's equation will normally be achieved. The normal *blackwater* depth may be converted to *whitewater* using the air voids ratio. For water flow with a relatively small head loss between upstream and downstream energy levels, normal depth may not be reached; a step-by-step method should be used to determine the depth of flow and maximum velocity.

(iii) Compare this velocity with the recommended velocity for the armor layer from Figure 3-5. If the recommended velocity is exceeded, it may be possible to decrease the discharge intensity or select a more erosion-resistant armor layer. If the velocity is less than that recommended, it may be possible to reduce the base width or select a less erosion-resistant armor layer.

5. Determine the tailwater conditions over a range of discharges and consider ways to dissipate energy at the toe of the waterway.

If the tailwater conditions cause a hydraulic jump to form on the slope (Figure 3-6, Case (a)), it may be advisable to provide heavier armor, stronger restraint, discharge, or anchorage than normally used to protect the waterway from erosion by high-velocity flow. The decision will depend on the energy loss and frequency of occurrence. The critical zone of potential erosion is at the front of the jump. Experience from field trials and embankment overtopping under high tailwater conditions has shown that high-velocity flow zones within the jump generally occur only at the front of the jump and that erosion is consequently restricted.

If Cases (b), (c), or (d) in Figure 3-6 apply, provided the slope reinforcement is terminated in a safe manner, limited erosion may be acceptable. Note that in all cases, the flow velocity decreases downstream of the toe. Erosion protection may be provided — either by continuing the slope reinforcement or by other means (*e.g.*, gabion mattress, rock armor).

If it is necessary to stabilize and contain the hydraulic jump — for example, to accommodate the short-term design discharge — then a control and/or armored stilling basin may be adopted.

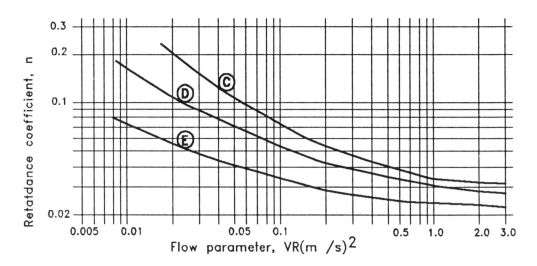

Grass Retardance Catagories

Average grass length	Retardance
150 mm to 250mm	C
50mm to 150mm	D
less than 50mm	E

(a) HYDRAULIC ROUGHNESS OF GRASSES FOR SLOPES FLATTER THAN 1 IN 10

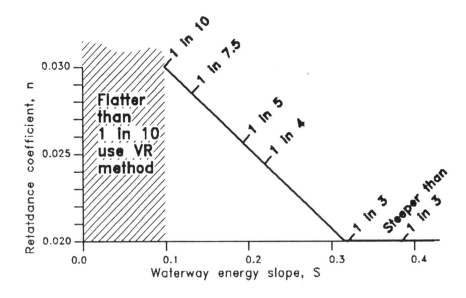

(b) RECOMMENDED RETARDANCE COEFFICIENTS FOR GRASSED SLOPES STEEPER THAN 1 ON 10

Figure 3-4   *Roughness and retardance coefficients n for grassed slopes (Hewlett et al., 1987).*

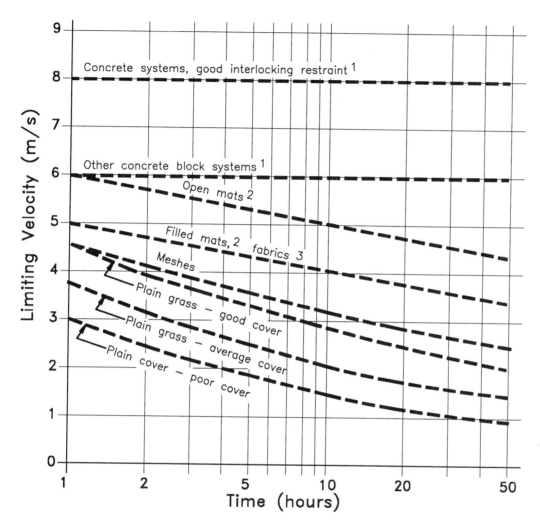

Notes:
1. Minimum superficial mass 135 kg/m^2.
2. Minimum nominal thickness 20 mm.
3. Installed within 20 mm of soil surface, or in conjunction with surface mesh.
4. See text for other criteria for geosynthetic reinforcement.
5. These graphs should only be used for erosion resistance to unidirectional flow. Values are based on available experiance and information as of 1987.
6. All reinforced grass values assume well established, good grass cover.
7. Other criteria (such as short term protection, ease of installation and management, susceptability to vandalism, etc) must be considered in choice of reinforcement.

EROSION RESISTANCE

Figure 3-5   Recommended limiting values for erosion resistance of plain and reinforced grass (Hewlett et al., 1987).

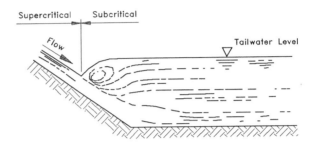

(a) HIGH TAILWATER — HYDRAULIC JUMP ON SLOPE

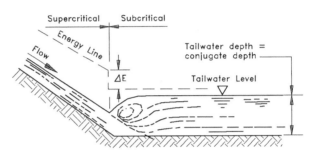

(b) HYDRAULIC JUMP AT TOE

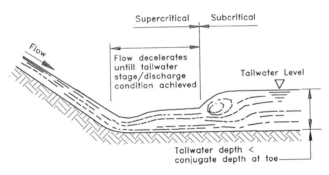

(c) LOW TAILWATER — HYDRAULIC JUMP IN DOWNSTREAM AREA

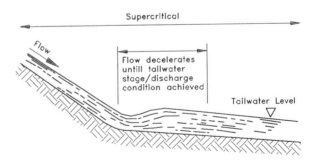

(c) STEEP SLOPE DOWNSTREAM — NO JUMP

Figure 3-6   Possible flow conditions at base of steep waterway (Hewlett et al., 1987).

<u>Geotechnical Considerations</u>: The principal geotechnical consideration is the effect that water entering the embankment (or excavation) will have on the subsoil. The procedure normally followed is listed below.

1. Conduct a site investigation to determine: (a) the nature of the subsoil, and (b) the existing and future groundwater level. This should include an in situ test to determine the rate at which water will infiltrate the subsoil. Samples of soil may be obtained for laboratory testing.
2. Perform laboratory tests on the subsoil to determine the soil strength and consolidation parameters necessary for designing the waterway.
3. Investigate the stability of the slope during normal *dry* conditions, as well as during and immediately following flow.
4. Consider whether any localized drainage should be provided beneath the waterway to provide relief of pore pressures for increased stability.
5. Consider whether there is likely to be any settlement of the subsoil and whether the armor layer is flexible enough to accommodate movement.

<u>Botanical Considerations</u>: Botanical considerations include the choice of grass mixture, and its establishment and management. Consider the following principal points.

1. Obtain samples of soil that will support the grass and carry out physical and chemical tests to determine its suitability.
2. Choose a grass mixture. The principal factors affecting this choice are soil conditions, climate, and management requirements.
3. Decide on the method of sowing and establishment of grass.

<u>Detailing and Specification</u>: A number of detailed points should be considered which combine the hydraulic, geotechnical, and botanical aspects, to complete the design process. These should be included on the drawings or in the specification and are listed below.

1. Anchorage: Requirements for anchorage (a) at the edges of all reinforcement systems and (b) within concrete systems.
2. Shear Restraint: Requirements for additional shear connection between concrete armor layers and the subsoil.
3. Underlayer: If concrete reinforcement is used, specify the details of any underlayer that is to be provided.
4. Crest Details: Complete a detailed design of the waterway or slope crest. The upstream end of the reinforcement system must be designed to avoid the risk of waterway erosion from the upstream area.
5. Channel Details: Cross-sections of the channel should be drawn. Estimate freeboard based on bulked depth of flow. Careful detailing is required at any transition between two or more plane surfaces.
6. Toe Details: Complete a detailed design of the toe of the waterway or slope.

7. Construction Details: Joints in geotextiles and concrete reinforcement, preparation of formation, temporary restraint of geotextile reinforcement, etc.

Details for each of these requirements are in Hewlett, et al. (1987). Remember to:
- check that the waterway will perform satisfactorily;
- produce the construction drawings;
- prepare a specification, including material and acceptance tests; and
- set up a framework for future construction, maintenance, and inspection.

It is important that adequate design and site supervision be exercised at all stages by the client or its representative to ensure that the work is constructed in accordance with good practice.

3.12   REFERENCES

AASHTO (1996) Standard Specification for Geotextiles, Designation: M 288-96 (draft), American Association of State Transportation and Highway Officials, Washington, D.C.

AASHTO (1991), Model Drainage Manual, 1st Ed., American Association of State Highway and Transportation Officials, Washington, D.C.

AASHTO (1990), Standard Specifications for Geotextiles - M 288, Standard Specifications for Transportation Materials and Methods of Sampling and Testing, American Association of State Transportation and Highway Officials, Washington, D.C., pp 689-692.

AASHTO (1990), Guide Specifications and Test Procedures for Geotextiles, Task Force 25 Report, Subcommittee on New Highway Materials, American Association of State Transportation and Highway Officials, Washington, D.C.

ASTM (1994), Soil and Rock (II): D 4393 - latest; Geosynthetics, Annual Book of ASTM Standards, Section 4, Volume 04.09, American Society for Testing and Materials, Philadelphia, PA, 516 p.

Cedergren, H.R. (1989), Seepage, Drainage, and Flow Nets, Third Edition, John Wiley and Sons, New York, 465 p.

Christopher, B.R. and Holtz, R.D. (1985), Geotextile Engineering Manual, Report No. FHWA-TS-86/203, Federal Highway Administration, Washington, D.C., March, 1044 p.

Dunham, J.W. and Barrett, R.J. (1974), Woven Plastic Cloth Filters for Stone Seawalls, Journal of the Waterways, Harbors, and Coastal Engineering Division, American Society of Civil Engineers, New York, February.

Hewlett, H.W.M., Boorman, L.A. and Bramley, M.E., Design of Reinforced Grass Waterways, Report 116, Construction Industry Research and Information Association, London, U.K., 1987, 116 p.

FHWA (1992), Evaluation Scour at Bridges, Hydraulic Engineering Circular No. 18, Federal Highway Administration IP-90-017.

FHWA (1989), Design of Riprap Revetment, Hydraulic Engineering Circular No. 11, Federal Highway Administration.

FHWA (1988), Design of Roadside Channels with Flexible Linings, Hydraulic Engineering Circular No. 15, Federal Highway Administration.

FHWA (1975), Hydraulic Design of Energy Dissipators for Culvert and Channels, Hydraulic Engineering Circular No. 14, Federal Highway Administration.

Keown, M.P. and Dardeau, E.A., Jr. (1980), Utilization of Filter Fabric for Streambank Protection Applications, TR HL-80-12, Hydraulics Laboratory, U.S. Army Engineer Waterways Experiment Station, Vicksburg, MS.

Koerner, R.M., Designing With Geosynthetics, 3rd Edition, Prentice-Hall Inc., Englewood Cliffs, NJ, 1994, 783p.

U.S. Army Corps of Engineers, Civil Works Construction Guide Specification for Plastic Filter Fabric, Corps of Engineer Specifications No. CW-02215, Office, Chief of Engineers, U.S. Army Corps of Engineers, Washington, D.C., 1977.

U.S. Department of the Navy, Foundations and Earth Structures, Design Manual 7.2, Naval Facilities Engineering Command, Alexandria, VA, 1982.

U.S. Department of the Navy, Soil Mechanics, Design Manual 7.1, Naval Facilities Engineering Command, Alexandria, VA, 1982.

4.0   TEMPORARY RUNOFF and SEDIMENT CONTROL

4.1   INTRODUCTION

Geotextiles, geosynthetic erosion control blankets, and other geosynthetic products can be used to temporarily control and minimize erosion and sediment transport during construction. Four specific application areas have been identified:

- Geotextile silt fences can be used as a substitute for hay bales or brush piles to remove suspended particles from sediment-laden runoff water.

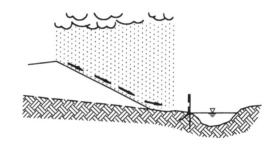

- Geotextiles can be used as a turbidity curtain placed within a stream, lake, or other body of water to retain suspended particles and allow sedimentation to occur.

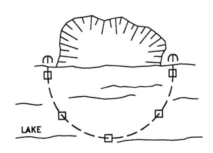

- Special soil retention blankets, made of both natural and synthetic grids, meshes, nets, fibers, and webbings, can be used to provide tractive resistance and resist water velocity on slopes. These products retain seeds and add a mulch effect to promote the establishment of a vegetative cover.

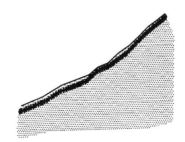

- Geotextiles held in place by pins or riprap can be used to temporarily control erosion in diversion ditches, culvert outfalls, embankment slopes, etc. Alternatively, soil retention blankets can be used for temporary erosion control until vegetation can be established in the ditch.

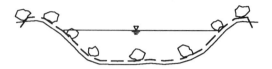

The main advantages of using geosynthetics over conventional techniques in sediment control applications include the following.

- In the case of a silt fence, the geotextile can be designed for the specific application, while conventional techniques are basically designed by trial-and-error.
- Geotextile silt fences in particular often prove to be very cost-effective, especially in comparison to hay bales, considering ease of installation and material costs.
- Control by material specifications is easier.

For runoff control, geosynthetic products are designed to help mitigate immediate erosion problems and provide long-term stabilization by promoting the establishment and sustainment of vegetative cover. The main advantages of using geosynthetics for erosion control applications include the following.

- Vegetative systems have desirable aesthetics.
- Products are lightweight and easy to handle.
- Temporary, degradable products improve establishment of vegetation.
- Continuity of protection is generally better over the entire protected area.
- Empirically predictable performance; traditional techniques such as seeding, mulch covers, and brush or hay bale barriers, are often less reliable.

The following sections review the function, selection specifications, and installation procedures for geosynthetics used as silt fences, turbidity curtains, and erosion control blankets. Design of geotextiles in temporary riprap-geotextile systems to control ditch erosion follows Chapter 3 design guidelines.

4.2   FUNCTION OF SILT FENCES

In most applications, a geotextile silt fence is placed downslope from a construction site or newly graded area to prevent eroded soils from being transported by runoff to the surrounding environment. Sometimes silt fences are used in permanent or temporary diversion ditches for the same purpose.

A silt fence primarily functions as a temporary dam (Mallard and Bell, 1981). It retains water long enough for suspended fine sand and coarse silt particles in the runoff to settle out before they reach the fence. Generally, a retention time of 20 to 25 minutes is sufficient to allow silts to settle out, so flow through the geotextile after the first charge must provide this retention time. Although smaller geotextile pore opening sizes and low permittivity can be selected to allow finer particles to settle out, some water must be able to pass through the fence to prevent possible overtopping of the fence.

Because not all the silt and clay in suspension will settle out before reaching the fence, water flowing through the fence will still contain some fines in suspension. Removal of fines by the geotextile creates a difficult filtration condition. If the openings in the geotextile (*i.e.*, AOS) are small enough to retain most of the suspended fines, the geotextile will blind and its permeability will be reduced so that bursting or overtopping of the fence could occur. Therefore, it is better to have some geotextile openings large enough to allow silt-sized particles to easily pass through. Even if some silt passes through the fence, the flow velocity will be small, and some fines may settle out. If the application is critical, *e.g.,* when the site is immediately adjacent to environmentally sensitive wetlands or streams, multiple silt fences should be used. The second fence is placed a short distance downstream of the first fence to capture silt that passed through the first fence.

In the past, the AOS and permittivity, ψ, have been used to design and specify the filtration requirements of the geotextile. However, Wyant (1980) and Allen (1994) indicate that these geotextile index properties are not directly related to silt fence performance. Experience indicates that, in general, most geotextiles have hydraulic characteristics that provide acceptable silt fence performance for even the most erodible silts (Wyant, 1980; Allen, 1994). Thus, geotextile selection and specification can be based on typical properties of silt fence geotextiles known to have performed satisfactorily in the past, or through the use of performance type tests such as ASTM D 514, Determining Filtering Efficiency and Flow Rate of a Geotextile for Silt Fence Applications Using Site-Specific Soil. Past experience is the basis for the AOS and permittivity values presented later in this chapter.

Most silt fence applications are temporary; the fence only must work until the site can be revegetated or otherwise protected from rainfall and erosion. According to Richardson and Middlebrooks (1991), silt fences are best limited to applications where sheet erosion occurs and where flow is not concentrated, though silt fences can be used in both ditch or swale applications by special design. Flow velocity should be less than about 0.3 m/s. Recommendations for allowable slope length versus slope angle to limit runoff velocity are presented in Table 4-1. Furthermore, the limiting slope angle and velocity requirements suggest that the drainage areas for overland flow to a fence should be less than about 1 ha per 30 m of fence.

Silt fence ends should be turned uphill to ensure they capture runoff water and prevent flow around the ends. The groundline at the fence ends should be at or above the elevation of the lowest portion of the fence top. Measures should be taken to prevent erosion along the fence backs that run downhill for a significant distance. Gravel check dams at approximately 2 to 3 m intervals along the back of the fence can be used.

Table 4-1

Limits of Slope Steepness and Length to Limit Runoff Velocity to 0.3 m/s (after Richardson and Middlebrooks, 1991)

SLOPE STEEPNESS (%)	MAXIMUM SLOPE LENGTH (m)
< 2	30
2 - 5	25
5 - 10	15
10 - 20	10
> 20	5

4.3   DESIGN OF SILT FENCES

4.3-1   Simplified Design Method

This section follows the simplified design method of Richardson and Middlebrooks (1991). See their paper for additional details.

STEP 1.   Estimate runoff volume.

Use the Rational Method (small watershed areas):

$$Q = 2.8 \times 10^{-3} \, C \, i \, A \qquad\qquad [4 - 1]$$

where:

Q = runoff (m³/s)

C = surface runoff coefficient

i = rainfall intensity (mm/hr)

A = area (ha)

Use $C = 0.2$ for rough surfaces, and $C = 0.6$ for smooth surfaces. A 10-year storm event is typically used for designing silt fences.

Use the appropriate rainfall intensity factor, i, for the locality. Assume a 10-year design storm, or use local design regulations. Neglect any concentration times (worst case). This calculation gives the total storage volume required of the silt fence.

STEP 2.   Estimate sediment volume.

Use the SCS (1977) Universal Soil Loss Equation (USLE)

$$A = 2.2 \ R \ K \ L \ S \ C \ P \qquad\qquad [4 - 2]$$

where:

A = soil loss (metric tons/ha/yr)

R = rainfall erosion index

K = soil erodibility factor

LS = slope length and steepness factor

C = vegetative cover factor ($C = 1$ for no cover)

P = erosion control practice factor ($P = 1$ for minimal practice)

Obtain rainfall erosion index from Figure 4-1. Note that this figure is not recommended for the western US (west of the 104° meridian), where locally or regionally mapped design rainfalls must be utilized. Use Figure 4-2 to obtain the values of KLS (limited slope lengths and steepness factors are applicable to most silt fence applications).

The USLE gives an estimate of predicted tons of sediment produced per hectare per year. This provides a reasonable safety factor for a 6-month design period, so sediment volume will not exceed the storage volume available behind the silt fence. Assume a density of about 800 kg/m³ for converting the soil loss in metric tons to a design volume.

STEP 3.   Select geotextile.

A. Hydraulic properties

Because site specific designs for retention and permittivity are not necessary for most soils (at least in a practical sense), use nominal AOS and permittivity values for geotextiles known to perform satisfactorily as silt fences. Suggested values are:

0.15 mm < AOS < 0.60 mm for woven silt films

0.15 mm < AOS < 0.30 mm for all other geotextiles

Permittivity, $\psi > 0.02$ s^{-1}

B. Physical and mechanical properties

The geotextile must be strong enough to support the pooled water and the sediments collected behind the fence. Minimum strength depends on height of impoundment and spacing between fence posts.

Use Figure 4-3 to determine required tensile strength for a range of impoundment heights and post spacings. For unreinforced geotextiles, limit impoundment heights to 0.6 m and post spacing to 2 m; for greater heights and spacings, use steel or plastic grid/mesh reinforcement to prevent burst failure of geotextile. Unsupported geotextiles must not collapse or deform, allowing silt-laden water to overtop the fence. Use Figure 4-4 to design the fence posts.

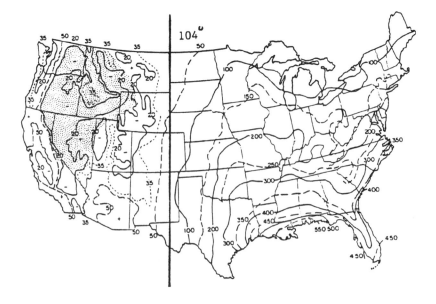

Figure 4-1   *Rainfall erosion index, R (Soil Conservation Service, 1977).*

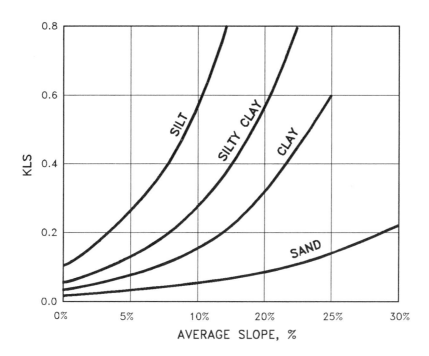

Figure 4-2   Universal soil loss KLS vs slope (Richardson and Middlebrooks, 1991).

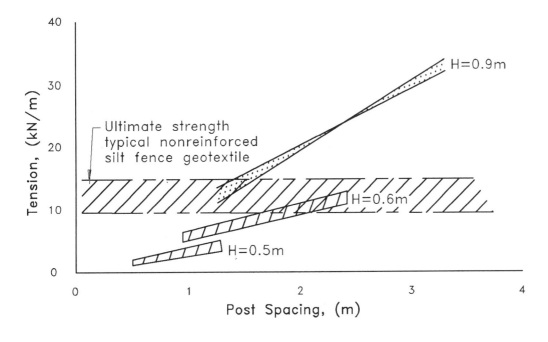

Figure 4-3   Geotextile strength versus post spacing.

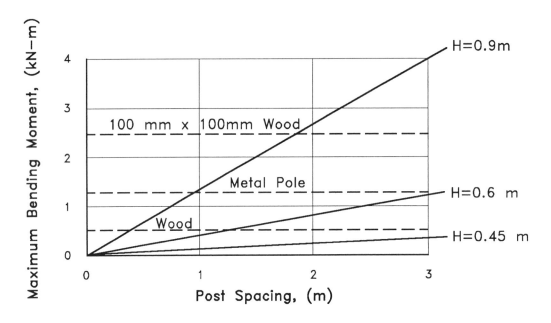

Figure 4-4   Post requirements vs post spacing (Richardson and Koerner, 1990).

4.3-2 Alternate Hydraulic Design Using Performance Tests

An alternate design approach for silt fences uses model studies to estimate filtration efficiency for specific site conditions. This method was developed by Wyant (1980) for the Virginia Highway and Transportation Research Council (VHTRC) and is based on observed field performance and laboratory testing. The procedures for this method are described in ASTM D 5141. The laboratory model consists of a flume with an outflow opening similar to the size of a hay bale and positioned at a fixed slope of 8%. The geotextile is strapped across the end of the flume. A representative soil sample from the site is then suspended in water to a concentration of about 3000 ppm (equivalent water content is 0.3 percent) and poured through the flume. Based on the performance of the geotextile, appropriate geotextiles can be selected to provide filtering efficiencies approximating of 75% or more and flow rates on the order of 0.1 L/min/m^2 after three test repetitions.

The model study approach provides a system performance evaluation by utilizing actual soils from the local area of interest. Thus, it cannot be performed by manufacturers. The approach lends itself to an approved list-type specification for silt fences. In this case, the agency or its representatives perform the flume test using their particular problem soils and prequalifies the geotextiles that meet filtering efficiency and flow criteria requirements. Qualifying geotextiles can be placed on an approved list that is then provided to contractors. Geotextiles on any approved list should be periodically retested because manufacturing changes often occur.

4.3-3 Constructability Requirements

The geotextile used as a silt fence must be strong enough to enable it to be properly installed. AASHTO-AGC-ARTBA Task Force 25 property recommendations are indicated in Table 4-2. Realize that these specifications are not based on research but on properties of existing geotextiles which have performed satisfactorily in silt fence applications. Also given are requirements for resistance to ultraviolet degradation. Although the applications are temporary (*e.g.*, 6 to 36 months), the geotextile must have sufficient UV resistance to function throughout its anticipated design life.

Table 4-2
Physical Requirements[1,2] For Temporary Silt Fence Geotextiles
(Modified from AASHTO, 1990 and 1996)

PROPERTY	TEST METHOD	WIRE FENCE SUPPORTED REQUIREMENTS	SELF SUPPORTED REQUIREMENTS
Tensile Strength(N)	ASTM D 4632	400 minimum[3]	550 minimum
Elongation at 200 N tensile strength	ASTM D 4632	N/A	50% maximum
Permittivity (sec[-1])	ASTM D 4491	0.05	0.05
AOS (mm)	ASTM D 4751	0.60 maximum	0.60 maximum
Ultraviolet Degradation[4]	ASTM D 4355	Minimum 70% strength retained	Minimum 70% strength retained

NOTES
1. Acceptance of geotextile material to be based on ASTM D 4759.
2   Contracting agency may require a letter from the supplier certifying that its geotextile meets specification requirements.
3. Minimum -- use value in weaker principal direction. All numeric values represent minimum (maximum) average roll values (*i.e.*, test results from any sampled roll in a lot shall meet or exceed the minimum (maximum) values in the table). Values are for less critical/less severe conditions. Lot should be sampled according to ASTM D 4534.
4. Strength retained after 500 hours of ultraviolet exposure when tested according to ASTM D 4355. This method specifies tensile testing of a 50 mm strip (or ravelled strip) for both control and exposed samples.

4.4   SPECIFICATIONS

The following specifications were developed by the Washington State Department of Transportation in 1994 and are included herein for your reference. They are meant to serve as guidelines for selecting and installing of geotextiles for routine (less critical) projects. They are not intended to replace site-specific evaluation, testing, and design.

WASHINGTON STATE DEPARTMENT OF TRANSPORTATION
MATERIALS LABORATORY
TUMWATER, WA
GEOTEXTILE FOR SILT FENCE
1994

Description

The Contractor shall furnish and place construction geotextile for silt fence in accordance with the details shown in the Plans.

Materials

Geotextile and Thread for Sewing

The material shall be a geotextile consisting only of long chain polymeric fibers or yarns formed into a stable network such that the fibers or yarns retain their position relative to each other during handling, placement, and design service life. At least 85 percent by weight of the material shall be polyolefins or polyesters. The material shall be free from defects or tears. The geotextile shall also be free of any treatment or coating which might adversely alter its hydraulic or physical properties after installation. The geotextile shall conform to the properties as indicated in Table 1.

Thread used for sewing shall consist of high strength polypropylene, polyester, or polyamide. Nylon threads will not be allowed. The thread used to sew permanent erosion control geotextiles must also be resistant to ultraviolet radiation.

Table 1:
Geotextile Property Requirements[1]
for Temporary Silt Fence

Geotextile Property	ASTM Test Method[2]	Unsupported Between Posts	Supported Between Posts with Wire or Polymeric Mesh
AOS	D 4751	0.60 mm max. for slit film wovens 0.30 mm max. for other geotextiles 0.15 mm min.	0.60 mm max. for slit film wovens 0.30 mm max. for other geotextiles 0.15 mm min.
Water Permittivity	D 4491	0.02 sec^{-1} min.	0.02 sec^{-1} min.
Grab Tensile Strength, min. in MD and CMD	D 4632	800 N min. in MD 450 N min. in CMD	450 N min.
Grab Failure Strain, min. in MD only	D 4632	30% max. at 800 N or more	---
Ultraviolet (UV) Radiation Stability	D 4355	70% Strength Retained min., after 500 hr in weatherometer	70% Strength Retained min., after 500 hr in weatherometer

NOTES
1. All geotextile properties in Table 1 are minimum average roll values (*i.e.,* the test result for any sampled roll in a lot shall meet or exceed the values shown in the table).
2. The test procedures used are essentially in conformance with the most recently approved ASTM geotextile test procedures, except for geotextile sampling and specimen conditioning, which are in accordance with WSDOT Test Methods 914 and 915, respectively. Copies of these test methods are available at the Headquarters Materials Laboratory in Tumwater.

Geotextile Approval and Acceptance

Source Approval

The Contractor shall submit to the Engineer the following information regarding each geotextile proposed for use:

Manufacturer's name and current address,

Full product name,

Geotextile structure, including fiber/yarn type, and

Proposed geotextile use(s).

If the geotextile source has not been previously evaluated, a sample of each proposed geotextile shall be submitted to the Headquarters Materials Laboratory in Tumwater for evaluation. After the sample and required information for each geotextile type have arrived at the Headquarters Materials Laboratory in Tumwater, a maximum of 14 calendar days will be required for this testing. Source approval will be based on conformance to the applicable values from Tables 1 through 6. Source approval shall not be the basis of acceptance of specific lots of material unless the lot sampled can be clearly identified and the number of samples tested and approved meet the requirements of WSDOT Test Method 914.

Geotextile Samples for Source Approval

Each sample shall have minimum dimensions of 1.5 meters by the full roll width of the geotextile. A minimum of 6 square meters of geotextile shall be submitted to the Engineer for testing. The geotextile machine direction shall be marked clearly on each sample submitted for testing. The machine direction is defined as the direction perpendicular to the axis of the geotextile roll. Source approval for temporary silt fences will be by manufacturer's certificate of compliance as described under "Acceptance Samples."

The geotextile samples shall be cut from the geotextile roll with scissors, sharp knife, or other suitable method which produces a smooth geotextile edge and does not cause geotextile ripping or tearing. The samples shall not be taken from the outer wrap of the geotextile roll nor the inner wrap of the core.

Acceptance Samples

Samples will be randomly taken by the Engineer at the job site to confirm that the geotextile meets the property values specified.

Approval will be based on testing of samples from each lot. A "lot" shall be defined for the purposes of this specification as all geotextile rolls within the consignment (*i.e.*, all rolls sent to the project site) which were produced by the same manufacturer during a continuous period of production at the same manufacturing plant and have the same product name. After the samples and manufacturer's certificate of compliance have arrived at the Headquarters Materials Laboratory in Tumwater, a maximum of 14 calendar days will be required for this testing. If the results of the testing show that a geotextile lot, as defined, does not meet the properties required for the specified use as indicated in Tables 1 through 6 the roll or rolls which were sampled will be rejected. Two additional rolls for each roll tested which failed from the lot previously tested will then be selected at random by the Engineer for sampling and retesting. If the retesting shows that any of the additional rolls tested do not meet the required properties, the entire lot will be rejected. If the test results from all the rolls retested meet the required properties, the entire lot minus the roll(s) which failed will be accepted. All geotextile which has defects, deterioration, or damage, as determined by the Engineer, will also be rejected. All rejected geotextile shall be replaced at no cost to the State.

Acceptance by Certificate of Compliance

When the quantities of geotextile proposed for use in each geotextile application are less than or equal to the following amounts, acceptance shall be by Manufacturer's Certificate of Compliance:

Application: Temporary Silt Fence   Geotextile Quantities: All quantities

The manufacturer's certificate of compliance shall include the following information about each geotextile roll to be used:

Manufacturer's name and current address,
Full product name,
Geotextile structure, including fiber/yarn type
Geotextile roll number,
Proposed geotextile use(s), and
Certified test results.

Approval of Seams

If the geotextile seams are to be sewn in the field, the Contractor shall provide a section of sewn seam before the geotextile is installed which can be sampled by the Engineer.

The seam sewn for sampling shall be sewn using the same equipment and procedures as will be used to sew the production seams. If production seams will be sewn in both the machine and cross-machine directions, the Contractor must provide sewn seams for sampling which are oriented in both the machine and cross-machine directions. The seams sewn for sampling must be at least 2 meters in length in each geotextile direction. If the seams are sewn in the factory, the Engineer will obtain samples of the factory seam at random from any of the rolls to be used. The seam assembly description shall be submitted by the Contractor to the Engineer and will be included with the seam sample obtained for testing. This description shall include the seam type, stitch type, sewing thread type(s), and stitch density.

Construction Geotextile (Installation Requirements)

Description

The Contractor shall furnish and place construction geotextile in accordance with the details shown in the Plans.

Identification, Shipment and Storage

Geotextile roll identification, storage, and handling shall be in conformance to ASTM D 4873. During periods of shipment and storage, the geotextile shall be kept dry at all times and shall be stored off the ground. Under no circumstances, either during shipment or storage, shall the material be exposed to sunlight, or other form of light which contains ultraviolet rays, for more than five calendar days.

Installation

The Contractor shall be fully responsible to install and maintain temporary silt fences at the locations shown in the Plans. A silt fence shall not be considered temporary if the silt fence must function beyond the life of the contract. The silt fence shall prevent soil carried by runoff water from going beneath, through, or over the top of the silt fence, but shall allow the water to pass through the fence. The minimum height of the top of the silt fence shall be 600 mm and the maximum height shall be 750 mm above the original ground surface. Damaged or otherwise improperly functioning portions of silt fences shall be repaired or replaced by the Contractor at no expense to the Contracting Agency, as determined by the Engineer.

The geotextile shall be attached on the up-slope side of the posts and support system with staples, wire, or in accordance with the manufacturer's recommendations. The staples or wire shall be installed through or around a 13 mm thick wood lath placed against the geotextile at the fence posts, or other method approved by the Engineer, to reduce potential for geotextile tearing at the staples or wire. Silt fence back-up support for the geotextile in the form of a wire or plastic mesh is optional, depending on the properties of the geotextile selected for use in Table 6. If wire or plastic back-up mesh is used, the mesh shall be fastened securely to the up-slope of the posts with the geotextile being up-slope of the mesh back-up support.

The geotextile shall be sewn together at all edges at the point of manufacture, or at an approved location as determined by the Engineer, to form geotextile lengths and widths as required. Alternatively, a geotextile seam may be formed by folding the geotextile from each geotextile section over on itself several times and firmly attaching the folded seam to the fence post, provided that the Contractor can demonstrate, to the satisfaction of the Engineer, that the folded geotextile seam can withstand the expected sediment loading.

The geotextile at the bottom of the fence shall be buried in a trench to a minimum depth of 150 mm below the ground surface. The trench shall be backfilled and the soil tamped in place over the buried portion of the geotextile as shown in the Plans such that no flow can pass beneath the fence nor scour occur. When wire or polymeric back-up support mesh is used, the wire or polymeric mesh shall extend into the trench a minimum of 80 mm. The fence posts shall be placed or driven a minimum of 600 mm into the ground. A minimum depth of 300 mm will be allowed if topsoil or other soft subgrade soil is not present, and the minimum depth of 600 mm cannot be reached. Fence post depths shall be increased by 150 mm if the fence is located on slopes of 3:1 or steeper and the slope is perpendicular to the fence. If the required post depths cannot be obtained, the posts shall be adequately secured by bracing or guying to prevent overturning of the fence due to sediment loading, as approved by the Engineer.

Silt fences shall be located on contour as much as possible, except at the ends of the fence, where the fence shall be turned uphill such that the silt fence captures the runoff water and prevents water from flowing around the end of the fence as shown in the Plans. If the fence must cross contours, with the exception of the ends of the fence, gravel check dams placed perpendicular to the back of the fence shall be used to minimize concentrated flow and erosion along the back of the fence. The gravel check dams shall be approximately 0.3 m deep at the back of the fence and be continued perpendicular to the fence at the same elevation until the top of the check dam intercepts the ground surface behind the fence as shown in the Plans. The gravel check dams shall consist of Crushed Surfacing Base Course (Section 9-03.9(3)), Gravel Backfill for Walls (Section 9-03.12(2)), or Shoulder Ballast (Section 9-03.9(2)). The gravel check dams shall be located every 3 m along the fence where the fence must cross contours. The slope of the fence line where contours must be crossed shall not be steeper than 3:1.

Either wood or steel posts shall be used. Wood posts shall have minimum dimensions of 40 mm by 40 mm by the minimum length shown in the Plans, and shall be free of defects such as knots, splits, or gouges. Steel posts shall consist of either size No. 8 rebar or larger, or shall consist of ASTM A 120 steel pipe with a minimum diameter of 25 mm. The spacing of the support posts shall be a maximum of 2.0 m as shown in the plans.

Fence backup support, if used, shall consist of steel wire with maximum a mesh spacing of 50 mm, or a prefabricated polymeric mesh. The strength of the wire or polymeric mesh shall be equivalent to or greater than that required in Table 6 for the geotextile (*i.e.*,

800 N grab tensile strength) if it is unsupported between posts. The polymeric mesh must be as resistant to ultraviolet radiation as the geotextile it supports.

Sediment deposits shall either be removed when the deposit reaches approximately one-third the height of the silt fence, or a second silt fence shall be installed, as determined by the Engineer.

Measurement

Construction geotextile, with the exception of temporary silt fence geotextile and underground drainage geotextile used in trench drains, will be measured by the square meter for the ground surface area actually covered. Temporary silt fence geotextile will be measured by the linear meter of silt fence installed. Underground drainage geotextile used in trench drains will be measured by the square meter for the perimeter of drain actually covered.

Payment

Payment will be made in accordance with Section 1-04.1, for each of the following bid items that are included in the
"Construction Geotextile For Temporary Silt Fence", per linear meter.
Sediment removal behind silt fences will be paid by force account under temporary water pollution/erosion control. If a new silt fence is installed in lieu of sediment removal, as determined by the Engineer, the silt fence will be paid for at the unit contract price per linear meter for "Construction Geotextile For Silt Fence".

4.5   INSTALLATION PROCEDURES

Silt fences are quite simple to construct; the normal construction sequence is shown in Figure 4-5. Installation of a prefabricated silt fence is shown is Figure 4-6.
1. Install wooden or steel fence posts or large wooden stakes in a row, with normal spacing between 0.5 to 3 m, center to center, and to a depth of 0.4 to 0.6 m. Most prefabricated fences have posts spaced approximately 2 to 3 m apart, which is usually adequate (Step 1).
2. Construct a small (minimum 0.15 m deep and 0.15 m wide) trench on the upstream side of the silt fence (Step 2).
3. Attach reinforcing wire, if required, to the posts (Step 3).
4. If a prefabricated silt fence is not being used, the geotextile must be attached to the posts using staples, reinforcing wire, or other attachments provided by the manufacturer. Geotextile should be extended at least 150 mm below the ground surface (Step 4 & 5).
5. Bury the lower end of the geotextile in the upstream trench and backfill with natural material, tamping the backfill to provide good anchorage (Step 6).
The field inspector should review the field inspection guidelines in Section 1.7.

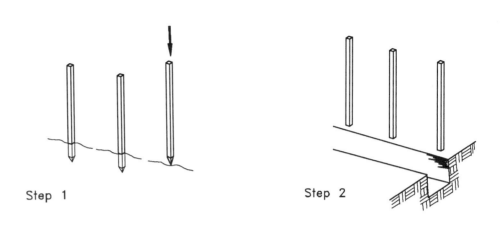

Step 1   Step 2

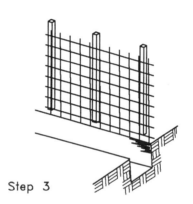

Step 3

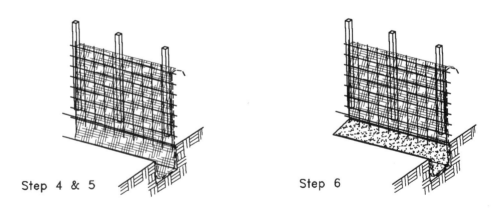

Step 4 & 5   Step 6

Figure 4-5   Typical silt fence installation.

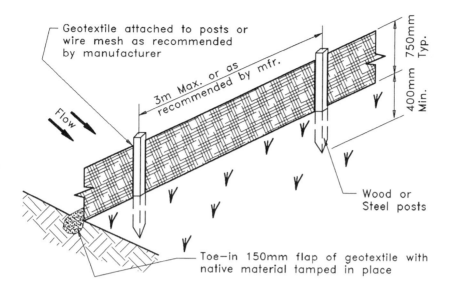

Geotextile attached to posts or
wire mesh as recommended
by manufacturer

3m Max. or as
recommended by mfr.

Flow

750mm
Typ.

400mm
Min.

Wood or
Steel posts

Toe-in 150mm flap of geotextile with
native material tamped in place

SILT FENCE INSTALLATION

Figure 4-6   *Installation of a prefabricated silt fence.*

4.6   MAINTENANCE

Silt fences should be checked periodically, especially after a rainfall or storm event. Excessive buildup of sediment must be removed so the silt fence can function properly. Generally, sediment buildup behind the fence should be removed when it reaches 1/3 to 1/2 of the fence height. The toe trench should also be checked to be ensure that runoff is not piping under the fence.

4.7   SILT AND TURBIDITY CURTAINS

Silt and turbidity curtains perform essentially the same function as silt fences; that is, the geotextile intercepts sediment-laden water while allowing clear water to pass. Thus, for maximum efficiency, a silt or turbidity curtain should pass a maximum amount of water while retaining a maximum amount of sediment. Unfortunately, such optimum performance is normally not possible because sediments will eventually blind or clog (Figure 2-3) the geotextile. To maximize the geotextile's efficiency, the following soil, site, and environmental conditions should be established, and the geotextile selected should provide a specific filtering efficiency while maintaining the required flow rate (Bell and Hicks, 1980).

1. Grain size distribution of soil to be filtered.
2. Estimate of the soil volume to be filtered during construction.
3. Flow conditions, anticipated runoff, and water level fluctuations.
4. Expected environmental conditions, including temperature and duration of sunlight exposure.
5. Velocity, direction, and quantity of discharge water.
6. Water depth and levels of turbidity.
7. Survey of the bottom sediments and vegetation at the site.
8. Wind conditions.

On the basis of these considerations, the geotextile can then be selected either according to the properties required to maximize particle retention and flow capacity while resisting clogging, or by performing filtration model studies such as ASTM D 5141. The first approach follows the criteria developed in Chapter 2 for drainage systems. Silt and turbidity curtains are generally concerned with fine-grained soils, therefore, the following criteria could be considered when selecting the geotextile.

A. Retention Criteria

$$AOS = D_{85} \text{ for woven geotextiles.}$$

$$AOS = 1.8 \times D_{85} \text{ for nonwovens.}$$

NOTE: The D_{85} is a characteristic large-grain size appropriate to the suspended sediment grain size distribution. It will be strongly influenced by items Nos. 1, 3, 5, 6, and 7 above.

B. Flow Capacity Criteria

$$\psi = (10\,q) \div A$$

where:

ψ = permittivity of geotextile (T^{-1})

q = flow rate (L^3/T)

A = cross-sectional area silt curtain (L^2)

10 = factor of safety

C. Clogging Resistance

Maximize AOS requirements using largest opening possible from criterion A above.

For silt and turbidity curtain construction, the geotextile forming the curtain is held vertical by flotation segments at the top and a ballast along the bottom (Bell and Hicks, 1980). A tension cable is often built into the curtain immediately above or below the flotation segments to absorb stress imposed by currents, wave action, and wind. Barrier sections are usually about 30 m long and of any required width. Curtains can also be constructed within shallow bodies of water using silt fence-type construction methods. Geotextiles have also been attached to soldier piles and draped across riprap barriers as semipermanent curtains.

The U.S. Army Corps of Engineers (1977) indicates that silt and turbidity curtains are not appropriate for certain conditions, such as:

- operations in open ocean;
- operations in currents exceeding 0.5 m/s;
- in areas frequently exposed to high winds and large breaking waves; and
- near hopper or cutter head dredges where frequent curtain movement would be necessary.

4.8   EROSION CONTROL BLANKETS

In freshly graded areas, the soil is susceptible to erosion by rainfall and runoff. Temporary, degradable blankets are used to enhance the establishment of vegetation. These products are used where vegetation alone provides sufficient site protection after the temporary products degrade. Such products are usually evaluated by field trial sections, and, therefore, are empirically designed. There are very few published records of comparative use, so the user must decide on the preferable system, usually based on local experience. You should be aware that a variety of products and systems exist. As an aid to selecting the best system, consult manufacturers and other agencies about their experiences.

Erosion protection must be provided for three distinct phases, namely:
1. prior to vegetation growth;
2. during vegetation growth; and
3. after vegetation is fully established.
Erosion control blankets provide protection during the first two phases. After vegetation is established protection can be provided by erosion control mats that reinforce the vegetation root mass, as discussed in Chapter 3.

Geosynthetic erosion control blankets are manufactured of light-weight polymer net(s) and a bedding of polymer webbing or organic materials such as straw or coconut. The bedding material helps retain moisture, seeds, and soil to promote growth. These polymer materials are typically not stabilized against ultraviolet light, and are designed to degrade over time. Erosion control blankets have design lives that vary between approximately 6 months to 5 years. Some blankets are provided with seeds encased in paper.

Erosion control blankets provide protection against moderate-flow velocities for short periods of time. They are typically used on moderate slopes and low velocity intermittent flow channels. Flows up to 1.5 m/s and durations of ½ to approximately 5 hours can be withstood, as illustrated in Figure 4-7. Again, design is empirical, and blanket product manufacturers should have actual flume test data and design recommendations available for their specific products. Duration of flume tests should be noted.

Since the design of erosion control blankets is empirical, specification by index properties is not easily accomplished. Also, relatively few test methods have been standardized for erosion control blankets. Therefore, it is recommended that specifications use an approved products list.

Construction plans and specifications should detail and note installation requirements. Details such as anchoring in trenches, use of pins, pin length, pin spacing, roll overlap requirements, and roll termination should be addressed.

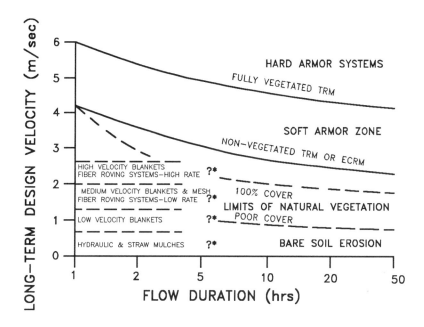

Figure 4-7   Recommended maximum design velocities and flow durations for various classes of erosion control materials (after Theisen, 1992). [Note: TRM is turf reinforcement mat, and ECRM is erosion control revegetation mat.]

4.9   REFERENCES

AASHTO (1996) Standard Specification for Geotextiles, Designation: M 288 (draft), American Association of State Transportation and Highway Officials, Washington, D.C.

AASHTO (1990), Standard Specifications for Geotextiles - M 288, Standard Specifications for Transportation Materials and Methods of Sampling and Testing, American Association of State Transportation and Highway Officials, Washington, D.C., pp 689-692.

Allen, T.M. (1994), personal communication.

ASTM (1994), Soil and Rock (II): D 4393 - latest; Geosynthetics, Annual Book of ASTM Standards, Section 4, Volume 04.09, American Society for Testing and Materials, Philadelphia, PA, 516 p.

Bell, J.R. and Hicks, R.G. (1980), Evaluation of Test Methods and Use Criteria for Geotechnical Fabrics in Highway Applications - Interim Report, Report No. FHWA/RD-80/021, Federal Highway Administration, Washington, D.C., June 190 p.

Mallard, P. and Bell, J.R. (1981), Use of Fabrics in Erosion Control, Transportation Research Report 81-4, Federal Highway Administration, Washington, D.C.

Richardson, G.N. and Middlebrooks, P. (1991), A Simplified Design Method for Silt Fences, Proceedings of the Geosynthetics '91 Conference, Industrial Fabrics Association International, St. Paul, MN, pp. 879-888.

Richardson, G.N. and Koerner, R.M., Editors (1990), A Design Primer: Geotextiles and Related Materials, Industrial Fabrics Association International, St. Paul, MN, 166 p.

Soil Conservation Service (1977), National Engineering Handbook, Sections 11 and 14, U.S. Department of Agriculture, Washington, D.C.

Theisen, M.S. (1992), Geosynthetics in Erosion Control and Sediment Control, Geotechnical Fabrics Report, Industrial Fabrics Association International, St. Paul, MN, May/June, pp. 26-35.

U.S. Army Corps of Engineers (1977), Civil Works Construction Guide Specification for Plastic Filter Fabric, Corps of Engineer Specifications No. CW-02215, Office, Chief of Engineers, U.S. Army Corps of Engineers, Washington, D.C.

Wyant, D.C. (1980), Evaluation of Filter Fabrics for Use as Silt Fences, Report No. VHTRC 80-R49, Virginia Highway and Transportation Research Council, Charlottesville, VA.

5.0   GEOSYNTHETICS in ROADWAYS

5.1   INTRODUCTION

The most common use of geosynthetics is in road and pavement construction. Geotextiles increase stability and improve performance of weak subgrade soils **primarily by separating the aggregate from the subgrade**. In addition, geogrids and some geotextiles can provide strength through friction or interlock developed between the aggregate and the geosynthetic. Geotextiles can also provide filtration and drainage by allowing excess pore water pressures in the subgrade to dissipate into the aggregate base course and, in cases of poor-quality aggregate, through the geotextile plane itself.

In this chapter, each of the geosynthetic functions will be discussed and related to design concepts and performance properties.   Selection, specification, and construction procedures will also be presented.

5.1-1   Functions of Geosynthetics in Roadways

A geosynthetic placed at the interface between the aggregate base course and the subgrade functions as a separator to prevent two dissimilar materials (subgrade soils and aggregates) from intermixing.   Geotextiles and geogrids perform this function by preventing penetration of the aggregate into the subgrade (localized bearing failures) (Figure 5-1).   In addition, geotextiles prevent intrusion of subgrade soils up into the base course aggregate.   Localized bearing failures and subgrade intrusion only occur in very soft, wet, weak subgrades.   It only takes a small amount of fines to significantly reduce the friction angle of select granular aggregate.   Therefore, separation is important to maintain the design thickness and the stability and load-carrying capacity of the base course.   Soft subgrade soils are most susceptible to disturbance during construction activities such as clearing, grubbing, and initial aggregate placement.   Geosynthetics can help minimize subgrade disturbance and prevent loss of aggregate during construction. Thus, the primary function of the geotextile in this application is separation (Table 1-1), and can in some cases be considered a secondary function for geogrids.

The system performance may also be influenced by secondary functions of **filtration**, **drainage**, and **reinforcement** (Table 1-1). The geotextile acts as a filter to prevent fines from migrating up into the aggregate due to high pore water pressures induced by dynamic wheel loads.   It also acts as a drain, allowing the excess pore pressures to dissipate through the geotextile and the subgrade soils to gain strength through consolidation and improve with time.

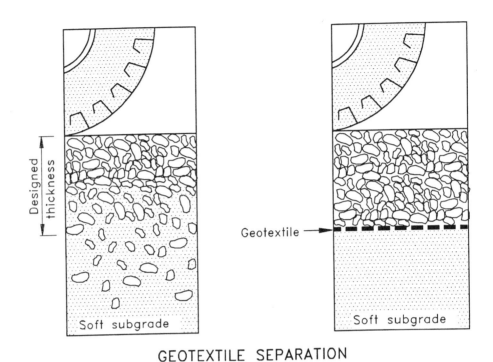

GEOTEXTILE SEPARATION

Figure 5-1   *Concept of geotextile separation in roadways (after Rankilor, 1981).*

Geogrids and geotextiles provide reinforcement through three possible mechanisms.

1. Lateral restraint of the base and subgrade through friction and interlock between the aggregate, soil and the geosynthetic (Figure 5-2a).
2. Increase in the system bearing capacity by forcing the potential bearing capacity failure surface to develop along alternate, higher shear strength surfaces (Figure 5-2b).
3. Membrane support of the wheel loads (Figure 5-2c).

When an aggregate layer is loaded by a wheel or track, the aggregate tends to move or shove laterally, as shown in Figure 5-2a, unless it is restrained by the subgrade or geosynthetic reinforcement. Soft, weak subgrade soils provide very little lateral restraint, so when the aggregate moves laterally, ruts develop on the aggregate surface and also in the subgrade. A geogrid with good interlocking capabilities or a geotextile with good frictional capabilities can provide tensile resistance to lateral aggregate movement. Another possible geosynthetic reinforcement mechanism is illustrated in Figure 5-2b.

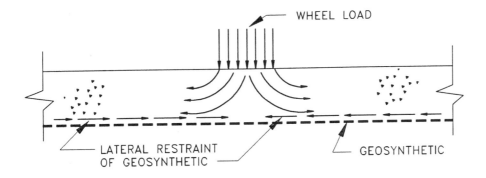

(a) LATERAL RESTRAINT

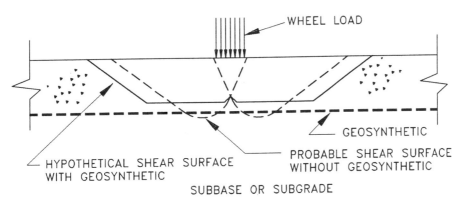

(b) BEARING CAPACITY INCREASE

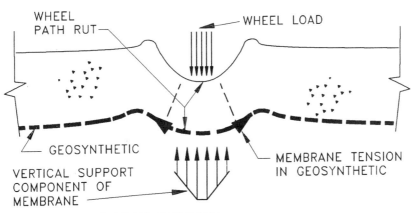

(c) MEMBRANE TENSION SUPPORT

Figure 5-2   *Possible reinforcement functions provided by geosynthetics in roadways: (a) lateral restraint, (b) bearing capacity increase, and (c) membrane tension support (after Haliburton, et al., 1981).*

Using the analogy of a wheel load to a footing, the geosynthetic reinforcement forces the potential bearing capacity failure surface to follow an alternate higher strength path. This tends to increase the bearing capacity of the roadway.

A third possible geosynthetic reinforcement function is membrane-type support of wheel loads, as shown conceptually in Figure 5-2c. In this case, the wheel load stresses must be great enough to cause plastic deformation and ruts in the subgrade. If the geosynthetic has a sufficiently high tensile modulus, tensile stresses will develop in the reinforcement, and the vertical component of this membrane stress will help support the applied wheel loads. As tensile stress within the geosynthetic cannot be developed without some elongation, wheel path rutting (in excess of 100 mm) is required to develop membrane-type support. Therefore, this mechanism is generally limited to temporary roads or the first aggregate lift in permanent roadways.

5.1-2 Subgrade Conditions in which Geosynthetics are Useful

Geotextile separators have a 20+ year history of successful use for the stabilization of very soft wet subgrades. Based on experience and several case histories summarized by Haliburton, Lawmaster, and McGuffey (1981) and Christopher and Holtz (1985), the following subgrade conditions are considered to be the most appropriate for geosynthetic use in roadway construction:

- Poor soils

  (USCS: SC, CL, CH, ML, MH, OL, OH, and PT)
  (AASHTO: A-5, A-6, A-7-5, and A-7-6)
- Low undrained shear strength

  $\tau_f = c_u < 90\,\text{kPa}$

  CBR < 3   {Note: CBR as determined with ASTM D 4429 Bearing

  $M_R \approx 30\,\text{MPa}$ Ratio of Soils in Place}
- High water table
- High sensitivity

Under these conditions, geosynthetics function primarily as separators and filters to stabilize the subgrade, improving construction conditions and allowing long-term strength improvements in the subgrade. If large ruts develop during placement of the first aggregate lift, then some reinforcing effect is also present. As a summary recommendation, the following geotextile functions are appropriate for the corresponding subgrade strengths:

Undrained Shear Strength (kPa)	Subgrade CBR	Functions
60 - 90	2 - 3	Filtration and possibly separation
30 - 60	1 - 2	Filtration, separation, and possibly reinforcement
< 30	< 1	All functions, including reinforcement

This table implicates that geotextiles do not provide a useful function when the undrained shear strength is greater than about 90 kPa (CBR about 3). From a foundation engineering point of view, clay soils with undrained shear strengths of 90 kPa are considered to be stiff clays (Terzaghi and Peck, 1967, p 30) and are generally quite good foundation materials. Allowable footing pressures on such soils equal 150 kPa or greater. Simple stress distribution calculations show that for static loads, such soils will readily support reasonable truck loads and tire pressures, even under relatively thin granular bases.

Dynamic loads and high tire pressures are another matter. Some rutting will probably occur in such soils, especially after a few hundred passes (Webster, 1993). If traffic is limited, as it is in many temporary roads, or if shallow (< 75 mm) ruts are acceptable, as in most construction operations, then a maximum undrained shear strength of about 90 kPa (CBR = 3) for geosynthetic use in highway construction seems reasonable.

An exception to this is in permeable base applications. Even on firm subgrades, a geotextile placed beneath the base functions as a separator and filter, as illustrated in Figure 5-3.

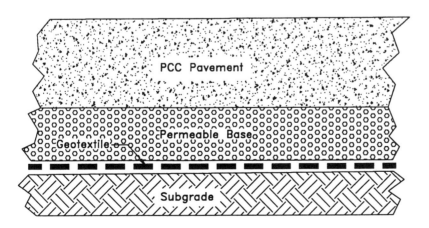

Figure 5-3   *Geotextile separator beneath permeable base (Baumgartner, 1994).*

5.2 APPLICATIONS

5.2-1 Temporary and Permanent Roads

Roads and highways are broadly classified into two categories: permanent and temporary, depending on their service life, traffic applications, or desired performance. Permanent roads include both paved and unpaved systems which usually remain in service 10 years or more. Permanent roads may be subjected to more than a million load applications during their design lives. On the other hand, temporary roads are, in most cases, unpaved. They remain in service for only short periods of time (often less than 1 year), and are usually subjected to fewer than 10,000 load applications during their services lives. Temporary roads include detours, haul and access roads, construction platforms, and stabilized working *tables* required for the construction of permanent roads, as well as embankments over soft foundations.

Geosynthetics allow construction equipment access to sites where the soils are normally too weak to support the initial construction work. This is one of the more important uses of geosynthetics. Even if the finished roadway can be supported by the subgrade, it may be virtually impossible to begin construction of the embankment or roadway. Such sites require stabilization by dewatering, demucking, excavation and replacement with select granular materials, utilization of stabilization aggregate, chemical stabilization, etc. Geosynthetics can often be a cost-effective alternate to these expensive foundation treatment procedures.

Furthermore, geosynthetic separators enable contractors to meet minimum compaction specifications for the first two or three aggregate lifts. This is especially true on very soft, wet subgrades, where the use of ordinary compaction equipment is very difficult or even impossible. Long term, a geosynthetic acts to maintain the roadway design section and the base course material integrity. Thus, the geosynthetic will ultimately increase the life of the roadway.

5.2-2 Benefits

Geosynthetics used in roadways on soft subgrades, may provide several and performance benefits, including the following.

1. Reducing the intensity of stress on the subgrade and preventing the base aggregate from penetrating into the subgrade (**function: separation**).
2. Preventing subgrade fines from pumping or otherwise migrating up into the base (**function: separation and filtration**).
3. Preventing contamination of the base materials which may allow more open-graded, free-draining aggregates to be considered in the design (**function: filtration**).

4. Reducing the depth of excavation required for the removal of unsuitable subgrade materials (**function: separation and reinforcement**).

5. Reducing the thickness of aggregate required to stabilize the subgrade (**function: separation and reinforcement**).

6. Reducing disturbance of the subgrade during construction (**function: separation and reinforcement**).

7. Allowing an increase in subgrade strength over time (**function: filtration**).

8. Reducing the differential settlement of the roadway, which helps maintain pavement integrity and uniformity (**function: reinforcement**). Geosynthetics will also aid in reducing differential settlement in transition areas from cut to fill. {NOTE: Total and consolidation settlements are not reduced by the use of geosynthetic reinforcement.}

9. Reducing maintenance and extending the life of the pavement (**functions: all**).

Geosynthetics are also used in permanent roadways to provide capillary breaks to reduce frost action in frost-susceptible soils, and to provide membrane-encapsulated soil layers (MESL) to reduce the effects of seasonal water content changes on roadways on swelling clays.

5.3   POSSIBLE FAILURE MODES OF PERMANENT ROADS

Yoder and Witczak (1975) define two types of pavement distress, or failure. The first is a structural failure, in which a collapse of the entire structure or a breakdown of one or more of the pavement components renders the pavement incapable of sustaining the loads imposed on its surface. The second type failure is a functional failure; it occurs when the pavement, due to its roughness, is unable to carry out its intended function without causing discomfort to drivers or passengers or imposing high stresses on vehicles. The cause of these failure conditions may be due to excessive loads, climatic and environmental conditions, poor drainage leading to poor subgrade conditions, and disintegration of the component materials. Excessive loads, excessive repetition of loads, and high tire pressures can cause either structural or functional failures.

Pavement failures may occur due to the intrusion of subgrade soils into the granular base, which results in inadequate drainage and reduced stability. Distress may also occur due to excessive loads that cause a shear failure in the subgrade, base course, or the surface. Other causes of failures are surface fatigue and excessive settlement, especially differential of the subgrade. Volume change of subgrade soils due to wetting and drying, freezing and thawing, or improper drainage may also cause pavement distress. Inadequate drainage of water from the base and subgrade is a major cause of pavement problems (Cedergren, 1987). If the subgrade is saturated, excess pore pressures will develop under traffic loads, resulting in subsequent softening of the subgrade.

Improper construction practices may also cause pavement distress. Wetting of the subgrade during construction may permit water accumulation and subsequent softening of the subgrade in the rutted areas after construction is completed. Use of dirty aggregates or contamination of the base aggregates during construction may produce inadequate drainage, instability, and frost susceptibility. Reduction in design thickness during construction due to insufficient subgrade preparation may result in undulating subgrade surfaces, failure to place proper layer thicknesses, and unanticipated loss of base materials due to subgrade intrusion. Yoder and Witczak (1975) state that a major cause of pavement deterioration is inadequate observation and field control by qualified engineers and technicians during construction.

After construction is complete, improper or inadequate maintenance may also result in pavement distress. Sealing of cracks and joints at proper intervals must be performed to prevent surface water infiltration. Maintenance of shoulders will also affect pavement performance.

As indicated in the list of possible benefits resulting from geosynthetic use in permanent roadway systems (section 5.2-2), properly designed geosynthetics can enhance pavement performance and reduce the likelihood of failures.

5.4   ROADWAY DESIGN USING GEOTEXTILES

Certain design principles are common to all types of roadways, regardless of the design method. Basically, the design of any roadway involves a study of each of the components of the system, (surface, aggregate base courses and subgrade) detailing their behavior under traffic load and their ability to carry that load under various climatic and environmental conditions. All roadway systems, whether permanent or temporary, derive their support from the underlying subgrade soils. Thus, the geotextile functions are similar for either temporary or permanent roadway applications. However, due to different performance requirements, **design methodologies for temporary roads should not be used to design permanent roads**. Temporary roadway design usually allows some rutting to occur over the design life, as ruts will not necessarily impair service. Obviously, ruts are not acceptable in permanent roadways. In the following two sections, recommended design procedures for both temporary and permanent roads are presented. Our permanent road and pavement design basically uses geotextiles for the construction or stabilization lift only; the base course thickness required to adequately carry the design traffic loads for the design life of the pavement is not reduced due to the use of a geotextile. There is some evidence, however, that suggests a geogrid placed at the bottom of the aggregate base may permit a 10 to 20% base thickness reduction (Barksdale, et al., 1989; Webster, 1993).

5.5   GEOTEXTILE SURVIVABILITY

Selecting a geotextile for either permanent or temporary roads depends upon one thing -- the survivability criteria. If the roadway system is designed correctly, then the stress at the top of the subgrade due to the weight of the aggregate and the traffic load should be less than the bearing capacity of the soil plus a safety factor. However, the stresses applied to the subgrade and the geotextile during construction may be much greater than those applied in service. **Therefore, selection of the geotextile in roadway applications is usually governed by the anticipated construction stresses**. This is the concept of geotextile survivability -- the geotextile must survive the construction operations if it is to perform its intended function.

Table 5-1 relates the elements of construction (*i.e.*, equipment, aggregate characteristics, subgrade preparation, and subgrade shear strength) to the severity of the loading imposed on the geotextile. If one or more of these items falls within a particular severity category (*i.e.*, moderate or high), then geotextiles meeting those survivability requirements should be selected. Variable combinations indicating a NOT RECOMMENDED rating suggests that one or more variables should be modified to assure a successful installation. Some judgment is required in using these criteria.

The geotextile strength required to survive the most severe conditions anticipated during construction can then be determined from Table 5-2, provided by AASHTO M288 (1990). Geotextiles that meet or exceed these survivability requirements can be considered acceptable for the project. These survivability requirements are not based on any systematic research but on the properties of geotextiles which have performed satisfactorily as separators in temporary roads and in similar applications. However, in the absence of any other information, they should be used as minimum property values. Judgment and experience are required to select final specification values. Geotextile survivability for major projects should be verified by conducting field tests under site-specific conditions. These field tests should involve trial sections using several geotextiles on typical subgrades at the project site and implementing various types of construction equipment. After placement of the geotextile and aggregate, the geotextiles are exhumed to see how well or how poorly they tolerated the imposed construction stresses.

Table 5-1
Construction Survivability Ratings
(after AASHTO, 1990 and 1996)

Site Soil CBR at Installation[1]	< 1		1 to 2		> 3	
Equipment Ground Contact Pressure (kPa)	> 350	< 350	> 350	< 350	> 550	< 550
Cover Thickness[2] (compacted, mm)						
100[3,4]	NR[6]	NR	H[6]	H	M[6]	M
150[5]	NR	NR	H	H	M	L[6]
300	NR	H	M	M	M.	L
450	H	M	M	M	M	L

NOTES:

1. Assume saturated CBR unless construction scheduling can be controlled.
2. Maximum aggregate size not to exceed one-half the compacted cover thickness
3. For low-volume, unpaved roads (ADT < 200 vehicles).
4. The 100 mm minimum cover is limited to existing road bases and is not intended for use in new construction.
5. Maximum aggregate size ≤ 30 mm
6. NR = NOT RECOMMENDED; L = LOW; M = MODERATE; and H = HIGH

These tests could be performed during design or after the contract was let, similar to the recommendations for riprap placement (Section 3.8-1, Item 6e). In the latter case, the contractor is required to demonstrate that the proposed subgrade condition, equipment, and aggregate placement will not damage the geotextile. If necessary, additional subgrade preparation, increased lift thickness, and/or possibly different construction equipment could be utilized. In rare cases, the contractor may even have to supply a different geotextile.

The selected geotextile must also retain the underlying subgrade soils, allowing the subgrade to drain freely, consolidate, and gain strength. Thus, the geotextile must be checked, using the drainage and filtration requirements discussed in Chapter 2 and summarized in Table 5-2.

Table 5-2
Physical Property Requirements[1]
(after AASHTO, 1990 and 1996)

Survivability Level	Grab Strength[4] ASTM D 4632 (N)		Puncture Resistance[4] ASTM D 4833 (N)		Tear Strength[4] ASTM D 4533 (N)	
	< 50% Geotextile Elongation[2,3]	> 50% Geotextile Elongation[2,3]	< 50% Geotextile Elongation	> 50% Geotextile Elongation	< 50% Geotextile Elongation	> 50% Geotextile Elongation
High (Class 1)	1400	900	500	350	500	350
Moderate (Class 2)	1100	700	400	250	400	250
Low (Class 3)	800	500	300	180	300	180

Additional Requirements	Test Method
Apparent Opening Size 1. < 50% soil passing 0.075 mm sieve, AOS < 0.6 mm 2. > 50% soil passing 0.075 mm sieve, AOS < 0.3 mm	ASTM D 4751
Permeability k of the geotextile > k of the soil (permittivity x the nominal geotextile thickness)	ASTM D 4491
Ultraviolet Degradation At 500 hours of exposure, 50% strength retained	ASTM D 4355
Geotextile Acceptance	ASTM D 4759

NOTES:

1. For the index properties, the first value of each set is for geotextiles which fail at less than 50% elongation, while the second value is for geotextiles which fail at greater than 50% elongation. Elongation is determined by ASTM D 4632
2. Values shown are minimum roll average values. Strength values are in the weakest principal direction.
3. The values of the geotextile elongation do not relate to the allowable consolidation properties of the subgrade soil. These must be determined by a separate investigation.
4. AASHTO classification.

5.6 DESIGN GUIDELINES FOR TEMPORARY AND UNPAVED ROADS

There are two main approaches to the design of temporary and unpaved roads. The first assumes no reinforcing effect of the geotextile; that is, the geotextile acts as a separator only. The second approach considers a possible reinforcing effect due to the geotextile. It appears that the separation function is more important for thin roadway sections with relatively small live loads where ruts, approximating 50 to 100 mm are anticipated. In these cases, a design which assumes no reinforcing effect is generally conservative. On the other hand, for large live loads on thin roadways where deep ruts (> 100 mm) may occur, and for thicker roadways on softer subgrades, the reinforcing function becomes increasingly more important if stability is to be maintained. It is for these latter cases that reinforcing analyses have been developed and are appropriate.

The design method presented in this manual considers mainly the separation function. It was selected because it has a long history of successful use, it is based on principles of soil mechanics, and it has been calibrated by full-scale field tests. It can also be adapted to a wide variety of conditions. Other methods considering reinforcement functions are described by Koerner (1994), Christopher and Holtz (1985) and Giroud and Noiray (1981). For roadways where stability of the embankment foundation is questionable, refer to Chapter 7 for information on reinforced embankments.

The following design method was developed by Steward, Williamson, and Mohney (1977) for the U.S. Forest Service (USFS). It allows the designer to consider:
- vehicle passes;
- equivalent axle loads;
- axle configurations;
- tire pressures;
- subgrade strengths; and
- rut depths.

The following limitations apply:
- the aggregate layer must be
  a) compacted to CBR 80,
  b) cohesionless (nonplastic);
- vehicle passes less than 10,000;
- geotextile survivability criteria must be considered; and
- subgrade undrained shear strength less than about 90 kPa (CBR < 3).

As discussed in Section 5.1-2, for subgrades stronger than about 90 kPa (CBR > 3), geotextiles are rarely required for stabilization, although they may provide some drainage and filtration, and will provide separation. In this case, the principles developed in Chapter 2 are applicable, just as they are for weaker subgrades where drainage and filtration are likely to be very important.

Based on both theoretical analysis and empirical (laboratory and full-scale field) tests on geotextiles, Steward, Williamson and Mohney (1977) determined that a certain amount of rutting would occur under various traffic conditions, both with and without a geotextile separator and for a given stress level acting on the subgrade. They present this stress level in terms of bearing capacity factors, similar to those commonly used for the design of shallow foundations on cohesive soils. These factors and conditions are given in Table 5-3.

Table 5-3
Bearing Capacity Factors for Different Ruts and Traffic Conditions both with and without Geotextile Separators
(after Steward, Williamson, and Mohney, 1977)

	Ruts (mm)	Traffic (Passes of 80 kN axle equivalents)	Bearing Capacity Factor, N_c
Without Geotextile:	<50 >100	>1000 <100	2.8 3.3
With Geotextile:	<50 >100	>1000 <100	5.0 6.0

The following design procedure is recommended:

STEP 1.   Determine soil subgrade strength.

Determine the subgrade soil strength in the field using the field CBR, cone penetrometer, vane shear, resilient modulus, or any other appropriate test. The undrained shear strength of the soil, c, can be obtained from the following relationships:

- for field CBR, c in kPa = 30 x CBR;
- for the WES cone penetrometer, c = cone index divided by 10 or 11; and
- for the vane shear test, c is directly measured.

Other in situ tests, such as the static cone penetrometer test (CPT) or dilatometer (DMT), may be used, provided local correlations with undrained shear strength exist. Use of the Standard Penetration Test (SPT) is not recommended for soft clays.

It is good practice to make strength determinations at several locations where the subgrade appears to be the weakest. Strengths should be evaluated at depths of 0 to 200 mm and from 200 to 500 mm; six to ten strength measurements are recommended at each location to obtain a good average value. Tests should also be performed when the soils are in their weakest condition (i.e., when the water table is the highest).

STEP 2.   Determine wheel loading.
Determine the maximum single wheel load, maximum dual wheel load, and the maximum dual tandem wheel load anticipated for the roadway during the design period. For example, an 8 m³ dump truck with tandem axles will have a dual wheel load of approximately 35 kN. A motor grader has a wheel load of 22 to 44 kN.

STEP 3.   Estimate amount of traffic.
Estimate the maximum amount of traffic anticipated for each design vehicle class.

STEP 4.   Establish tolerable rutting.
Establish the amount of tolerable rutting during the design life of the roadway. For example, 50 to 75 mm of rutting is generally acceptable during construction.

STEP 5.   Obtain bearing capacity factor.
Obtain appropriate subgrade stress level in terms of the bearing capacity factors in Table 5-3.

STEP 6.   Determine required aggregate thickness.
Determine the required aggregate thickness from the USFS design charts (Figures 5-4, 5-5, and 5-6) for each maximum loading. Enter the curve with appropriate bearing capacity factors (N_c) multiplied by the design subgrade undrained shear strength (c) to evaluate each required stress level (cN_c).

STEP 7.   Select design thickness.
Select the design thickness based on the design requirements. The design thickness should be given to the next higher 25 mm.

STEP 8.   Check geotextile drainage and filtration characteristics.

Check the geotextile drainage and filtration requirements. Use the gradation and permeability of the subgrade, the water table conditions, and the retention and permeability criteria given in Chapter 2. In high water table conditions with heavy traffic, filtration criteria may also be required. From Chapter 2, that criteria is:

$$AOS \leq D_{85} \quad \text{(Wovens)} \tag{Eq. 2-3}$$

$$AOS \leq 1.8\, D_{85} \quad \text{(Nonwovens)} \tag{Eq. 2-4}$$

$$k_{geotextile} \geq k_{soil} \tag{Eq. 2-7a}$$

$$\psi \geq 0.1 \text{ sec}^{-1} \tag{Eq. 2-8c}$$

STEP 9.   Determine geotextile survivability requirements.

Check the geotextile survivability strength requirements as discussed in Section 5.5.

STEP 10.   Specify geotextile property requirements.

Specify geotextiles that meet or exceed these survivability criteria.

STEP 11.   Specify construction requirements.

Follow the construction recommendations in Section 5.12.

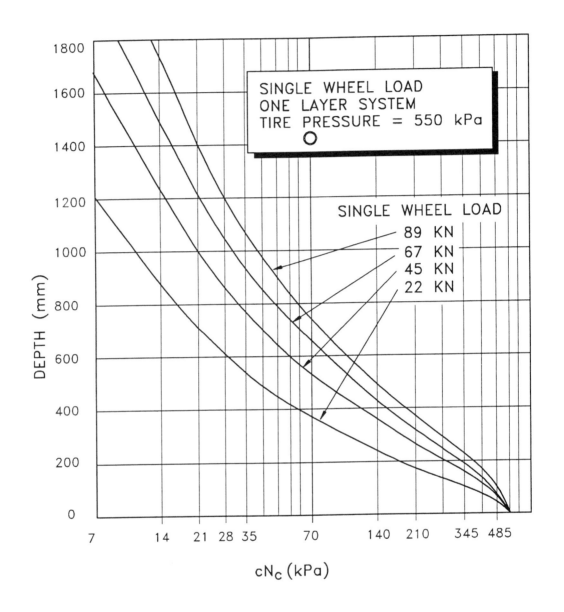

Figure 5-4   *U.S. Forest Service thickness design curve for single wheel load (Steward et al., 1977).*

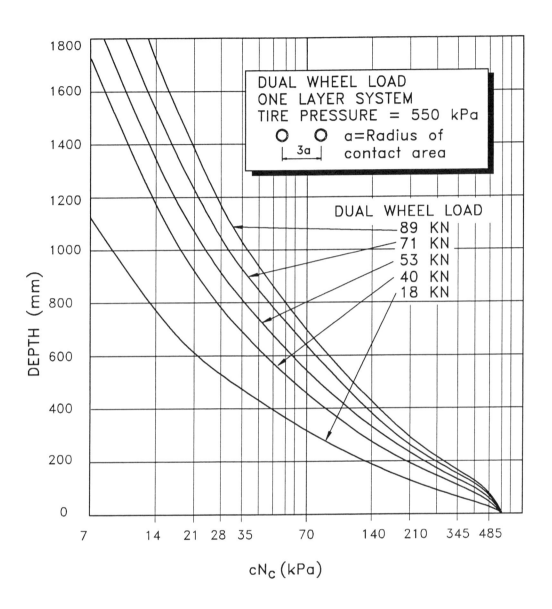

Figure 5-5   *U.S. Forest Service thickness design curve for dual wheel load (Steward et al., 1977).*

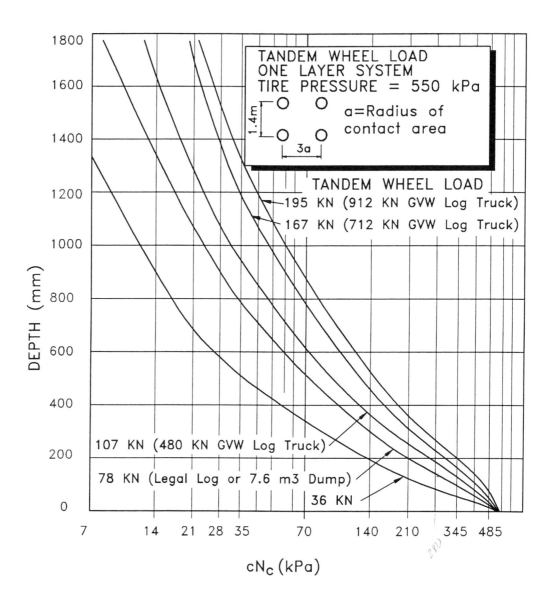

Figure 5-6   *U.S. Forest Service thickness design curve for tandem wheel load (Steward et al., 1977).*

5.7   TEMPORARY ROAD DESIGN EXAMPLE

DEFINITION OF DESIGN EXAMPLE

Project Description: A haul road over wet, soft soils is required for a highway
construction project.

Type of Structure: temporary unpaved road

Type of Application: geotextile separator

Alternatives: i) conventional design without geotextile
                     ii) geotextile separator between aggregate and subgrade

GIVEN DATA

subgrade   -   cohesive subgrade soils
            -    high water table
            -    average undrained shear strength about 30 kPa or
                 CBR = 1

traffic   -   approximately 5000 passes
            -    90 kN single axle truck
            -    550 kPa tire pressure

ruts   -   maximum of 50 to 100 mm

REQUIRED

Design the roadway section.

Consider: 1) design without a geotextile; and 2) alternate with geotextile.

DEFINE

A. Geotextile function(s)

B. Geotextile properties required

C. Geotextile specification

SOLUTION

A. Geotextile function(s):
   Primary   -   separation
   Secondary -   filtration, drainage, reinforcement

B. Geotextile properties required:
   survivability
   apparent opening size (AOS)

DESIGN

Design roadway with and without geotextile inclusion.   Compare options.

STEP 1.   DETERMINE SOIL SUBGRADE STRENGTH

Determine subgrade strength at several locations. Assume that CBR ≈ 1 is taken from area(s) where the subgrade appears to be the weakest.

   given   -   CBR ≈ 1

STEP 2.   DETERMINE WHEEL LOADING

   given   -   90 kN single-axle truck, with 550 kPa tire pressure
            -    therefore, 45 kN single wheel load

STEP 3.   ESTIMATE AMOUNT OF TRAFFIC

   given   -   5,000 passes

STEP 4.   ESTABLISH TOLERABLE RUTTING

   given   -   50 to 100 mm

STEP 5.   OBTAIN BEARING CAPACITY FACTOR

   without a geotextile: - $2.8 < N_c < 3.3$
                          -    assume $N_c ≈ 3.0$ for 5,000 passes and 50 to 100 mm ruts

   with a geotextile: - $5.0 < N_c < 6.0$
                          -    assume $N_c ≈ 5.5$ for 5,000 passes and 50 to 100 mm ruts

STEP 6.   DETERMINE REQUIRED AGGREGATE THICKNESSES

without a geotextile
- c N_c = 30 kPa x 3.0 = 90 kPa
- depth of aggregate $\approx$ 475 mm

with a geotextile
- c Nc = 30 kPa x 5.5 = 165 kPa
- depth of aggregate $\approx$ 325 mm

STEP 7. SELECT DESIGN THICKNESS

Use 325 mm and a geotextile

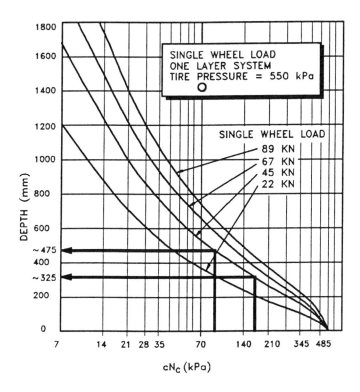

STEP 8.   CHECK GEOTEXTILE DRAINAGE AND FILTRATION CHARACTERISTICS

Use AOS < 0.3 mm, per requirement of Table 5-2 since soil has > 50% passing the
0.075 mm sieve
Permeability of geotextile must be greater than soil permeability.

STEP 9. DETERMINE GEOTEXTILE SURVIVABILITY REQUIREMENTS

Use Table 5-1: with CBR = 1, dump truck contact pressure > 350 kPa, and 325 mm cover thickness, and find a MODERATE survivability to NOT RECOMMENDED rating.

Use a HIGH survivability geotextile, or greater.

STEP 10. SPECIFY GEOTEXTILE PROPERTY REQUIREMENTS

From Table 5-2; geotextile separator shall meet or exceed the minimum average roll values, with elongation at failure determined with the ASTM D 4632 test method, of:

Property	Test Method	Elongation $< 50\%$	Elongation $\geq 50\%$
Grab Strength	ASTM D 4632	1400	900
Puncture Resistance	ASTM D 4833	500	350
Tear Resistance	ASTM D 4533	500	350

The geotextile shall have an AOS < 0.3 mm, $\psi \geq 0.1$ sec^{-1}, and the permeability shall be _____.

STEP 11. SPECIFY CONSTRUCTION REQUIREMENTS

See Section 5.12

5.8   DESIGN GUIDELINES FOR PERMANENT ROADWAYS

5.8-1   Recommended Method

The recommended design method for using geotextiles in permanent pavements is that developed by Christopher and Holtz (1985; 1991). It is based on the following concepts:

1. Standard methods are used to design the overall pavement system (i.e., AASHTO, CBR, R-value, resilient modulus, etc.).
2. **The geotextile is assumed to provide no structural support, therefore, no reduction is allowed in aggregate thickness required for structural support.**
3. The recommended method is used to design the first construction lift, which is called the *stabilizer lift* since it sufficiently stabilizes the subgrade to allow access by normal construction equipment.
4. Aggregate savings is achieved through a reduction in the thickness of stabilization aggregate required for construction.

5. Once the stabilizer lift is completed, construction proceeds using usual methods.

The design method assumes that the stabilizer lift is an unpaved road that will be exposed to relatively few vehicle passes (*i.e.*, construction equipment only) and that can tolerate 50 to 75 mm of rutting under the equipment loads. The design consists of the following steps:

STEP 1.   Assess need for geotextile.
Estimate the need for a geotextile based on the subgrade strength and by past performance in similar types of soils.

STEP 2.   Design pavement without geotextile.
Design the roadway for structural support using normal pavement design methods; provide no allowance for the geotextile.

STEP 3.   Determine need for additional aggregate.
See Figure 5-7 to determine if additional aggregate above that required for structural support is needed due to susceptibility of soils to pumping and base course intrusion. If so, reduce that aggregate thickness and include a geotextile at the base/subgrade interface. Note that a thickness reduction of approximately 50% is normally cost effective.

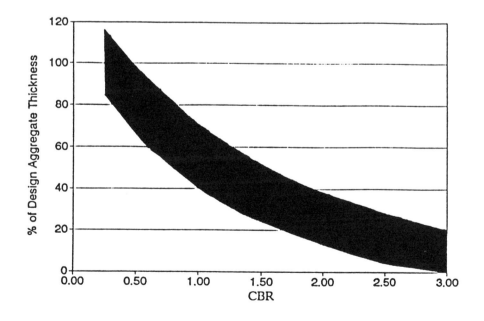

Figure 5-7   *Aggregate loss to weak subgrades (FHWA, 1989; in Christopher and Holtz, 1991).*

STEP 4.   Determine aggregate depth required to support construction equipment.
Determine the additional aggregate required for stabilization of the subgrade during construction activities. Use a 50 to 75 mm rutting criteria for construction equipment, and refer to the procedures outlined in Section 5.6.

STEP 5.   Compare thicknesses.
Compare the aggregate-geotextile system thicknesses determined in Steps 3 and 4. Use the system with the greater thickness.

STEP 6.   Check geotextile filtration.
Check the geotextile filtration characteristics using the gradation and permeability of the subgrade, the water table conditions, and the retention and permeability criteria. From Chapter 2, that criteria is:

$$AOS \leq D_{85} \text{ (Wovens)} \qquad \text{(Eq. 2-3)}$$

$$AOS \leq 1.8\, D_{85} \text{ (Nonwovens)} \qquad \text{(Eq. 2-4)}$$

$$k_{geotextile} \geq k_{soil} \qquad \text{(Eq. 2-7a)}$$

$$\psi \geq 0.1 \text{ sec}^{-1} \qquad \text{(Eq. 2-8c)}$$

STEP 7.   Determine geotextile survivability requirements.
Check the geotextile strength requirements for survivability as discussed in Section 5.5.

STEP 8.   Specify geotextiles that meet or exceed those survivability criteria.

STEP 9.   Follow the construction recommendations in Section 5.12.

5.8-2   Other Design Methods

Design methods for improving the structural capacity of permanent roads using geotextiles (e.g., Hamilton and Pearce, 1981) and geogrids (e.g., Haas, 1986; Haas, et. al., 1988; Barksdale, et al., 1989; Webster, 1993) have been proposed and may also be used. If a geogrid is used, either the base material should be sufficiently well graded to provide subgrade filtration and prevent soil intrusion, or for more open bases, a geotextile filter should be used with the geogrid.

5.9   PERMANENT ROAD DESIGN EXAMPLE (Christopher and Holtz, 1991)

DEFINITION OF DESIGN EXAMPLE

Project Description: New public street and service drive for a suburban Washington, D.C., townhouse development. Commonwealth of Virginia DOT regulations apply.

Type of Structure: Category IV street (permanent road)

Type of Application: geotextile separator

Alternatives:

i) excavate unsuitable material and increase subgrade aggregate thickness; or

ii) geotextile separator between aggregate and subgrade

GIVEN DATA

subgrade
- surficial soils: micaceous silts (CBR ≈ 2)
- local areas of very poor soils (CBR ≈ 0.5)
- low-lying topography
- poor drainage
- other nearby streets and roads require frequent maintenance

traffic
- maximum 300 vehicles per day
- 96% passenger, 5% single-axle, 1 multiaxle
- equivalent daily 90 kN single-axle load (EAL) applications = 10

REQUIRED

Design the pavement section.

Consider: 1) standard AASHTO design; and 2) alternate with geotextile

DEFINE

A. Geotextile function(s)

B. Geotextile properties required

C. Geotextile specification

SOLUTION

A. Geotextile function(s):
   Primary   -   separation
   Secondary -   filtration

B. Geotextile properties required:
   survivability
   apparent opening size (AOS)
   permeability

DESIGN

Design pavement with and without geotextile inclusion. Compare options.

STEP 1. ESTIMATE NEED FOR GEOTEXTILE

Ideal conditions for considering a geotextile; *e.g.*, low CBR, saturated subgrade, and poor performance history with conventional design.

STEP 2. DESIGN WITHOUT GEOTEXTILE

The structural design for the pavement section is based on the AASHTO Guide for Design of Pavement Structures (1977) using an equivalent design structural number for the anticipated loading and soil support conditions. (NOTES: i) AASHTO design uses English units; and ii) this case history used the AASHTO guide, 1977, which was current at time of design.)

traffic - as given

Determine structural number, SN:
from AASHTO design charts and with 20 years, CBR = 2, EAL = 10, and Regional Factor = 2
SN is equal to 2.9

Compute pavement thickness for structural support on a stable subgrade (*i.e.*, no fines pumped into aggregate subbase and no aggregate loss down into the subgrade):

Assume 2.5 inches asphaltic concrete surface and 8 inches aggregate base course

	surface	+	base	+	subbase
SN =	$a_1 D_1$	+	$a_2 D_2$	+	$a_3 D_3$
2.9 =	0.4 x 2.5"	+	0.14 x 8"	+	0.13 x D_3

Therefore, $D_3 = 6$ inches required for subbase.

Structural design: 2.5-in.   asphaltic concrete
                    8-in.   aggregate base
                    6-in.   aggregate subbase

STEP 3. ADDITIONAL AGGREGATE FOR PUMPING AND INTRUSION

By local experience in this area, and for subgrades of CBR ≤ 2, an additional 8 inches of aggregate subbase is required (stabilization aggregate).

For the geotextile separator alternate, this entire stabilization layer could be eliminated. However, some very poor soils are anticipated, and some conservatism can be applied. Therefore, reduce subbase aggregate thickness to 4 inches (100 mm) and use a geotextile separator.

STEP 4. DESIGN FOR CONSTRUCTABILITY USING A GEOTEXTILE

Use temporary road design procedures.

Assume:
   CBR = 2
            loaded dump trucks
            < 100 passes
            50 mm rut depth acceptable

Use Figure 5-5 (use 40 kN load) and
   $c \approx 30 \times$ CBR ≈ 60 kPa
   N_c $= 6$
   cN_c $= 360$ kPa
   aggregate depth $= 100$ mm

(See diagram on following page)

STEP 4. DESIGN FOR CONSTRUCTABILITY USING A GEOTEXTILE

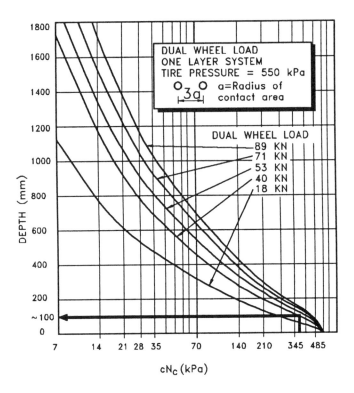

STEP 5. COMPARE THE THICKNESSES DETERMINED IN STEPS 3 AND 4

The two thicknesses are equal; therefore, use 100 mm of stabilization aggregate with a geotextile separator. Note that the minimum thickness of aggregate for a construction haul road is 100 mm, though the contractor will likely use a greater thickness.

STEP 6. CHECK GEOTEXTILE FILTRATION CHARACTERISTICS

Use AOS < 0.3 mm per Table 5-2 because > 50% passes the 0.075 mm sieve.

Permeability of geotextile must be greater than soil permeability per Table 5-2. Estimate soil permeability and determine geotextile requirement.

STEP 7. DETERMINE GEOTEXTILE SURVIVABILITY REQUIREMENTS

Use Table 5-1: with CBR = 2, dump truck contact pressure > 350 kPa, and 150 mm cover thickness (note that 100 mm cover is limited to existing road bases, therefore 150 mm minimum compacted lift thickness is recommended), and determine that a geotextile with a HIGH survivability rating is required.

STEP 8. SPECIFY GEOTEXTILE PROPERTY REQUIREMENTS

Geotextile separator shall meet or exceed the minimum average roll values, with elongation at failure determined with the ASTM D 4632 test method, of:

Property	Test Method	Elongation $\leq 50\%$	Elongation $> 50\%$
Grab Strength	ASTM D 4632	1400	900
Puncture Resistance	ASTM D 4833	500	350
Tear Resistance	ASTM D 4533	500	350

The geotextile shall have an AOS < 0.3 mm, $\psi \geq 0.1$ sec^{-1}, and the permeability shall be 0.1 cm/sec.

STEP 9. SPECIFY CONSTRUCTION REQUIREMENTS

See Section 5.12.

5.10   COST CONSIDERATIONS

Estimation of construction costs and benefit-cost ratios for geosynthetic-stabilized road construction is straight-forward and basically the same as that required for alternative pavement designs. Primary factors include the following:

1. cost of the geosynthetic;
2. cost of constructing the conventional design versus a geosynthetic design (*i.e.*, stabilization requirements for conventional design versus geosynthetic design), including
   a) stabilization aggregate requirements,
   b) excavation and replacement requirements,
   c) operational and technical feasibility, and
   d) construction equipment and time requirements;
3. cost of conventional maintenance during pavement service life versus improved service anticipated by using geosynthetic (estimated through pavement management programs); and
4. regional experience.

Annual cost formulas, such as the Baldock method (Illinois Department of Transportation, 1982), can be applied with an appropriate present worth factor to obtain the present worth of future expenditures.

Cost tradeoffs should also be evaluated for different construction and geosynthetic combinations. This should include subgrade preparation and equipment control versus geosynthetic survivability. In general, higher-cost geosynthetics with a higher survivability on the existing subgrade will be less expensive than the additional subgrade preparation necessary to use lower-survivability geosynthetics.

Research is ongoing to quantify the cost-benefit life cycle ratio of using geosynthetics in permanent roadway systems. In any case, the cost of a geosynthetic is generally $1.25/m^2$ while the cost of the pavement section is generally $25/m^2$. **The life extension of the roadway section will more than make up for the cost of the geosynthetic. The ability of a geosynthetic to prevent premature failure provides an extremely low-cost performance insurance.**

5.11   SPECIFICATIONS

Specifications should generally follow the guidelines in Section 1.6. The main considerations include the minimum geotextile requirements for design and those obtained from the survivability, retention, and filtration requirements in (Sections 5.5 and 5.8), as well as the construction requirements covered in Section 5.12. As with other applications, it is very important that an engineer's representative be on site during placement to observe that the correct geotextile has been delivered, that the specified construction sequence is being followed in detail, and that no damage to the geotextile is occurring. The following draft specification, slightly modified, from AASHTO-AGC-ARTBA Task Force 25 (1990) is an example.

AASHTO-AGC-ARTBA TASK FORCE 25
SPECIFICATION GUIDE FOR GEOTEXTILES USED IN SEPARATION
AND FILTER APPLICATIONS
(May, 1987, with slight modifications)

1.   *Description*
    This work shall consist of furnishing, and placing a geotextile primarily for use as a separator and filter to prevent mixing of dissimilar materials such as subgrades and surfaced and unsurfaced pavement structures, zones in embankments, foundations and select fill materials. The geotextile shall be designed to allow passage of water while retaining in situ soil without clogging. This specification does not address reinforcement applications which require an engineered project specific design.

2.	*Materials*

Fibers used in the manufacture of geotextile, and the threads used in joining geotextiles by sewing, shall consist of long chain synthetic polymers, composed of at least 85% by weight polyolefins, polyesters, or polyamides. Both the geotextile and threads shall be resistant to chemical attack, mildew and rot. These materials, based on construction survivability conditions defined in Table 5-1, shall conform to the physical requirements of Table 5-2.

3.	*Construction Methods/Requirements*

3.1	Geotextiles Packaging and Storing - Geotextile rolls shall be furnished with suitable wrapping for protection against moisture, and extended ultraviolet exposure prior to placement. Each roll shall be labeled or tagged to protect product identification sufficient for field identification as well as inventory and quality control purposes. Rolls shall be stored in a manner which protects them from the elements. If stored outdoors, they shall be elevated and protected with a waterproof cover.

3.2	Geotextile Exposure Following Placement - Exposure of geotextiles to the elements between lay down and cover shall be as soon as possible but not more than 3 days to minimize damage potential.

3.3	Site Preparation - The installation site shall be prepared by cleaning, grubbing, and excavation or filling to the design grade.

NOTE: Soft spots and unsuitable areas will be identified during site preparation or subsequent proof rolling. These areas shall be excavated and backfilled with select material compacted to normal procedures.

3.4	Installation - Geotextile installation shall proceed in the direction of construction. The geotextile shall be laid and overlapped in the direction as shown on the plans and shall be as wrinkle free as possible. On curves, the fabric may be folded or cut to accommodate the curve. As shown in Figure 5-7, the fold or overlap of cut pieces shall be in the direction of construction, and pinned, stapled, or weighted with cover material. The minimum initial cover will comply with the plans and specifications or shall be selected with aid of Table 5-3. Placement and grading of fill, subbase, or base material shall proceed in the direction of construction. Ruts that occur in placed material during construction shall be filled with the appropriate material which should be subsequently compacted.

3.5	Joints, Seams and Overlays - Where seams are required, they shall be joined by either sewing, sealing, or overlapping. All seams shall be subject to approval of the Engineer. Both factory and field sewn or sealed seams shall conform to the requirements of Table 5-2. Optionally, overlapped seams shall have a minimum overlap of 300 mm or as shown on the plans.

3.6	The contractor shall patch rips or tears in the geotextile as approved by the Engineer (repairs shall be performed by placing a new layer of fabric extending beyond the defect in all directions a minimum of the overlap required for parallel rolls. Alternatively, the defective section shall be replaced as directed by the Engineer).

4.	*Method of Measurement*

4.1	The geotextile shall be measured by the number of square meters computed from the payment lines shown on the plans or from payment lines established in writing by the

Engineer. This excludes seam overlaps.

4.2 Excavation, backfill, bedding and cover material are separate pay items.

5. *Basis of Payment*

5.1 The accepted quantities of geotextile shall be paid for at the contract unit price square meter in place.

5.2 Payment will be made under:

Pay Item	Pay Unit
Separator Geotextile	square meter

5.12   INSTALLATION PROCEDURES

5.12-1 Roll Placement

Successful use of geotextiles in pavements requires proper installation, and Figure 5-8 shows the proper sequence of construction. Even though the installation techniques appear fairly simple, most geotextile problems in roadways occur as the result of improper construction techniques. If the geotextile is ripped or punctured during construction activities, it will not likely perform as desired. If the geotextile is placed with a lot of wrinkles or folds, it will not be in tension, and, therefore, cannot provide a reinforcing effect. Other problems occur due to insufficient cover over the geotextile, rutting of the subgrade prior to placing the geotextile, and thin lifts that exceed the bearing capacity of the soil. The following step-by-step procedures should be followed, along with careful observations of all construction activities.

1. The site should be cleared, grubbed, and excavated to design grade, stripping all topsoil, soft soils, or any other unsuitable materials (Figure 5-8a). If moderate site conditions exist, *i.e.*, CBR greater than 1, lightweight proofrolling operations should be considered to help locate unsuitable materials. Isolated pockets where additional excavation is required should be backfilled to promote positive drainage. Optionally, geotextile-wrapped trench drains could be used to drain isolated areas.

2. During stripping operations, care should be taken not to excessively disturb the subgrade. This may require the use of lightweight dozers or grade-alls for low-strength, saturated, noncohesive and low-cohesive soils. For extremely soft ground, such as peat bog areas, do not overexcavate surface materials so you may take advantage of the root mat strength, if it exists. In this case, all vegetation should be cut at the ground surface. Sawdust or sand can be placed over stumps or roots that extend above the ground surface to cushion the geotextile. Remember, the subgrade preparation must correspond to the survivability properties of the geotextile.

3. Once the subgrade along a particular segment of the road alignment has been prepared, the geotextile should be rolled in line with the placement of the

new roadway aggregate (Figure 5-8b). Field operations can be expedited if the geotextile is pre-sewn to design widths in the factory or on firm ground so it can be unrolled on site in one continuous sheet. The geotextile should not be dragged across the subgrade. The entire roll should be placed and rolled out as smoothly as possible. Wrinkles and folds in the fabric should be removed by stretching and staking as required.

4. Parallel rolls of geotextiles should be overlapped, sewn, or joined as required. (Specific requirements are given in Sections 5.12-2 and 5.12-3.)

5. For curves, the geotextile should be folded or cut and overlapped in the direction of the turn (previous fabric on top) (Figure 5-9). Folds in the geotextile should be stapled or pinned approximately 0.6 m on centers.

6. When the geotextile intersects an existing pavement area, the geotextile should extend to the edge of the old system. For widening or intersecting existing roads where geotextiles have been used, consider anchoring the geotextile at the roadway edge. Ideally, the edge of the roadway should be excavated down to the existing geotextile and the existing geotextile sewn to the new geotextile. Overlaps, staples, and pins could also be utilized.

7. Before covering, the condition of the geotextile should be checked for excessive damage (*i.e.*, holes, rips, tears, etc.) by an inspector experienced in the use of these materials. If excessive defects are observed, the section of the geotextile containing the defect should be repaired by placing a new layer of geotextile over the damaged area. The minimum required overlap required for parallel rolls should extend beyond the defect in all directions. Alternatively, the entire defective section can be replaced.

8. The base aggregate should be end-dumped on the previously placed aggregate (Figure 5-8c). For very soft subgrades, pile heights should be limited to prevent possible subgrade failure. The maximum placement lift thickness for such soils should not exceed the design thickness of the road.

9. The first lift of aggregate should be spread and graded to 300 mm, or to the design thickness if less than 300 mm, prior to compaction (Figure 5-8d). At no time should traffic be allowed on a soft roadway with less than 200 mm (150 mm for CBR $\geq$ 3) of aggregate over the geotextile. Equipment can operate on the roadway without aggregate for geotextile installation under permeable bases, if the subgrade is of sufficient strength. For extremely soft soils, lightweight construction vehicles will likely be required for access on the first lift. Construction vehicles should be limited in size and weight so rutting in the initial lift is limited to 75 mm. If rut depths exceed 75 mm, it will be necessary to decrease the construction vehicle size and/or weight or to increase the lift thickness. For example, it may be necessary to reduce the size of the dozer required to blade out the fill or to deliver the fill in half-loaded rather than fully loaded trucks.

a. Prepare the ground by removing
stumps, boulders, etc.; fill in low spots

PREPARE THE GROUND

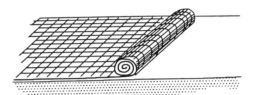

b. Unroll the geotextile directly over
the ground to be stabilized. If more
than one roll is required, overlap rolls.
Inspect geotextile.

UNROLL THE GEOTEXTILE

c. Back dump aggregate onto previously
placed aggregate. Do not drive on the
geotextile. Maintain 150 mm to 300 mm
cover between truck tires and Geotextile.

BACK DUMP AGGREGATE

d. Spread the aggregate over the
geotextile to the design thickness.

SPREAD THE AGGREGATE

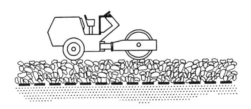

e. Compact the aggregate using dozer
tracks or smooth drum vibratory roller.

COMPACT THE AGGREGATE

Figure 5-8   Construction sequence using geotextiles.

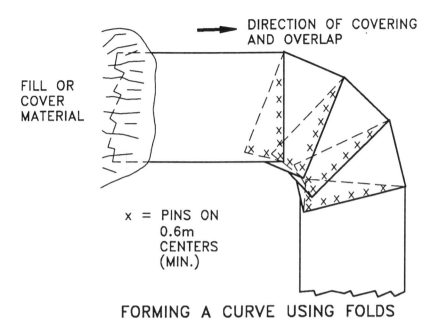

FORMING A CURVE USING FOLDS

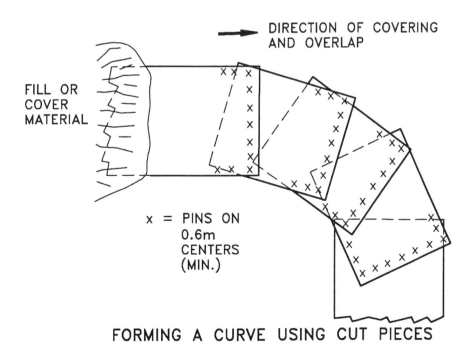

FORMING A CURVE USING CUT PIECES

Figure 5-9   *Forming curves using geotextiles.*

10. The first lift of base aggregate should be compacted by tracking with the dozer, then compacted with a smooth-drum vibratory roller to obtain a minimum compacted density (Figure 5-8e). For construction of permeable bases, compaction shall meet specification requirements. For very soft soils, design density should not be anticipated for the first lift and, in this case, compaction requirements should be reduced. One recommendation is to allow compaction of 5% less than the required minimum specification density for the first lift.

11. Construction should be performed parallel to the road alignment. Turning should not be permitted on the first lift of base aggregate. Turn-outs may be constructed at the roadway edge to facilitate construction.

12. On very soft subgrades, if the geotextile is to provide some reinforcing, pretensioning of the geotextile should be considered. For pretensioning, the area should be proofrolled by a heavily loaded, rubber-tired vehicle such as a loaded dump truck. The wheel load should be equivalent to the maximum expected for the site. The vehicle should make at least four passes over the first lift in each area of the site. Alternatively, once the design aggregate has been placed, the roadway could be used for a time prior to paving to prestress the geotextile-aggregate system in key areas.

13. Any ruts that form during construction should be filled in, as shown in Figure 5-10 to maintain adequate cover over the geotextile. In no case should ruts be bladed down, as this would decrease the amount of aggregate cover between the ruts.

14. All remaining base aggregate should be placed in lifts not exceeding 250 mm in loose thickness and compacted to the appropriate specification density.

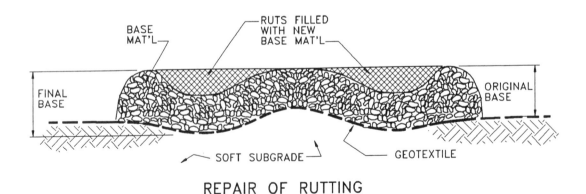

REPAIR OF RUTTING

Figure 5-10   Repair of rutting with additional material.

5.12-2 Overlaps

Overlaps can be used to provide continuity between adjacent geotextile rolls through frictional resistance between the overlaps. Also, a sufficient overlap is required to prevent soil from squeezing into the aggregate at the joint. The amount of overlap depends primarily on the soil conditions and the potential for equipment to rut the soil. If the subgrade does not rut under construction activities, only a minimum overlap is required to provide some pullout resistance. As the potential for rutting and squeezing of soil increases, the required overlap increases. Since rutting potential can be related to CBR, it can be used as a guideline for the minimum overlap required, as shown in Table 5-4.

Table 5-4
Recommended Minimum Geotextile Overal Requirements

CBR	Minimum Overlap
> 2	300 - 450 mm
1 - 2	600 - 900 mm
0.5 - 1	900 mm or sewn[1]
< 0.5	sewn[1]
All roll ends	900 mm or sewn[1]
NOTE:	1. See Section 5.12-3.

The geotextile can be stapled or pinned at the overlaps to maintain their position during construction activities. Nails 250 to 300 mm long should be placed at a minimum of 15 m on centers for parallel rolls and 1.5 m on centers for roll ends.

Geotextile roll widths should be selected so overlaps of parallel rolls occur at the roadway centerline and at the shoulders. Overlaps should not be placed along anticipated primary wheel path locations. Overlaps at the end of rolls should be in the direction of the aggregate placement (previous roll on top).

5.12-3 Seams

When seams are required for separation applications, they should meet the same tensile strength requirements for survivability (Table 5-1) as those of the geotextile perpendicular to the seam (as determined by the same testing methods). Seaming is discussed in detail in Section 1.8. All factory or field seams should be sewn with thread as strong and durable as the material in the fabric. J-seams with interlocking stitches are recommended. Alternatively, if bag-type stitches, which can easily unravel, or butt-type seams are used, seams should be double-sewn with parallel stitching spaced no more than 5 to 10 mm apart. Double sewing is required to safeguard against undetected missed stitches. The geotextile strength may actually have to exceed the specifications in order to provide seam strengths equal to the specified tensile strength.

For geogrids, overlap joints, tying or interlocking with wire cables, plastic pipe, hog rings, or bodkin joints may be required. Geotextile seam strength requirements should also be applied to overlapped or mechanically fastened geogrids. Consult the manufacturer for specific recommendations and strength test data.

5.13 FIELD INSPECTION

The field inspector should review the field inspection guidelines in Section 1.7. Particular attention should be paid to factors that affect geotextile survivability: subgrade condition, aggregate placement, lift thickness, and equipment operations.

5.14 SELECTION CONSIDERATIONS

For a geotextile to perform its intended function as a separator in a roadway, it must be able to tolerate the stresses imposed on it during construction; *i.e.,* the geotextile must have sufficient survivability to tolerate the anticipated construction operations. Geotextile selection for roadways is usually controlled by survivability, and the guidelines given in Section 5.5 are important in this regard. As mentioned, the specific geotextile property values given in Table 5-2 are minimums and are subject to revision. In the meantime, for important projects, you are strongly encouraged to conduct your own field trials, as described in Section 5.5.

5.15 REFERENCES

References quoted within this section are listed below.

AASHTO (1996), Standard Specification for Geotextiles, Designation: M 288 (draft). American Association of State Transportation and Highway Officials, Washington, D.C.

AASHTO (1990), Standard Specifications for Geotextiles - M 288, Standard Specifications for Transportation Materials and Methods of Sampling and Testing, American Association of State Transportation and Highway Officials, Washington, D.C., pp 689-692.

AASHTO (1972), Interim Guide for the Design of Pavement Structures, American Association of State Transportation and Highway Officials, Washington, D.C.

AASHTO-AGC-ARTBA (1990), Guide Specifications and Test Procedures for Geotextiles, Task Force 25 Report, Subcommittee on New Highway Materials, American Association of State Transportation and Highway Officials, Washington, D.C.

ASTM (1994), Soil and Rock (II): D 4393 - latest; Geosynthetics, Annual Book of ASTM Standards, Section 4, Volume 04.09, American Society for Testing and Materials, Philadelphia, PA, 516 p.

Barksdale, R.D., Brown, S.F. and Chan, F. (1989), Potential Benefits of Geosynthetics in Flexible Pavement Systems, National Cooperative Highway Research Program Report 315, Transportation Research Board, Washington, D.C., 56 p.

Baumgartner, R.H. (1994), Geotextile Design Guidelines for Permeable Bases, Federal Highway Administration, Washington, D.C., June, 33 p.

Cedergren, H.R. (1987), Drainage of Highway and Airfield Pavements, Krieger, 289 p.

Chan, F., Barksdale, R.D. and Brown, S.F. (1989), Aggregate Base Reinforcement of Surface Pavements, Geotextiles and Geomembranes, Vol. 8, No. 3, pp. 165-189.

Christopher, B.R. and Holtz, R.D. (1985), Geotextile Engineering Manual, Report No. FHWA-TS-86/203, Federal Highway Administration, Washington, D.C., 1044 p.

Christopher, B.R. and Holtz, R.D. (1991), Geotextiles for Subgrade Stabilization in Permanent Roads and Highways, Proceedings of Geosynthetics '91, Atlanta, GA, Vol. 2, pp. 701-713.

FHWA (1989), Geotextile Design Examples, Geoservices, Inc., Federal Highway Administration, Contract No. DTFH-86-R-00102, Washington, D.C.

Giroud, J.P. and Noiray, L. (1981), Geotextile-Reinforced Unpaved Roads, Journal of the Geotechnical Engineering Division, American Society of Civil Engineers, Vol 107, No GT 9, September, pp. 1233-1254.

Haas, R. (1986), Granular Base Reinforcement of Flexible Pavements Using Tensar Geogrids, Technical Note BR1, The Tensar Corporation, Atlanta, GA, 22 p.

Haas, R., Walls, J. and Carroll, R.G. (1988), Geogrid Reinforcement of Granular Base in Flexible Pavements, presented at the 67th Annual Meeting, Transportation Research Board, Washington, D.C.

Haliburton, T.A., Lawmaster, J.D. and McGuffey, V.C. (1981), Use of Engineering Fabrics in Transportation Related Applications, FHWA DTFH61-80-C-00094.

Hamilton, J.S. and Pearce, R.A. (1981), Guidelines for Design of Flexible Pavements using Mirafi Woven Stabilization Fabrics, Law Engineering Testing Co. Report to Celanese Corp., 47 p.

Illinois Department of Transportation (1982), Design Manual - Section 7: Pavement Design, I-82-2, Bureau of Design, Springfield.

Koerner, R.M. (1994), Designing With Geosynthetics, 3rd Edition, Prentice-Hall Inc., Englewood Cliffs, NJ, 783 p.

Rankilor, P.R. (1981), Membranes in Ground Engineering, John Wiley & Sons, Inc., Chichester, England, 377 p.

Steward, J., Williamson, R. and Nohney, J. (1977), Guidelines for Use of Fabrics in Construction and Maintenance of Low-Volume Roads, USDA, Forest Service, Portland, OR. Also reprinted as Report No. FHWA-TS-78-205.

Terzahgi, K. and Peck, R.B. (1967), Soil Mechanics in Engineering Practice, John Wiley & Sons, New York, 729 p.

Webster, S.L. (1993), Geogrid Reinforced Base Courses for Flexible Pavements for Light Aircraft: Test Section Construction, Behavior Under Traffic, Laboratory Tests, and Design Criteria, USAE Waterways Experiment Station, Vicksburg, MS, Technical Report GL-93-6, 100 p.

Yoder, E.J. and Witczak, M.W. (1975), Principles of Pavement Design, Second Edition, Wiley, 711 p.

6.0   PAVEMENT OVERLAYS

6.1   BACKGROUND

The second largest application of geotextiles in North America is in asphalt overlays of asphalt concrete (AC) and Portland cement concrete (PCC) pavement structures. (The largest single application is separation/stabilization, which utilizes an estimated 100 million square meters of geotextile.)   An estimated 85 million square meters of geotextiles were used in overlays in 1993 (Jagielski, 1993). This is 26% of the estimated 330 million square meters used in North America in 1993. This is indeed an impressive statistic. Notwithstanding, use of geotextiles in overlays may be described as a love-hate relationship with user agencies.

Many engineers are thoroughly convinced of the performance and cost benefit of geotextiles incorporated into overlays. Many other engineers are thoroughly convinced that geotextiles are either not beneficial or not economical in overlay construction. And, still other engineers are confused by the claims of performance and cost benefits. These divergent opinions regarding performance and benefits are addressed in this chapter.

The history of geotextiles in this application accounts for much of the confusion and skepticism. Promotion of geotextiles in overlays in the 1970s and early 1980s claimed that the geotextile reinforced the pavement and that the reinforcement prevented cracks in the old pavement from reflecting up through the new overlay. These claims are rarely, if ever, presented today. The tensile moduli of the light-weight nonwoven geotextiles typically used in this application are too low to mobilize significant tension under acceptable pavement deflections for the geotextile to act as a reinforcement. It is also commonly accepted that geotextiles do not prevent reflection cracking from occurring. How, then, are geotextiles beneficial in overlay pavement construction?

6.2   GEOTEXTILE FUNCTIONS

Geotextiles can be used as alternatives to stress-relieving granular layers, seal coats, rubberized asphalts, etc., for controlling surface moisture infiltration and retarding reflection cracks in pavement overlays. Properly installed, asphalt-saturated geotextiles function as a moisture barrier that protects the underlying pavement structure from further degradation due to ingress of surface water. In addition, geotextiles can provide cushioning for the overlay, thus functioning as a stress-relieving interlayer.   When properly installed, both functions combine to extend the life of the overlay.

6.3   APPLICATIONS

Pavement rehabilitation is required where structural deficiencies adversely affect the load-carrying capability of the existing pavement structure. Impregnated geotextiles are used beneath asphalt concrete (AC) overlays for rehabilitation and as preventive measures to slow the deterioration rate of existing roadways. Preventive measures such as AC overlays and moisture protection are used to extend pavement life and are often economical when life-cycle costs are computed.

6.3-1   Asphalt Concrete Pavements

When geotextiles are used with AC overlays of AC pavements, they can be effective in controlling (retarding) reflection cracking of low- and medium-severity alligator-cracked pavements. They also may be useful for controlling reflection of thermal cracks, although they are not as effective in retarding reflection of cracks due to significant horizontal or vertical movements. (AASHTO, 1993)

The variable performance of geotextile overlays, and the divergent opinions regarding benefits, is strongly influenced by the following factors (Barksdale, 1991):

- type and extent of existing pavement distress, including crack widths;
- extent of remedial work performed on the old pavement, such as crack sealing and/or filling, pothole repair, and replacement of failed base and subgrade areas;
- overlay thickness;
- variability of pavement structural strength from one section to another; and
- climate.

Obviously, an additional factor is the geotextile. These factors are summarized below.

Distress Type (Barksdale, 1991): Geotextiles generally have performed best when used for load-related fatigue distress (*e.g.*, closely spaced alligator cracking). Fatigue cracks should be less than 3 mm wide for best results. Cracks greater than 10 mm wide require a stiff filler. Geotextiles used to retard thermal cracking have, in general, been found to be ineffective.

Remediation of Old Pavement (AASHTO, 1993): Much of the deterioration that occurs in overlays is the result of unrepaired distress in the existing pavement prior to the overlay. Distressed areas of the existing pavement should be repaired if the distress is likely to affect performance of the overlay within a few years. The amount of preoverlay repair is related to the overlay design. The engineer should consider the cost implications of preoverlay repair versus overlay design. Guidelines on preoverlay repair techniques are available from the FHWA (FHWA, {current edition}; FHWA, 1987).

Effect of Overlay Thickness: Pavement performance is quite sensitive to the overlay thickness, either with or without a geotextile interlayer. Correspondingly, the benefits of a geotextile in retarding reflective cracking will increase with increasing thickness of the overlay. Geotextiles are most effective in retarding reflective cracking in thin (e.g., 40 mm) overlays. These observations, as reported by Barksdale (1991), are based upon extensive research conducted by Caltrans (Predoehl, 1990). The California results imply that a relatively thin geotextile interlayer is structurally adequate and equivalent to 30 mm of asphalt concrete.

Variability of Pavement Structural Strength (Barksdale, 1991): The structural strength of existing pavement, and, therefore, required overlay thickness, often varies greatly along a roadway. The significant effect of such variation on overlay performance has not often been considered in the past. This oversight likely contributes to some of the diverse opinions regarding geotextile benefits in asphalt overlays. (Variation of pavement strength along a roadway should be addressed for future demonstration or test sections of overlays.)

Climate: It has been observed (Aldrich, 1986) geotextile interlayers have generally performed better in warm and mild climates than in cold climates. However, the beneficial effects of reducing water infiltration - a principal function of the geotextile - were not considered in Aldrich's (1986) study. Successful installation and beneficial performance of geotextile interlayers in cold regions, such as Alaska, challenge the generality regarding climate.

Geotextile: Lightweight (e.g., 135 g/m^2) nonwoven geotextiles are typically used for asphalt overlays. These asphalt-impregnated geotextiles primarily function as a moisture barrier. Use of heavier, nonwoven geotextiles can provide cushioning or stress-relieving membrane interlayer-like benefits, in addition to moisture-barrier functions.

6.3-2 Portland Cement Concrete Pavements

Geotextiles are used with AC overlays of crack/seat-fractured plain Portland cement concrete (PCC) pavement to help control reflection cracking (AASHTO, 1993).

Geotextiles also may be used for AC overlays of jointed plain concrete pavement (JPCP) and of jointed reinforced concrete pavement (JRCP) to control reflective cracking. However, the effectiveness with these pavements is listed as questionable in the AASHTO Guide for Design of Pavement Structures (1993).

Important factors in assessing applicability and potential benefits of using a geotextile interlayer with PCC pavements include:

- existing structural strength of the pavement;
- slab preparation;
- geotextile installation;
- required overlay thickness;
- climate; and
- economics of geotextile overlay versus other design alternatives (Barksdale, 1991).

6.3-3   AC-Overlaid PCC Pavements

Geotextiles are also used for new AC overlays of AC-overlaid Portland cement concrete pavements (AC/PCC), where the original pavements may be JPCP, JRCP or continuously reinforced concrete pavement (CRCP). Some pavements are constructed as AC/PCC, although most are PCC pavements that have already been overlaid with AC. In addition to controlling reflecting cracking, an impregnated geotextile can help control surface water infiltration into the pavement, which can result in loss of bond between AC and PCC, stripping in the AC layers, progression of D cracking or reactive aggregate distress (in pavements with these problems), and weakening of the base and subgrade materials.

6.4   ADVANTAGES AND POTENTIAL DISADVANTAGES

6.4-1   Advantages

An asphalt-impregnated geotextile functioning as a moisture barrier and a stress-relieving interlayer provides several possible benefits and, therefore, advantages to their use. Retardation of reflection cracks will:

- increase the overlay and the roadway life;
- decrease roadway maintenance costs; and
- increase pavement serviceability.

Reflection cracking can have a considerable, often controlling, influence on the life of an AC overlay (AASHTO, 1993). After a pavement cracks, its longevity is quickly reduced. Deteriorated reflection cracks require more frequent maintenance, such as sealing and patching. Reflection cracks also permit water to enter the pavement structure, which can weaken the base layers and subgrade, and decrease the structural capacity of the pavement. Base and subgrade will be weakened by ingress of water if the base does not have excellent (*i.e.*, water removed within 2 hours) or good (*i.e.*, water removed within 1 day) drainage. Water infiltration causes a reduction in shear strength of the subgrade, which in turn leads to a rapid deterioration of the pavement system. The sealing function of the asphalt-impregnated geotextile is intended to reduce surface water infiltration through reflection cracks (when they eventually reappear at the surface of the overlay) and through thermal-induced cracks.

Reduction in surface water entering PCC pavements potentially provides additional benefits of:

- reduction or elimination of pumping (*i.e.*, no water, no pumping);
- decreased slab movements through reduced erosion of fines from beneath the slab (lower moisture gradients might also reduce slab warping); and
- increased subgrade strength through a decrease in moisture (Barksdale, 1991).

6.4-2 Potential Disadvantages

Correct construction of the asphalt-impregnated geotextile and AC overlay is paramount to its functioning as designed. Too little asphalt in the tack coat can result in a partially saturated geotextile, which in turn can absorb moisture and lead to spalling or popping off of the surface treatment due to freeze-thaw action within the geotextile. Bleeding occurs with too much asphalt which can result in overlay slippage, as well as potential pavement slippage planes. Bleeding also can cause difficulty with installation, as it can result in the geotextile sticking to and being pulled up by the tires and tracks of the asphalt trucks and paving vehicle. The AC overlay must be placed below the specified temperature, which requires inspection and control. AC placement significantly above the specified temperature can result in the asphalt tack coat being drawn out of the geotextile which can result in shrinkage or even melting of the geotextile. Shrinkage and melting is a concern for a polypropylene geotextile which has a typical melt temperature of 165°C, it is not a concern for a polyester geotextile which has a typical melt temperature of 225°C. Improper pavement preparation and crack filling can also decrease the effectiveness of the geotextile moisture barrier.

Geotextiles cannot be expected to perform well when the roadway being overlaid is structurally inadequate. Nor will such surface treatments do anything to solve ground water problems, subgrade softening, base course contamination, or freeze-thaw problems. These problems must be corrected before resurfacing, independent of geotextile used.

Geotextiles have also been found ineffective in reducing thermal cracking. Pavement overlay systems have also had limited success in areas of heavy rainfall and regions with significant freeze-thaw (FHWA Manual, 1982).

6.5 DESIGN

6.5-1 General

Design of AC overlays is thoroughly presented in the AASHTO Guide for Design of Pavement Structures (1993). To have a high probability of success, a carefully planned and executed study is required to develop an engineered overlay design using a geotextile (Barksdale, 1991). A carefully planned and executed study also is required for successful, alternative (*i.e.*, non-geotextile) overlay designs.

The steps required to develop an overlay design for flexible pavements with a geotextile, as summarized from Barksdale (1991), are as follows.

STEP 1. Pavement condition evaluation.

The results of a general pavement condition survey are valuable in establishing the type, severity, and extent of pavement distress. Candidate pavements should be divided into segments, and a thorough visual evaluation made of each segment to determine the type, extent, and severity of cracking, and to classify the present distress as: alligator cracking, block cracking, transverse cracks, joint cracking, patching, potholes, widening drop-offs, etc. Crack widths should be measured. See AASHTO Guide for Design of Pavement Structures (1993) for guidance.

STEP 2. Structural strength.

The overall structural strength of the pavement should be evaluated along its length, using suitable nondestructive techniques, such as the Benkelman beam, falling-weight deflectometer, Dynaflect, or Road Rater.

STEP 3. Base/subgrade failure.

Areas with base or subgrade failures should be identified. Benkelman beam pavement deflections greater than approximately 0.6 mm are indicative of failure, as is excessive rutting.

STEP 4. Remedial pavement treatment.

The results of the pavement condition survey and deflection measurements should be used to develop a pavement repair strategy for each segment.

STEP 5. Overlay design.

A realistic overlay thickness must be selected to ensure a reasonable overlay life. Design methodologies are presented in the AASHTO Guide for Design of Pavement Structures (1993). The overlay thickness with a geotextile should be determined as if the interlayer is not present.

STEP 6. Geotextile selection.

STEP 7. Performance monitoring.

Performance monitoring during the life of the overlay is highly desirable for developing a local data bank of performance histories using geotextiles in overlays. Using a control section without a geotextile interlayer, with all other items equal, will yield valuable comparative data.

The steps in developing an overlay design for PCC pavements where a geotextile may be used is generally similar to that for flexible pavements (Barksdale, 1991). Vertical joint deflection surveys should be performed. Full-width geotextiles should not be used when vertical joint deflections are greater than about 0.2 mm, unless corrective measures such as undersealing, are taken to reduce joint movement. Horizontal thermal joint movement should be less than about 1.3 mm. As before, the thickness of the overlay is not reduced with the use of a geotextile interlayer.

6.5-2 Drainage Considerations

As noted previously, the primary function of the geotextile in an overlay is to minimize infiltration of surface water into the pavement structure. The benefits of this are normally not quantified and, if incorporated into design, are only subjectively treated. To objectively quantify the benefits of a moisture barrier a potential design approach is to estimate the effects of the asphalt-impregnated geotextile barrier on the drainage characteristics of the pavement structure. The 1993 AASHTO Guide for Design of Pavement Structures presents provisions for modifying pavement design equations to take advantage of performance improvements due to good drainage. Although not discussed in the design guide, a geotextile overlay could be considered a method to improve drainage via reduced infiltration.

From the AASHTO Guide (1993), modified layer coefficients determine the treatment for the expected drainage level for flexible pavements. The factor for modifying the layer coefficient is referred to as an m_i value, thus the structural number (SN) becomes:

$$SN = a_1 D_1 + a_2 D_2 m_2 + a_3 D_3 m_3$$

where:

D_1, D_2, D_3 = thicknesses of existing pavement surface, base, and subbase layers;

a_1, a_2, a_3 = corresponding structural layer coefficients; and

m_2, m_3 = drainage coefficients for granular base and subbase.

The recommended m_i values are presented in Table 6-1 as a function of the drainage quality and the percent of time during the year the pavement structure is near saturation. Definitions of quality of drainage are presented in Table 6-2.

From the AASHTO guide (1993), the drainage coefficient, C_d, in the performance equation determines the treatment for the expected drainage level for rigid pavements. The performance equation is used to calculate a design slab thickness for a rigid pavement. Recommended values for C_d are presented in Table 6-3.

Again, these modification factors are not discussed for use with geotextile overlays in the AASHTO design guide. They are presented here, as it is hypothesized that these values could aid in objectively estimating the structural benefit of a geotextile moisture barrier.

Table 6-1

Recommended m_i Values for Modifying Structural Layer Coefficients of Untreated Base and Subbase Materials in Flexible Pavements
(from AASHTO, 1993)

Quality of Drainage	Percent of Time Pavement Structure is Exposed to Moisture Levels Approaching Saturation			
	< 1 %	1 to 5 %	5 to 25 %	> 25 %
Excellent	1.40 - 1.35	1.35 - 1.30	1.30 - 1.20	1.20
Good	1.35 - 1.25	1.25 - 1.15	1.15 - 1.00	1.00
Fair	1.25 - 1.15	1.15 - 1.05	1.00 - 0.80	0.80
Poor	1.15 - 1.05	1.05 - 0.80	0.80 - 0.60	0.60
Very Poor	1.05 - 0.95	0.95 - 0.75	0.75 - 0.40	0.40

Table 6-2
Quality of Pavement Drainage
(from AASHTO, 1993)

Quality of Drainage	Water Removed Within
Excellent	2 hours
Good	1 day
Fair	1 week
Poor	1 month
Very Poor	(water will not drain)

Table 6-3
Recommended Values of Drainage Coefficient, C_d,
for Pavement Design
(from AASHTO, 1993)

Quality of Drainage	Percent of Time Pavement Structure is Exposed to Moisture Levels Approaching Saturation			
	< 1 %	1 to 5 %	5 to 25 %	> 25 %
Excellent	1.25 - 1.20	1.20 - 1.15	1.15 - 1.10	1.10
Good	1.20 - 1.15	1.15 - 1.10	1.10 - 1.00	1.00
Fair	1.15 - 1.10	1.10 - 1.00	1.00 - 0.90	0.90
Poor	1.10 - 1.00	1.00 - 0.90	0.90 - 0.80	0.80
Very Poor	1.00 - 0.90	0.90 - 0.80	0.80 - 0.70	0.70

6.6 COST CONSIDERATIONS

As previously stated, the design thickness of an AC overlay with a geotextile interlayer should be determined as if the geotextile is not present. The economic justification of geotextile use is then derived from:

- an increase in pavement life; a decrease in pavement maintenance costs; and an increase in pavement serviceability due to retardation and possible reduction of reflection cracks;
- an increased structural capacity due to drier base and subgrade materials; or
- a combination of the above items.

The old pavement surface condition and overall installation play a very important role in the performance of the paving geotextile. The deteriorated pavement should be repaired, including filling joints and cracks and replacing sections with potholes and faults in their base or subgrade. Under favorable conditions, reflection cracks can be *retarded* for approximately 2 to 5 years as compared to the overlay without the paving grade geotextile. **The anticipated life improvement, under favorable conditions, is approximately 100 to 200% that of an overlay of the same design thickness without a geotextile.** Favorable conditions for the use of a paving grade geotextile with pavement repaving include:

- the presence of fatigue-related pavement failure, evidenced by alligator cracks;
- pavement cracks no wider than 3 mm; and
- the thickness of the new overlay designed to meet the structural requirements of the pavement.

The economic benefit of the geotextile interlayer functioning as a moisture barrier is currently not quantified in cost analyses. The effect of the geotextile on the quality of drainage might be used to objectively estimate an increase in pavement structural capacity. This increased capacity then can be used to estimate increased pavement life or to design a thinner AC overlay.

These potential economic benefits can be combined for a particular project. Other cost benefits, currently not quantified, include potential improvement of aesthetics and improved ride quality.

Alternatively, some engineers may reduce the overlay thickness based upon an equivalent performance thickness to justify economics. Extensive research conducted by Caltrans (Predoehl, 1989), implies that a geotextile interlayer is equivalent to 30 mm of asphalt concrete for relatively thin (*i.e.*, ≤ 120 mm) overlays that are structurally adequate. A useful rule of thumb, based upon typical in-place costs, is that a geotextile interlayer is roughly equivalent to the cost of about 15 mm of asphalt concrete. Cost of installed geotextile generally decreases with increased quantities and an experienced local installer.

Considerable insight into the economics of overlay design with geotextiles can be gained from historic cost and performance data (Barksdale, 1991). This data may be locally, regionally, nationally, or internationally generated.

A final economic analysis issue is the probability of success. Geotextile interlayers, as well as other rehabilitation techniques, are not always effective in improving pavement performance. Therefore, an estimate of the probability of success should be included in all economic analysis (Barksdale, 1991). The probability for success will obviously increase with thoroughness of rehabilitation design, local experience with geotextile interlayers, and thoroughness of construction inspection.

6.7   SPECIFICATIONS

Specific geotextile considerations, property requirements, and construction procedures are detailed in a specification prepared by the AASHTO-AGC-ARTBA Task Force 25 and are included herein. The specification was based on the combined experience of the Texas and California Departments of Transportation, which have had the greatest success using geotextiles in pavement overlays.

SPECIFICATIONS FOR PAVING FABRICS
(after AASHTO-AGC-ARTBA TASK FORCE No. 25, 1990)

1. *Description*

Work shall consist of furnishing and placing a fabric between pavement layers for the purpose of incorporating a waterproofing and stress relieving membrane within the pavement structure. This specification guide is applicable to fabric membranes used for full coverage of the pavement, or as strips over transverse and longitudinal pavement joints. It is not intended to describe membrane systems specifically designed for pavement joints and localized (spot) repairs.

2. *Materials*

2.1 Paving Fabric - The fabric used with this specification shall be constructed of nonwoven synthetic fibers; resistant to chemical attack, mildew, and rot; and shall meet the following physical requirements:

Property	Requirements	Task Method
Tensile Strength, N	350 minimum*	ASTM D 4632
Elongation-at-break, %	50 minimum	ASTM D 4632
Asphalt Retention, L/m^2	0.9 minimum	Task Force 25 Method 8
Melting Point, °C	150 or greater	ASTM D 276

*Minimum - Value in weaker principal direction. All numerical values represent minimum average roll values (i.e., any roll in a lot shall meet or exceed the minimum values in the table).

2.2 Asphalt Sealant - The material used to impregnate and seal the fabric, as well as bond it to both the base pavement and overlay, shall be a paving grade asphalt recommended by the fabric manufacturer and approved by the Engineer.

Uncut asphalt cements are the preferred sealant; however, cationic and anionic emulsions may be used provided the precautions outlined in Section 4.4 are followed. Cutbacks and emulsions which contain solvents shall not be used.

REMARKS
The grade of asphalt cement specified for hot-mix design in each geographic location is generally the most acceptable material.

2.3 Aggregate - Washed concrete sand may be spread over asphalt-saturated fabric to facilitate movement of equipment during construction or to prevent tearing or delamination of the fabric. Hot-mix broadcast in front of construction vehicle tires may also be used to serve this purpose. If sand is applied, excess quantities shall be removed from the fabric prior to placing the surface course.

REMARKS

Sand is not usually required. However, ambient temperatures are occasionally sufficiently high to cause bleed-through of the asphalt sealant resulting in undesirable fabric adhesion to construction vehicle tires.

3. *Equipment*

    3.1 <u>Asphalt Distributor</u> - The distributor shall be capable of spraying the asphalt sealant at the prescribed uniform application rate. No streaking, skipping or dripping will be permitted. The distributor shall be also equipped with a hand spray having a single nozzle and positive shut-off valve.

    3.2 <u>Fabric Handling Equipment</u> - Mechanical or manual laydown equipment shall be capable of laying the fabric smoothly.

    3.3 <u>Miscellaneous Equipment</u> - Stiff bristle brooms or squeegees to smooth the fabric, scissors or blades to cut the fabric, and brushes for applying asphalt sealant at fabric overlaps shall be provided. Pneumatic rolling equipment to smooth the fabric into the sealant and sanding equipment may be required for certain jobs.

REMARKS

Rolling is especially required on jobs where thin lifts or chip seals are being placed. Rolling helps ensure fabric bond to adjoining pavement layers in the absence of the heat and weight associated with thicker lifts of asphaltic pavement. An example of when rolling is extremely important is when the ambient temperature is so low that the normal wicking of the asphalt sealant into the fabric does not occur.

4. *Construction Methods/Requirements*

    4.1 <u>Fabric Packaging and Storing</u> - Fabric rolls shall be furnished with suitable wrapping for protection against moisture and extended ultraviolet exposure prior to placement. Each roll shall be labeled or tagged to provide product identification sufficient for inventory and quality control purposes. Rolls shall be stored in a manner which protects them from the elements. If stored outdoors, they shall be elevated and protected with a waterproof cover.

    4.2 <u>Weather Limitations</u> - Neither the asphalt sealant nor fabric shall be placed when weather conditions, in the opinion of the Engineer, are not suitable. Air and pavement temperatures shall be sufficient to allow the asphalt sealant to hold the fabric in place. For asphalt cements, air temperatures shall be 10 °C and rising. When using asphalt emulsions, air temperature shall be 16 °C and rising.

    4.3 <u>Surface Preparation</u> - The surface on which the fabric is to be placed shall be reasonably free of dirt, water, vegetation, or other debris. Cracks exceeding 3 mm in width shall be filled with a suitable crack filler and potholes shall be properly repaired as directed by the Engineer. The crack fillers shall be allowed to cure prior to fabric placement.

REMARKS

If the condition of the existing pavement is such that a simple crack fill operation is not adequate for surface preparation, then it may be more economical to place a leveling course prior to placing the fabric.

4.4 <u>Application of Asphalt Sealant</u> - The asphalt shall be uniformly spray applied to the prepared dry pavement surface at the rate 0.9 to 1.35 liters per square meter or as recommended by the fabric manufacturer and approved by the Engineer.

Application of the sealant shall be the distributor spray bar, with hand spraying kept to a minimum. Temperature of the asphalt sealant shall be sufficiently high to permit a uniform spray pattern. For asphalt cements, the minimum temperature shall be 140 °C. To avoid damage to the fabric, however, distributor tank temperatures shall not exceed 160 °C. Spray patterns for asphalt emulsion are improved by heating. Temperatures in the 55 to 70 °C range are desirable. A temperature of 70 °C shall not be exceeded since higher temperatures may break the emulsion.

The target width of asphalt sealant application shall be fabric width plus 150 mm. The asphalt sealant shall not be applied any farther in advance of fabric placement than the contractor can maintain free of traffic.

Asphalt spills shall be cleaned from the road surface to avoid flushing and fabric movement.

When asphalt emulsions are used, the emulsion shall be cured (essentially no moisture remaining) prior to placing the fabric and final wearing surface.

REMARKS

The rate specified must be sufficient to satisfy the asphalt retention properties of the fabric and bond the fabric and overlay to the old pavement. In order to account for the variables in pavement texture and precision of distributor truck operation, a rate of at least 0.9 liters per square meter should be specified. Rough and raveled surfaces may require a higher application rate. Within street intersections, on steep grades or in other zones where vehicle speed changes are commonplace, the normal application rate should be reduced by about 20 percent, but no less than 0.9 liters per square meter or as specified by the manufacturer. NOTE: When using emulsions, the application rate must be increased to offset water content of the emulsion.

4.5 <u>Fabric Placement</u> - The fabric shall be placed into the asphalt sealant with minimum wrinkling prior to the time the asphalt has cooled and lost tackiness. As directed by the Engineer, wrinkles or folds in excess of 25 mm shall be slit and laid flat. Brooming and/or pneumatic rolling will be required to maximize fabric contact with the pavement surface.

Overlap of the fabric joints shall be sufficient to ensure full closure of the joint, but should not exceed 150 mm. Transverse joints shall be lapped in the direction of paving to prevent edge pickup by the paver. A second application of asphalt sealant to fabric overlaps will be required if in the judgment of the Engineer additional asphalt sealant is needed to ensure proper bonding of the double fabric layer.

Removal and replacement of fabric that is damaged will be the responsibility of the contractor.

REMARKS

The problems associated with wrinkles are related to thickness of the asphalt lift being placed over the fabric. When wrinkles are large enough to be folded over, there usually is not enough asphalt available from the tack coat to satisfy the requirement of the multiple layers of fabric. Therefore, wrinkles should be slit and laid flat. Sufficient asphalt sealant should be sprayed on the top of the fabric to satisfy the requirement of the lapped fabric. In overlapping adjacent rolls of fabric, it is desirable to keep the lapped dimension as small as possible and still provide a positive overlap. If the lapped dimension becomes too large, the problem of inadequate tack to satisfy the two lifts of fabric and the old pavement may occur. If this problem does occur, then additional asphaltic sealant should be added to the lapped areas. In the application of additional asphalt sealant, care should be exercised not to apply too much since an excess will cause flushing.

4.6 Fabric Trafficking - Trafficking the fabric will be permitted for emergency or construction equipment only.

4.7 Asphalt Overlay - Placement of the hot mix overlay should closely follow fabric laydown. The temperature of the mix shall not exceed 160 °C. In the event asphalt bleeds through the fabric causing construction problems before the overlay is placed, the affected areas shall be blotted by spreading sand or hot mix. To avoid movement or damage to the fabric membrane, turning of the paver and other vehicles shall be gradual and kept to a minimum.

4.8 Seal Coats - Prior to placing a seal cost (or thin overlay such as an open-graded fiction course), lightly sand the fabric at a spread rate of 0.8 to 1.1 kg/m^2 and pneumatically roll the fabric tightly into the sealant.

REMARKS

The task force believes that trafficking of the fabric should not be allowed due to safety considerations. If the contracting agency policy allows trafficking of the fabric, then the following verbiage is recommended: If approved by the Engineer, the membrane may be opened to traffic for 24 to 48 hours prior to installing the surface course. Warning signs shall be placed which advise the motorist that the surface may be slippery when wet. The signs shall also post the appropriate safe speed. Excess sand shall be broomed from the fabric surface prior to placing the overlay. If, in the judgment of the Engineer, the fabric surface appears dry and lacks tackiness following exposure to traffic, a light tack cost shall be applied prior to the overlay.

5.   Method of Measurement

5.1 The paving fabric will be measured by the square meter.

5.2 Asphalt sealant for the paving fabric will be measured by the liter.

6.   Basis of Payment

6.1 The accepted quantities of paving fabric will be paid for at the contract unit price per square meters in place.

6.2 The accepted quantities of asphalt sealant for the paving fabric will be paid for at the contract unit price per liter complete in place.

6.3 Payment will be made under:

Pay Item   **Pay Unit**
Paving Fabric   square meter
Asphalt Sealant for Paving Fabric   liter

6.8   FIELD INSPECTION

Proper construction procedures are essential for a successful AC overlay project with geotextiles, and therefore, good field inspection is very important. Prior to construction, the field inspector should review the guidelines in Section 1.7. Most geotextile manufacturers and suppliers will provide technical assistance during the initial stages of a fabric overlay project. This assistance may be particularly beneficial to inexperienced inspectors and contractors.

One construction problem observed in some installations is the placement of insufficient tack coat. Tack coat should be a separate pay item, and field inspection should monitor the quantity of tack coat placed. Monitoring can be done by gauging or by weighing.

6.9   GEOTEXTILE SELECTION

The selected paving grade geotextile must have the ability to absorb and retain the asphalt tack coat to effectively form a waterproofing and stress-relief interlayer. The most common paving grade geotextiles are needlepunched, nonwoven materials, with a mass per unit area of 120 to 200 g/m^2. These types of geotextiles are very porous and have a high asphalt retention property that benefits the waterproofing and stress-reducing properties of the paving geotextile layer. Thinner, heat-bonded geotextiles have also been used. However, a significant variation in constructability and performance has been found between different paving geotextiles.

Although lighter-weight (*i.e.,* 120 to 135 g/m^2) geotextiles are typically used, both theory and a limited amount of field evidence indicates that geotextile with a greater mass per unit area (and a greater retention of asphalt), perform better than lighter-weight geotextiles by further reducing stress at the tip of the underlying pavement crack (Graf and Werner, 1993; Grzybowska et al., 1993; Walsh, 1993). Numerical analysis indicates that at some level of thickness, the bonding of the overlay would be reduced due to shear on the geotextile (Grzybowska et al., 1993; Walsh, 1993).

For overlay design, the appropriate paving grade geotextile should be selected with consideration given to pavement conditions, pavement deflection measurements, and the overlay design traffic (EAL), as presented in Table 6-4.

Table 6-4
Paving Grade Geotextile Selection

Paving Grade Geotextile		Pavement Conditions Rating[1]	Deflections (mm)	Design Traffic (EAL)
Class 1 (lighter grade)[2]		65 - 80	< 2.5	≤ 50,000
Class 2 (medium grade)		40 - 50	> 1.5	≤ 2,000,000
Class 3 (heaviest grade)		20 - 30	> 1.5	> 2,000,000

NOTES:
1. From The Asphalt Institute (1983), and adopted from TRB Record No. 700
   - 65 - 80   Fairly good, slight longitudinal and alligator cracking. Few slightly rough and uneven
   - 40 - 50   Poor to fair, moderate longitudinal and alligator crackings. Surface is slightly rough and uneven
   - 20 - 30   Poor conditions with extensive alligator and longitudinal crackings. Surface is very rough and uneven.
2. The suggested Class 1 weight is a low-cost solution. For optimum performance use Class 2.

6.10   RECYCLING

AC overlays used with a geotextile can be recycled. The most common practice is to mill down to just above the geotextile interlayer. This process maintains the benefits of the interlayer when the recycled overlay is replaced (Marienfeld and Smiley, 1994). Alternatively, milling may include the geotextile interlayer. A detailed study of recycling a nonwoven polypropylene geotextile (Christman, 1981) concluded that overlay with geotextile interlayers does not pose any problem to the milling operation. Additionally, no apparent differences were noted in the properties of dryer-drum recycled mixtures (50-50 blend) containing or lacking geotextiles.

6.11   OTHER GEOSYNTHETIC MATERIALS

6.11-1 Membrane Strips

A variety of commercially available, heavy-duty membrane strips are used over cracks and joints of PCC pavements that are overlaid with AC. Typically, these materials are composites of woven or nonwoven geotextiles and modified asphalt membranes. Materials of single-layer geotextiles with rubber-asphalt membranes are typically used for strip waterproofing. Materials of double-layer geotextiles that sandwich a modified asphalt membrane are typically used to reduce and retard reflective cracking. Crack reduction interlayers are typically 3.5 mm thick and are capable of maintaining 95% of their thickness during installation and in-service use. Interlayer strips are typically 0.3 to 1 m wide, and usually weigh 1600 to 3300 g/m^2 -- which is heavier than the typical geotextile that weighs about 1300 g/m^2 with asphalt impregnation (Barksdale, 1991).

The installation of heavy-duty membranes is relatively easy. Usually the manufacturer's installation recommendations are followed because of the wide variation of products and installation requirements. Single-, two-, or three-step installation processes are required for the various products.

Advantages of strips include:

- limited area of installation, and, therefore, less potential installation problems;
- factory-applied asphalt, and, therefore, less field variances; and
- heavier weight, and possible function as a stress-relieving membrane interlayer of the material.

The moderate amount of documented field performance data developed to date has been summarized by Barksdale (1991).

6.11-2 Geogrids

High-strength and high-stiffness polymer and fiberglass geogrids have been used on a relatively limited basis for full-width and strip overlay applications. These grids primarily function as a reinforcing element, provided they are sufficiently stiff. The limited use of geogrids in this application shows promise. However, additional well-planned, monitored installations are needed.

6.11-3 Geocomposites

Geogrid-geotextile composites are available; however, there is limited experience with these products to date. The intent of such a composite is to have a material that installs similarly to geotextile overlays. A properly installed composite should function as stress-relieving interlayer, moisture barrier, and reinforcement (provided the geogrid has a high modulus).

6.12   REFERENCES

References quoted within this section are listed below.

AASHTO (1993), AASHTO Guide for Design of Pavement Structures, American Association of State Highway and Transportation Officials, Washington, D.C.

AASHTO-AGC-ARTBA (1990), Guide Specifications and Test Procedures for Geotextiles, Task Force 25 Report, Subcommittee on New Highway Materials, American Association of State Transportation and Highway Officials, Washington, D.C.

Aldrich, R.C. (1986), Evaluation of Asphalt Rubber and Engineering Fabrics as Pavement Interlayers, Misc. Paper GL-86-34 (untraced series N-86), November, 7 p.

Asphalt Institute (1983), Asphalt Overlays for Highway and Street Rehabilitation, Manual Series No. 17 (MS-17), June.

Barksdale, R.D. (1991), Fabrics in Asphalt Overlays and Pavement Maintenance, National Cooperative Highway Research Program Report 171, Transportation Research Board, National Research Council, Washington. D.C., July 72 p.

Christman, R. (1981), Material Properties and Equipment Capabilities Resulting From Recycling Bituminous Concrete Pavements Containing Petromat®, Pavement Resource Managers, Newington, CT, February, 139 p.

FHWA {current edition}, Pavement Rehabilitation Manual, Pavement Division, Office of Highway Operations, Federal Highway Administration, Washington, D.C.

FHWA (1987), Techniques for Pavement Rehabilitation, Training Course Participants Notes, National Highway Institute, Federal Highway Administration, 3rd Edition.

FHWA (1982), Report on Performance of Fabrics in Asphalt Overlays, Experimental Application and Evaluation, Demonstration Projects Division, Office of Highway Operations, Federal Highway Administration, September, 39p.

Graf, B. and Werner, G. (1993), Design of Asphalt Overlay/Fabric System Against Reflective Cracking, Reflective Cracking in Pavements -- Proceedings of the Second International RILEM Conference, Liege, Belgium, pp. 159-168.

Grzybowska,, W.J., Wojtowicz, J. and Fonferko, L. (1993), Application of Geosynthetics to Overlays in Cracow Region of Poland, Reflective Cracking in Pavements -- Proceedings of the Second International RILEM Conference, Liege, Belgium, pp. 290-298.

Jagielski,K. (1993), North American Geotextile Industry Faces Many Challenges, Geotechnical Fabrics Report, Industrial Fabrics Association International, Vol. 7, No. 11, St. Paul, MN, October, pp. 18-20.

Marienfeld, M.L. and Smiley, D. (1994), Paving Fabrics: The Why and the How-To, Geotechnical Fabrics Report, Industrial Fabrics Association International, Vol. 12, No. 4, St. Paul, MN, June/July, pp 24-29.

Predoehl, N.H. (1990), Evaluation of Paving Fabric Test Installations in California - Final Report, FHWA/CA/TL-90/02, Office of Transportation Materials and Research, California Department of Transportation, Sacramento, CA, February.

Walsh, I.D. (1993), Thin Overlay to Concrete Carriageway to Minimise Reflective Cracking, Reflective Cracking in Pavements -- Proceedings of the Second International RILEM Conference, Liege, Belgium, pp. 464-481.

7.0 REINFORCED EMBANKMENTS ON SOFT FOUNDATIONS

7.1 BACKGROUND

Embankments constructed on soft foundation soils have a tendency to spread laterally because of horizontal earth pressures acting within the embankment. These earth pressures cause horizontal shear stresses at the base of the embankment which must be resisted by the foundation soil. If the foundation soil does not have adequate shear resistance, failure can result. Properly designed horizontal layers of high-strength geotextiles or geogrids can provide reinforcement which increase stability and prevent such failures. Both materials can be used equally well, provided they have the requisite design properties. There are some differences in how they are installed, especially with respect to seaming and field workability. Also, at some very soft sites, especially where there is no root mat or vegetative layer, geogrids may require a lightweight geotextile separator to provide filtration and prevent contamination of the first lift if it is an open-graded or similar type soil. A geotextile is not required beneath the first lift if it is sand, which meets soil filtration criteria.

The reinforcement may also reduce horizontal and vertical displacements of the underlying soil and thus reduce differential settlement. *It should be noted that the reinforcement will not reduce the magnitude of long-term consolidation or secondary settlement of the embankment.*

The use of reinforcement in embankment construction may allow for:
- an increase in the design factor of safety;
- an increase in the height of the embankment;
- a reduction in embankment displacements during construction, thus reducing fill requirements; and
- an improvement in embankment performance due to increased uniformity of post-construction settlement.

This chapter assumes that all the common foundation treatment alternatives for the stabilization of embankments on soft or problem foundation soils have been carefully considered during the preliminary design phase. Holtz (1989) discusses these treatment alternatives and provides guidance about when embankment reinforcement is feasible. In some situations, the most economical final design may be some combination of a conventional foundation treatment alternative together with geosynthetic reinforcement.

Examples include preloading and stage construction with prefabricated *(wick)* vertical drains, the use of stabilizing berms, or pile-supported bridge approach embankments - each used with geosynthetic reinforcement at the base of the embankment.

7.2   APPLICATIONS

Reinforced embankments over weak foundations typically fall into one of two situations - construction over uniform deposits, and construction over local anomalies (Bonaparte, Holtz, and Giroud, 1987). The more common is embankments, dikes, or levees constructed over very soft, saturated silt, clay, or peat layers (Figure. 7-1a). In this situation, the reinforcement is usually placed with its strong direction perpendicular to the centerline of the embankment, and plane strain conditions are assumed to prevail. Additional reinforcement with its strong direction oriented parallel to the centerline may also be required at the ends of the embankment.

The second reinforced embankment situation includes foundations below the embankment that are locally weak or contain voids. These zones or voids may be caused by sink holes, thawing ice (thermokarsts), old stream beds, or pockets of silt, clay, or peat (Figure 7-1b). In this application, the role of the reinforcement is to bridge over the weak zones or voids, and tensile reinforcement may be required in more than one direction. Thus, the strong direction of the reinforcing must be placed in proper orientation with respect to the embankment centerline (Bonaparte and Christopher, 1987).

Geotextiles may also be used as separators for displacement-type embankment construction (Holtz, 1989). In this application, the geotextile does not provide any reinforcement but only acts as a separator to maintain the integrity of the embankment as it displaces the subgrade soils. In this case, geotextile design is based upon constructability and survivability, and a high elongation material may be selected.

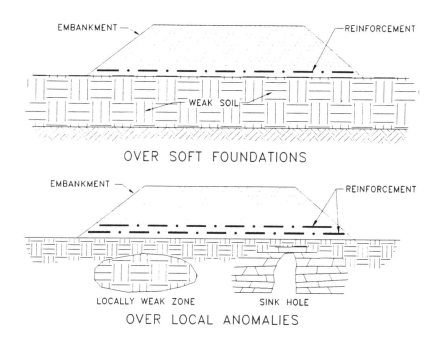

EMBANKMENT ─── ┌──────────── ─── REINFORCEMENT

── WEAK SOIL ──

OVER SOFT FOUNDATIONS

EMBANKMENT ─── ┌──────────── ─── REINFORCEMENT

LOCALLY WEAK ZONE   SINK HOLE

OVER LOCAL ANOMALIES

Figure 7-1   *Reinforced embankment applications (after Bonaparte and Christopher, 1987).*

7.3   DESIGN GUIDELINES FOR REINFORCED EMBANKMENTS ON SOFT SOILS

7.3-1   Design Considerations

As with ordinary embankments on soft soils, the basic design approach for reinforced embankments is to design against failure. The ways in which embankments constructed on soft foundations can fail have been described by Terzaghi and Peck (1967); Haliburton, Anglin and Lawmaster (1978 a and b); Fowler (1981); Christopher and Holtz (1985); and Koerner (1990), among others. Figure 7-2 shows unsatisfactory behavior that can occur in reinforced embankments. The three possible modes of failure indicate the types of stability analyses that are required. In addition, settlement of the embankment and potential creep of the reinforcement must be considered, although creep is only a factor if the creep rate in the reinforcement is greater than the strength gain occurring in the foundation due to consolidation. Because the most critical condition for embankment stability is at the end of construction, the reinforcement only has to function until the foundation soils gain sufficient strength to support the embankment.

The calculations required for stability and settlement utilize conventional geotechnical design procedures modified only for the presence of the reinforcement.

The stability of an embankment over soft soil is usually determined by the *total stress* method of analysis, which is conservative since the analysis generally assumes that no strength gain occurs in the compressible soil. The stability analyses presented in this text uses the *total stress* approach, because it is simple and appropriate for reinforcement design (Holtz, 1989).

It is always possible to calculate stability in terms of the effective stresses using the *effective stress* shear strength parameters. However, this calculation requires an accurate estimate of the field pore pressures to be made during the project design phase. Additionally, high-quality, undisturbed samples of the foundation soils must be obtained and K_o consolidated-undrained triaxial tests conducted in order to obtain the required design soil parameters. Because the prediction of in situ pore pressures in advance of construction is not easy, it is essential that field pore pressure measurements using high quality piezometers be made during construction to control the rate of embankment filling. Preloading and staged embankment construction are discussed in detail by Ladd (1991). Note that by taking into account the strength gain that occurs with staged embankment construction, lower strength and therefore lower cost reinforcement can be utilized. However; the time required before the facility is available may be significantly increased and the costs of the site investigation, laboratory testing, design analyses, field instrumentation, and inspection are much greater.

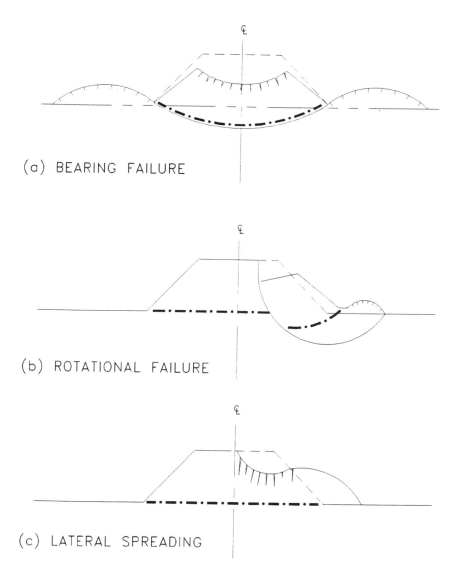

(a) BEARING FAILURE

(b) ROTATIONAL FAILURE

(c) LATERAL SPREADING

Figure 7-2   *Reinforced embankments failure modes (after Haliburton et al.,*
                 1978a).

7.3-2 Design Steps

The following is a step-by-step procedure for design of reinforced embankments. Additional comments on each step can be found in Section 7.3-3.

STEP 1.   Define embankment dimensions and loading conditions.

            A. Embankment height, H

B. Embankment length

C. Width of crest

D. Side slopes, b/H

E. External loads
    1. surcharges
    2. temporary (traffic) loads
    3. dynamic loads

F. Environmental considerations
    1. frost action
    2. shrinkage and swelling
    3. drainage, erosion, and scour

G. Embankment construction rate
    1. project constraints
    2. anticipated or planned rate of construction

STEP 2.   Establish the soil profile and determine the engineering properties of the foundation soil.

A. From a subsurface soils investigation, determine
    1. subsurface stratigraphy and soil profile
    2. groundwater table (location, fluctuation)

B. Engineering properties of the subsoils
    1. Undrained shear strength, c_u, for end of construction
    2. Drained shear strength parameters, c' and ϕ', for long-term conditions
    3. Consolidation parameters (C_c, C_r, c_v, σ_p')
    4. Chemical and biological factors that may be detrimental to the reinforcement

C. Variation of properties with depth and areal extent

STEP 3.   Obtain engineering properties of embankment fill materials.

A. Classification properties

B. Moisture-density relationships

C. Shear strength properties

D. Chemical and biological factors that may be detrimental to the reinforcement

STEP 4. Establish minimum appropriate factors of safety and operational settlement criteria for the embankment. Suggested minimum factors of safety are as follows.

A. Overall bearing capacity: 1.5 to 2

B. Global (rotational) shear stability at the end of construction: 1.3

C. Internal shear stability, long-term: 1.5

D. Lateral spreading (sliding): 1.5

E. Dynamic loading: 1.1

F. Settlement criteria: dependent upon project requirements

STEP 5. Check bearing capacity.

A. When the thickness of the soft soil is much greater than the width of the embankment, use classical bearing capacity theory:

$$q_{ult} = \gamma_{fill} H = c_u N_c \qquad [7\text{-}1]$$

where N_c, the bearing capacity factor, is usually taken as 5.14 -- the value for a strip footing on a cohesive soil of constant undrained shear strength, c_u, with depth. This approach underestimates the bearing capacity of reinforced embankments, as discussed in Section 7.3-3.

B. When the soft soil is of limited depth, perform a *lateral squeeze* analysis (Section 7.3-3).

STEP 6. Check rotational shear stability.

Perform a rotational slip surface analysis on the unreinforced embankment and foundation to determine the critical failure surface and the factor of safety against local shear instability.

A. If the calculated factor of safety is greater than the minimum required, then reinforcement is not needed. Check lateral embankment spreading (Step 7).

B. If the factor of safety is less than the required minimum, then calculate the required reinforcement strength, T_g, to provide an adequate factor of safety using Figure 7-3 or alternative solutions (Section 7.3-3), where:

$$T_g = \frac{FS(M_D) - M_R}{R \cos(\theta - \beta)}$$

STEP 7. Check lateral spreading (sliding) stability.

Perform a lateral spreading or sliding wedge stability analysis (Figure 7-4).

A. If the calculated factor of safety is greater than the minimum required, then reinforcement is not needed for this failure possibility.

B. If the factor of safety is inadequate, then determine the lateral spreading strength of reinforcement, T_{ls}, required -- see Figure 7-4b. Soil cohesion should be assumed equal to 0 for extremely soft soils and low embankments. A cohesion value should be included with placement of the second and subsequent fills in staged embankment construction.

C. Check sliding above the reinforcement. See Figure 7-4a.

STEP 8. Establish tolerable geosynthetic deformation requirements and calculate the required reinforcement modulus, J, based on wide width (ASTM D 4595) tensile testing.

Recommendations, based on type of fill soil materials and for construction over peats, are:

Reinforcement Modulus:	$J = T_{ls} / \varepsilon_{geosynthetic}$	[7-2]
Cohesionless soils:	$\varepsilon_{geosynthetic} = 5$ to 10%	[7-3]
Cohesive soils:	$\varepsilon_{geosynthetic} = 2\%$	[7-4]
Peats:	$\varepsilon_{geosynthetic} = 2$ to 10%	[7-5]

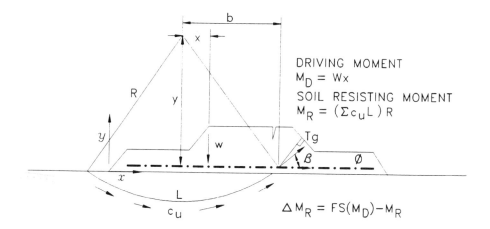

DRIVING MOMENT
$M_D = Wx$
SOIL RESISTING MOMENT
$M_R = (\Sigma c_u L)R$

$\Delta M_R = FS(M_D) - M_R$

(a) ROTATIONAL FAILURE MODEL ($\beta=0$)

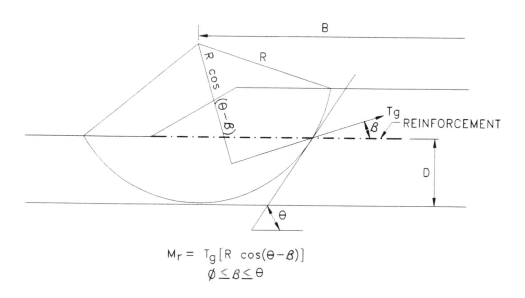

$M_r = T_g[R\,\cos(\theta - \beta)]$
$\phi \le \beta \le \theta$

(b) ROTATIONAL FAILURE MODEL ($\beta \ne 0$)

Figure 7-3   *Reinforcement required to provide rotational stability: (a)*
Christopher and Holtz (1985) after Wager (1981);   (b)
Bonaparte and Christopher (1987) for the case in which the
reinforcement does not increase soil strength.

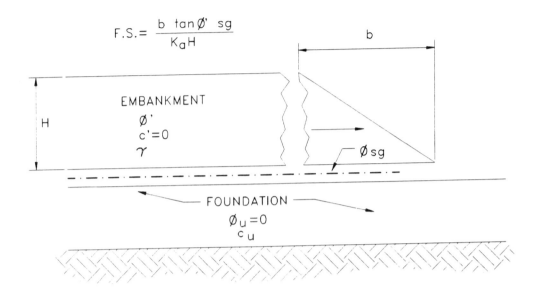

(a) SLIDING

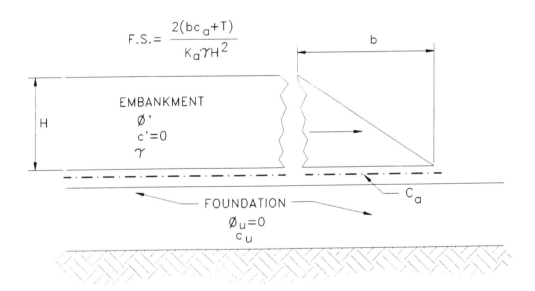

(b) RUPTURE

Figure 7-4   Reinforcement required to limit lateral embankment spreading: (a) embankment sliding on reinforcement; (b) rupture of reinforcement and embankment sliding on foundation soil (from Bonaparte and Christopher, 1987).

STEP 9.   Establish geosynthetic strength requirements in the longitudinal direction.

    A. Check bearing capacity and rotational slope stability at the ends of the embankment (Steps 5 and 6).

    B. Use strength and elongation determined from Steps 7 and 8 to control embankment spreading during construction and to control bending following construction.

    C. As the strength of the transverse seams control, seam strength requirements are the higher of the strengths determined from Steps 6 or 7.

STEP 10.   Establish geosynthetic properties (Section 7.4).

    A. Design strengths and modulus based on ASTM D 4595, wide width method.

    B. Seam strength is also based on ASTM D 4884, and is equal to the strength required in the longitudinal direction.

    C. Soil-geosynthetic friction, ϕ_{sg}, based on ASTM D 5321 with on-site soils. For preliminary estimates, assume $\phi_{sg} = 2/3\phi$; for final design, testing is recommended.

    D. Geotextile stiffness based on site conditions and experience. See Section 7.4-5.

    E. Select survivability and constructability requirements for the geosynthetic based on site conditions, backfill materials, and equipment, using Tables 7-1, 7-2, and 7-3.

STEP 11.   Estimate magnitude and rate of embankment settlement.

Use conventional geotechnical procedures and practices for this step.

STEP 12.   Establish construction sequence and procedures.

See Section 7.8.

STEP 13.   Establish construction observation requirements.

See Sections 7.8 and 7.9.

STEP 14.   Hold preconstruction meetings.

Consider a *partnering type* contract with a disputes resolution board.

STEP 15.   Observe construction and build with confidence (if the procedures outlined in these guidelines are followed!)

7.3-3   Comments on the Design Procedure

STEPS   1 and 2 need no further elaboration.

STEP 3.   Obtain embankment fill properties.

Follow traditional geotechnical practice, except that the first few lifts of fill material just above the geosynthetic should be free-draining granular materials. This requirement provides the best frictional interaction between the geosynthetic and fill, as well as providing a drainage layer for excess pore water to dissipate from the underlying soils. Other fill materials may be used above this layer as long as the strain compatibility of the geosynthetic is evaluated with respect to the backfill materials (Step 8).

STEP 4.   Establish design factors of safety.

The minimum factors of safety previously stated are recommended for projects with modern state-of-the-practice geotechnical site investigations and laboratory testing. Those factors may be adjusted depending on the method of analysis, type and use of facility being designed, the known conditions of the subsurface, the quality of the samples and soils testing, the cost of failure, the probability of unusual events occurring, and the engineer's previous experience on similar projects and sites. In short, all of the uncertainties in loads, analyses, and soil properties influence the choice of appropriate factors of safety. Typical factors of safety for unreinforced embankments also seem to be appropriate for reinforced embankments.

When the calculated factor of safety is greater than 1 but less than the minimum allowable factor of safety for design, say 1.3 or 1.5, then the geosynthetic provides an additional factor of safety or a *second line of defense* against failure. On the other hand, when the calculated factor of safety for the unreinforced embankment is significantly less than 1, the geosynthetic reinforcement is the difference between success and failure. In this latter case, **construction considerations (Section 7.8) become crucial to the project success.**

Maximum tolerable post-construction settlement and embankment deformations, which depend on project requirements, must also be established.

STEP 5.   Check overall bearing capacity.

Reinforcement does not increase the overall bearing capacity of the foundation soil. If the foundation soil cannot support the weight of the embankment, then the embankment cannot be built. Thus, the overall bearing capacity of the entire embankment must be satisfactory before considering any possible reinforcement. As such, the vertical stress due to the embankment can be treated as an average stress over the entire width of the embankment, similar to a semi-rigid mat foundation.

The bearing capacity can be calculated using classical soil mechanics methods (Terzaghi and Peck, 1967; Vesic, 1975; Perloff and Baron, 1976; and U.S. Navy, 1982) which use limiting equilibrium-type analyses for strip footings, assuming logarithmic spiral failure surfaces on an infinitely deep foundation. These analyses are not appropriate if the thickness of the underlying soft deposit is small compared to the width of the embankment. In this case, high lateral stresses in the confined soft stratum beneath the embankment could lead to a lateral squeeze-type failure. The shear forces developed under the embankment should be compared to the corresponding shear strength of the soil. Approaches discussed by Jürgenson (1934), Silvestri (1983), and Bonaparte, Holtz and Giroud (1987), Rowe and Soderman (1987a), Hird and Jewell (1990), and Humphrey and Rowe (1991) are appropriate. The designer should be aware that the analysis for lateral squeeze is only approximate, and no single method is completely accepted by geotechnical engineers at present.

In a review of 40 reinforced embankment case histories, Humphrey and Holtz (1986) and Humphrey (1987) found that in many cases, the failure height predicted by classical bearing capacity theory was significantly less than the actual constructed height, especially if high strength geotextiles and geogrids were used as the reinforcement. Figure 7-5 shows the reinforced embankment height versus average undrained shear strength of the foundation. Significantly, four embankments failed at heights 2 m greater than predicted by Equation 7-1 (line B in Figure 7-5). The two reinforced embankments that failed below line B were either on peat or under reinforced (Humphrey, 1987). It appears that in many cases, the reinforcement enhances the beneficial effect the following factors have on stability:

- limited thickness or increasing strength with depth of the soft foundation soils (Rowe and Soderman, 1987 a and b; Jewell, 1988);
- the dry crust (Humphrey and Holtz, 1989);
- flat embankment side slopes (e.g., Humphrey and Holtz, 1987); or
- dissipation of excess pore pressures during construction.

If the factor of safety for bearing capacity is sufficient, then continue with the next step. If not, consider increasing the embankment's width, flattening the slopes, adding toe berms, or improving the foundation soils by using stage construction and drainage enhancement or other alternatives, such as relocating the alignment or placing the roadway on an elevated structure.

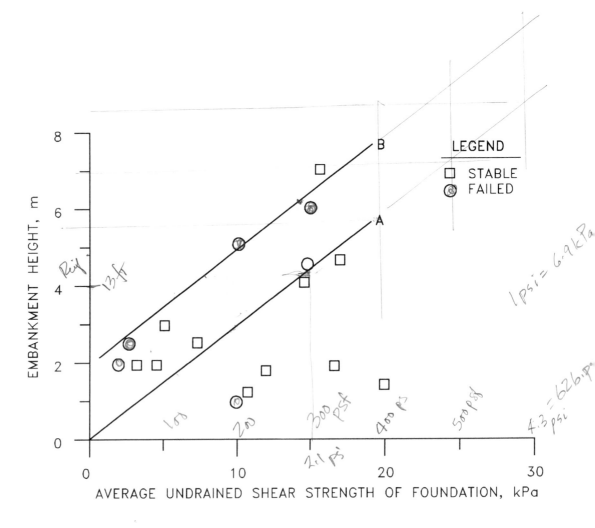

Figure 7-5   *Embankment height versus undrained shear strength of foundation; line A: classical bearing capacity theory (Eq.7-1); line B: line A + 2 m (after Humphrey, 1987).*

STEP 6.   Check rotational shear stability.

The next step is to calculate the factor of safety against a circular failure through the embankment and foundation using classical limiting equilibrium-type stability analyses. If the factor of safety does not meet the minimum design requirements (Step 4), then the reinforcing tensile force required to increase the factor of safety to an acceptable level must be estimated.

This is done by assuming that the reinforcement acts as a stabilizing tensile force at its intersection with the slip surface being considered. The reinforcement thus provides the additional resisting moment required to obtain the minimum required factor of safety. The analysis is shown in Figure 7-3.

The analysis consists of determining the most critical failure surface(s) using conventional limiting equilibrium analysis methods. For each critical sliding surface, the driving moment (M_D) and soil resisting moment (M_R) are determined as shown in Figure 7-3a. The additional resisting moment ΔM_R to provide the required factor of safety is calculated as shown in Figure 7-3b. Then one or more layers of geotextiles or geogrids with sufficient tensile strength at tolerable strains (Step 7) are added at the base of the embankment to provide the required additional resisting moment. If multiple layers are used, they must be separated by a granular layer and they must have compatible stress-strain properties (*e.g.*, the same type of reinforcement must be used for each layer).

A number of procedures have been proposed for determining the required additional reinforcement, and these are summarized by Christopher and Holtz (1985), Bonaparte and Christopher (1987), Holtz (1990), and Humphrey and Rowe (1991). The basic difference in the approaches is in the assumption of the reinforcement force orientation at the location of the critical slip surface (the angle ß in Figures 7-3a and 7-3b). It is conservative to assume that the reinforcing force acts horizontally at the location of the reinforcement (ß = 0). In this case, the additional reinforcing moment is equal to the geosynthetic strength, T_g, times the vertical distance, y, from the plane of the reinforcement to the center of rotation, or:

$$\Delta M_R = T_g y \qquad\qquad\qquad [7\text{-}6a]$$

as determined for the most critical failure surface, shown in Figure 7-3a. This approach is conservative because it neglects any possible reinforcement reorientation along the alignment of the failure surface, as well as any confining effect of the reinforcement.

A less-conservative approach assumes that the reinforcement bends due to local displacements of the foundation soils at the onset of failure, with the maximum possible reorientation located tangent to the slip surface ($\beta = \theta$ in Figure 7-3b). In this case,

$$\Delta M_R = T_g [R \cos (\theta - \beta)] \qquad\qquad [7\text{-}6b]$$

where,

θ = angle from horizontal to tangent line as shown in Figure 7-3.

Limited field evidence indicates that it is actually somewhere in between the horizontal and tangential (Bonaparte and Christopher, 1987) depending on the foundation soils, the depth of soft soil from the original ground line in relation to the width of the embankment (D/B ratio), and the stiffness of the reinforcement. Based on the minimal information available, the following suggestions are provided for selecting the orientation:

$\beta = 0$ for brittle, strain-sensitive foundations soils (*e.g.*, leached marine clays) or where a crust layer is considered in the analysis for increased support;

$\beta = \theta/2$ for D/B < 0.4 and moderate to highly compressible soils (*e.g.*, soft clays, peats);

$\beta = \theta$ for D/B ≥ 0.4 highly compressible soils (*e.g.*, soft clays, peats); and reinforcement with high elongation potential ($\varepsilon_{design} \geq 10\%$), and large tolerable deformations; and

$\beta = 0$ when in any doubt!

Other approaches, as discussed by Bonaparte and Christopher (1987), require a more rigorous analysis of the foundation soils deformation characteristics and the reinforcement strength compatibility.

In each method, the depth of the critical failure surface must be relatively shallow, *i.e.*, y in Figure 7-3a must be large, otherwise the geosynthetic contribution toward increasing the resisting moment will be small. On the other hand, Jewell (1988) notes that shallow slip surfaces tend to underestimate the driving force in the embankment, and both he and Leshchinsky (1987) have suggested methods to address this problem.

STEP 7. Check lateral spreading (sliding) stability.

A simplified analysis for calculating the reinforcement required to limit lateral embankment spreading is illustrated in Figure 7-4. For unreinforced as well as reinforced embankments, the driving forces result from the lateral earth pressures developed within the embankment and which must, for equilibrium, be transferred to the foundation by shearing stresses (Holtz, 1990). Instability occurs in the embankment when either:

1. the embankment slides on the reinforcement (Figure 7-4a); or
2. the reinforcement fails in tension and the embankment slides on the foundation soil (Figure 7-4b).

In the latter case, the shearing resistance of the foundation soils just below the embankment is insufficient to maintain equilibrium. Thus, in both cases, the reinforcement must have sufficient friction to resist sliding on the reinforcement plane, and the geosynthetic tensile strength must be sufficient to resist rupture as the potential sliding surface passes through the reinforcement.

The forces involved in the analysis of embankment spreading are shown in Figure 7-4 for the two cases above. The lateral earth pressures, usually assumed to be active, are of a maximum at the crest of the embankment. The factor of safety against embankment spreading is found from the ratio of the resisting forces to the actuating (driving) forces. The recommended factor of safety against sliding is 1.5 (Step 4). If the required soil-geosynthetic friction angle is greater than that reasonably achieved with the reinforcement, embankment soils and subgrade, then the embankment slopes must be flattened or berms must be added. Sliding resistance can be increased by the soil improvement techniques mentioned above. Generally, however, there is sufficient frictional resistance between geotextiles and geogrids commonly used for reinforcement and granular fill. If this is the case, then the resultant lateral earth pressures must be resisted by the tension in the reinforcement.

STEP 8.   Establish tolerable deformation requirements for the geosynthetic.

Excessive deformation of the embankment and its reinforcement may limit its serviceability and impair its function, even if total collapse does not occur. Thus, an analysis to establish deformation limits of the reinforcement must be performed. The most common way to limit deformations is to limit the allowable strain in the geosynthetic. This is done because the geosynthetic tensile forces required to prevent failure by lateral spreading are not developed without some strain, and some lateral movement must be expected. Thus, geosynthetic modulus is used to control lateral spreading (Step 7). The distribution of strain in the geosynthetic is assumed to vary linearly from zero at the toe to a maximum value beneath the crest of the embankment. This is consistent with the development of lateral earth pressures beneath the slopes of the embankment.

For the assumed linear strain distribution, the maximum strain in the geosynthetic will be equal to twice the average strain in the embankment. Fowler and Haliburton (1980) and Fowler (1981) found that an average lateral spreading of 5% was reasonable, both from a construction and geosynthetic property standpoint. If 5% is the average strain, then the maximum expected strain would be 10%, and the geosynthetic modulus would be determined at 10% strain (Equation 7-3). However, it has been suggested that a modulus at 10% strain would be too large, and that smaller maximum values at, say, 2 to 5% are more appropriate.

If cohesive soils are used in the embankment, then the modulus should be determined at 2% strain to reduce the possibility of embankment cracking (Equation 7-4). Of course, if embankment cracking is not a concern, then these limiting reinforcement strain values could be increased. Keep in mind, however, that if cracking occurs, no resistance to sliding is provided. Further, the cracks could fill with water, which would add to the driving forces.

Additional discussion of geosynthetic deformation is given in Christopher and Holtz (1985 and 1989), Bonaparte, Holtz and Giroud (1985), Rowe and Mylleville (1989 and 1990), and Humphrey and Rowe (1991).

STEP 9.   Establish geosynthetic strength requirements in the longitudinal direction.

Most embankments are relatively long but narrow in shape. Thus, during construction, stresses are imposed on the geosynthetic in the longitudinal direction, *i.e.*, in the along direction the centerline. Reinforcement may be also required for loadings that occur at bridge abutments, and due to differential settlements and embankment bending, especially over nonuniform foundation conditions and at the edges of soft soil deposit.

Because both sliding and rotational failures are possible, analyses procedures discussed in Steps 6 and 7 should be applied, but in the direction along the alignment of the embankment. This determines the longitudinal strength requirements of the geosynthetic. Because the usual placement of the geosynthetic is in strips perpendicular to the centerline, the longitudinal stability will be controlled by the strength of the transverse seams.

STEP 10.   Establish geosynthetic properties.

See Section 7.4 for a determining the required properties of the geosynthetic.

STEP 11.   Estimate magnitude and rate of embankment settlement.

Although not part of the stability analyses, both the magnitude and rate of settlement of the embankment should be considered in any reinforcement design. There is some evidence from finite element studies that differential settlements may be reduced somewhat by the presence of geosynthetic reinforcement. Long-term or consolidation settlements are not influenced by the geosynthetic, since compressibility of the foundation soils is not altered by the reinforcement, although the stress distribution may be somewhat different. Present recommendations provide for reinforcement design as outlined in Steps 6 - 10 above. Then use conventional geotechnical methods to estimate immediate, consolidation, and secondary settlements, as if the embankment was unreinforced (Christopher and Holtz, 1985).

Possible creep of reinforced embankments on soft foundations should be considered in terms of the geosynthetic creep rate versus the consolidation rate and strength gain of the foundation. If the foundation soil consolidates and gains strength at a rate faster than (or equal to) the rate the geosynthetic loses strength due to creep, there is no problem. Many soft soils such as peats, silts, and clays with sand lenses have high permeability, therefore, they gain strength rapidly, but each case should be analyzed individually.

STEP 12.   Establish construction sequence and procedures.

The importance of proper construction procedures for geosynthetic reinforced embankments on very soft foundations cannot be over emphasized. A specific construction sequence is usually required to avoid failures during construction.

See Section 7.8 for details on site preparation, special construction equipment, geosynthetic placement procedures, seaming techniques, and fill placement and compaction procedures.

STEP 13.   Establish construction observation requirements

See Sections 7.8 and 7.9.

A. Instrumentation. As a minimum, install piezometers, settlement points, and surface survey monuments. Also consider inclinometers to observe lateral movement with depth.

Note that the purpose of the instrumentation in soft ground reinforcement projects is not for research but to verify design assumptions and to control and, usually, expedite construction.

B. Geosynthetic inspection. Be sure field personnel understand:
- geosynthetic submittal for acceptance prior to installation;
- testing requirements;
- fill placement procedures; and
- seam integrity verification.

STEP 14. Hold preconstruction meetings

It has been our experience that the more potential contractors know about the overall project, the site conditions, and the assumptions and expectations of the designers, the more realistically they can bid; and, the project is more successful. Prebid and preconstruction information meetings with contractors have been very successful in establishing a good, professional working relationship between owner, design engineer, and contractor. Partnering type contracts and a disputes resolution board can also be used to reduce problems, claims, and litigation.

STEP 15. Observe construction

Inspection should be performed by a trained and knowledgeable inspector, and good documentation of construction should be maintained.

7.4   SELECTION OF GEOSYNTHETIC AND FILL PROPERTIES

Once the design strength requirements have been established, the appropriate geosynthetic must be selected. In addition to its tensile and frictional properties, drainage requirements, construction conditions, and environmental factors must also be considered. Geosynthetic properties required for reinforcement applications are given in Table 7-1. The selection of appropriate fill materials is also an important aspect of the design. When possible, granular fill is preferred, especially for the first few lifts above the geosynthetic.

7.4-1   Geotextile and Geogrid Strength Requirements

The most important mechanical properties are the tensile strength and modulus of the reinforcement, seam strength, soil-geosynthetic friction, and system creep resistance.

The tensile strength and modulus values should preferably be determined by an in-soil tensile test. From research by McGown, Andrawes, and Kabir (1982) and others, we know that in-soil properties of many geosynthetics are markedly different than those from tests conducted in air. However, in-soil tests are not yet routine nor standardized, and the test proposed by Christopher, Holtz, and Bell (1986) needs additional work. The practical alternate is to use a wide strip tensile test (ASTM D 4595) as a measure of the in-soil strength. This point is discussed by Christopher and Holtz (1985) and Bonaparte, Holtz, and Giroud (1987). Traditional grab or narrow-strip tensile tests are not appropriate for obtaining design properties of reinforcing geosynthetics.

Table 7-1
Geosynthetic Properties Required for Reinforcement Applications

Criteria and Parameter	Property[1]
Design Requirements:	
a. Mechanical	
Tensile strength	Wide width strength
Tensile modulus	Wide width strength
Seam strength	Wide width strength
Tension creep	Tension creep
Soil-geosynthetic friction	Soil-geosynthetic friction angle
b. Hydraulic	
Piping resistance	Apparent opening size
Permeability	Permeability
Constructability Requirements:	
Tensile strength	Grab strength
Puncture resistance	Puncture resistance
Tear resistance	Trapezoidal tear
Longevity:	
UV stability (if exposed)	UV resistance
Soil compatibility (where required)	Chemical; Biological
NOTE:	
1, See Table 1-3 for specific test procedures.	

The following minimum criteria for tensile strength of geosynthetics are recommended.

1. For ordinary cases, determine the design tensile strength (the larger of T_g and T_{ls}) and the required secant modulus at 2 to 10% strain.

2. The ultimate tensile strength T_{ult} obviously must be greater that the design tensile strength, T_d. Note that T_g includes an inherent safety factor against overload and sudden failure that is equal to the rotational stability safety factor. The tensile strength requirements should be increased to account for installation damage, depending on the severity of the conditions.

3. The strain of the reinforcement at failure should be at least 1.5 times the secant modulus strain to avoid brittle failure. For exceptionally soft

foundations where the reinforcement will be subjected to very large tensile stresses during construction, the geosynthetic must have either sufficient strength to support the embankment itself, or the reinforcement and the embankment must be allowed to deform. In this case, an elongation at rupture of up to 50% may be acceptable. In either case, high tensile strength geosynthetics and special construction procedures (Section 7.8) are required.

4. If there is a possibility of tension cracks forming in the embankment or high strain levels occurring during construction (such as might occur, for example, with cohesive embankments), the lateral spreading strength, T_{ls}, at 2% strain should be required.

5. The required lateral spreading strength, T_{ls}, should be increased to account for creep and installation damage as the creep potential of the geosynthetic depends on the creep potential of the foundation. If significant creep is expected in the foundation, the creep potential of the geosynthetic at design stresses should be evaluated, recognizing that strength gains in the foundation will reduce the creep potential. Installation damage potential will depend on the severity of the conditions.

6. Strength requirements must be evaluated and specified for both the transverse and longitudinal directions of the embankment. Usually, the transverse seam strength controls the longitudinal strength requirements.

Depending on the strength requirements, geosynthetic availability, and seam efficiency, more than one layer of reinforcement may be necessary to obtain the required tensile strength. If multiple layers are used, a granular layer of 200 to 300 mm must be placed between each successive geosynthetic layer or the layers must be mechanically connected (*e.g.*, sewn) together. Also, the geosynthetics must be strain compatible; that is, the same type of geosynthetic should be used for each layer.

For soil-geosynthetic friction values, either direct shear or pullout tests should be utilized. If test values are not available, Bell (1980) recommends that for sand embankments, the soil-geosynthetic friction is from $2/3\phi$ up to the full ϕ of the sand. Pullout tests by Holtz (1977) have shown that soil-geotextile friction is approximately equal to the ϕ of the sand. For clay soils, friction tests are definitely warranted and should be performed under all circumstances.

The creep properties of geosynthetics in reinforced soil systems are not well established. In-soil creep tests are possible but are far from routine today. For design, it is recommended that the working stress be kept much lower than the creep limit of the geosynthetic. Values of 40 to 60% of the ultimate stress are typically satisfactory for this purpose. A polyester will probably have less creep than a polypropylene or a polyethylene. Live loads versus dead loads also must be taken into account. Short-term live loadings are much less detrimental in terms of creep than sustained dead loads. And finally, as discussed in Section 7.3-3 Step 11, the relative rates of deformation of the geosynthetic versus the consolidation and strength gain of the foundation soil must be considered. In most cases, creep is not an issue in reinforced embankment stability.

7.4-2 Drainage Requirements

The geosynthetic must allow for free vertical drainage of the foundation soils to reduce pore pressure buildup below the embankment. Pertinent geosynthetic hydraulic properties are piping resistance and permeability (Table 7-1). It is recommended that the permeability of the geosynthetic be at least 10 times that of the underlying soil. The opening size should be selected based on the requirements of Section 2.3. The opening size should be a maximum to reduce the risk of clogging, while still providing retention of the underlying soil.

7.4-3 Environmental Considerations

For most embankment reinforcement situations, geosynthetics have a high resistance to chemical and biological attack; therefore, chemical and biological compatibility is usually not a concern. However, in unusual situations such as very low (*i.e.*, < 3) or very high (*i.e.*, > 9) pH soils, or other unusual chemical environments -- such as in industrial areas or near mine or other waste dumps -- the chemical compatibility of the polymer(s) in the geosynthetic should be checked to assure it will retain the design strength at least until the underlying subsoil is strong enough to support the structure without reinforcement.

7.4-4 Constructability (Survivability) Requirements

In addition to the design strength requirements, the geotextile or geogrid must also have sufficient strength to survive construction. If the geotextile is ripped, punctured, or torn during construction, support strength for the embankment structure will be reduced and failure could result. Constructability property requirements are listed in Table 7-1. (These are also called survivability requirements.) Tables 7-2 and 7-3 were developed by Haliburton, Lawmaster, and McGuffey (1982) specifically for reinforced embankment construction with varying subgrade conditions, construction equipment, and lift thicknesses (see also Christopher and Holtz, 1985). The specific property values in Table 7-4 are interim values provided by Task Force 25 of AASHTO-AGC-ARTBA, and are subject to revision. For all critical applications, high to very high survivability geotextiles and geogrids are recommended.

As the construction of the first lift of the embankment is analogous to construction of a temporary haul road, survivability requirements discussed in Section 5.9 are also appropriate here.

7.4-5 Stiffness and Workability

For extremely soft soil conditions, geosynthetic stiffness or workability may be an important consideration. The workability of a geosynthetic is its ability to support workpersons during initial placement and sewing operations and to support construction equipment during the first lift placement. Workability is generally related to geosynthetic stiffness; however, stiffness evaluation techniques and correlations with field workability are very poor (Tan, 1990). In the absence of any other stiffness information, ASTM Standard D 1388, Option A using a 50 x 300 mm specimen is recommended (see Christopher and Holtz, 1985). The values obtained should be compared with actual field performance to establish future design criteria. The workability guidelines based on subgrade CBR as suggested by Task Force 25 (Christopher and Holtz, 1985) are satisfactory for CBR > 1.0. For very soft subgrades, much stiffer geosynthetics are required. Other aspects of field workability such as water absorption and bulk density, should also be considered, especially on very soft sites.

Table 7-2

Required Degree of Geosynthetic Survivability as a Function of Subgrade Conditions and Construction Equipment

SUBGRADE CONDITIONS	Construction Equipment and 150 to 300 mm Cover Material Initial Lift Thickness		
	Low Ground Pressure Equipment (≤ 30 kPa)	Medium Ground Pressure Equipment (> 30 kPa ≤ 60 kPa)	High Ground Pressure Equipment (>60 kPa)
Subgrade has been cleared of all obstacles except grass, weeds, leaves, and fine wood debris. Surface is smooth and level, and shallow depressions and humps do not exceed 150 mm in depth and height. All larger depressions are filled. Alternatively, a smooth working table may be placed.	Low	Moderate	High
Subgrade has been cleared of obstacles larger than small- to moderate-sized tree limbs and rocks. Tree trunks and stumps should be removed or covered with a partial working table. Depressions and humps should not exceed 450 mm in depth and height. Larger depressions should be filled.	Moderate	High	Very high
Minimal site preparation is required. Trees may be felled, delimbed, and left in place. Stumps should be cut to project not more than 150 mm ± above subgrade. Geosynthetic may be draped directly over the tree trunks, stumps, large depressions and humps, holes, stream channels, and large boulders. Items should be removed only if, where placed, the geosynthetic and cover material over them will distort the finished road surface.	High	Very high	Not Recommended

NOTES:
1. Recommendations are for 150 to 300 mm initial thickness. For other initial lift thickness:
    - 300 to 450 mm: Reduce survivability requirement one level
    - 450 to 600 mm: Reduce survivability requirement two levels
    - > 600 mm: Reduce survivability requirement three levels
2. For special construction techniques such as prerutting, increase survivability requirement one level.
3. Placement of excessive initial cover material thickness may cause bearing failure of soft subgrades.

Table 7-3
Required Degree of Geosynthetic Survivability as a Function of Cover Material and Construction Equipment

CONSTRUCTION		COVER MATERIAL		
		Fine sand to +50 mm diameter gravel, rounded to subangular	Coarse aggregate with diameter up to one-half proposed lift thickness, may be angular	Some to most aggregate with diameter greater than one-half proposed lift thickness, angular and sharp-edged, few fines
150 to 300 mm Initial Lift Thickness	Low ground pressure equipment (≤30 kPa)	Low	Moderate	High
	Medium ground pressure equipment (>30 kPa, ≤60 kPa)	Moderate	High	High
300 to 450 mm Initial Lift Thickness	Medium ground pressure equipment (>30 kPa, ≤60 kPa)	Low	Moderate	High
	High ground pressure equipment (>60 kPa)	Moderate	High	High
450 to 600 mm Initial Lift Thickness	High ground pressure equipment (>60 kPa)	Low	Moderate	High
> 600 mm Initial Lift Thickness	High ground pressure equipment (>60 kPa)	Low	Low	Moderate

NOTES:
1. For special construction techniques such as prerutting, increase geosynthetic survivability requirement one level.
2. Placement of excessive initial cover material thickness may cause bearing failure of soft subgrades.

Table 7-4

Physical Property Requirements[1] for Geotextile Survivability (after AASHTO, 1990)

Survivability Level	Grab Strength[4] ASTM D 4632 (N)		Puncture Resistance[4] ASTM D 4833 (N)		Tear Strength[4] ASTM D 4533 (N)	
	< 50% Geotextile Elongation[2,3]	> 50% Geotextile Elongation[2,3]	< 50% Geotextile Elongation	> 50% Geotextile Elongation	< 50% Geotextile Elongation	> 50% Geotextile Elongation
High (Class 1)	1400	900	500	350	500	350
Moderate (Class 2)	1100	700	400	250	400	250
Low (Class 3)	800	500	300	180	300	180

Additional Requirements	Test Method
Apparent Opening Size	ASTM D 4751
1. < 50% soil passing 0.075 mm sieve, AOS < 0.6 mm	
2. > 50% soil passing 0.075 mm sieve, AOS < 0.3 mm	
Permeability	ASTM D 4491
k of the geotextile > k of the soil (permittivity x the nominal geotextile thickness)	
Ultraviolet Degradation	ASTM D 4355
At 500 hours of exposure, 50% strength retained	
Geotextile Acceptance	ASTM D 4759

NOTES:
1.   For the index properties, the first value of each set is for geotextiles which fail at less than 50% elongation, while the second value is for geotextiles which fail at greater than 50% elongation. Elongation is determined by ASTM D 4632.
2.   Values shown are minimum roll average values. Strength values are in the weakest principal direction.
3.   The values of the geotextile elongation do not relate to the allowable consolidation properties of the subgrade soil. These must be determined by a separate investigation.
4.   AASHTO classification.

7.4-6   Fill Considerations

When possible, the first few lifts of fill material just above the geosynthetic should be free-draining granular materials. This requirement provides the best frictional interaction between the geosynthetic and fill, as well as a drainage layer for excess pore water dissipation of the underlying soils. Other fill materials may be used above this layer as long as the strain compatibility of the geosynthetic is evaluated with respect to the backfill material, as discussed in Section 7.3-3, Step 8.

Most reinforcement analyses assume that the fill material is granular. In fact, in the past the use of cohesive soils together with geosynthetic reinforcement has been discouraged. This may be an unrealistic restriction, although there are problems with placing and compacting cohesive earth fills on especially soft subsoils. Furthermore, the frictional resistance between geosynthetics and cohesive soils is problematic. It may be possible to use composite embankments. Cohesionless fill could be used for the first 0.5 to 1 m; then the rest of the embankment could be constructed to grade with locally available materials.

7.5   DESIGN EXAMPLE

DEFINITION OF DESIGN EXAMPLE

- Project Description: A 4-lane highway is to be constructed over a peat bog. Alignment and anticipated settlement require construction of an embankment with an average height of 2 m. See project cross section figure.

- Type of Structure: embankment supporting a permanent paved road

- Type of Application: geosynthetic reinforcement

- Alternatives:
  - i) excavate and replace - wetlands do not allow;
  - ii) lightweight fill - high cost;
  - iii) stone columns - soils too soft;
  - iv) drainage and surcharge - yes; or
  - v) very flat (8H:1V) slope - right-of-way restriction

GIVEN DATA

- Geometry   -   as shown in project cross section figure

- Soils   -   subsurface exploration indicates $c_u = 5$ kPa in weakest areas
  - soft soils are underlain by firmer soils of $c_u = 25$ kPa
  - embankment fill soil will be sands and gravel
  - lightweight fill costs $250,000 more than sand/gravel

- Stability   -   Stability analyses of the unreinforced embankment were conducted with the STABL computer program. The most critical condition for embankments on soft soils is end-of-construction case; therefore, UU (unconsolidated, undrained) soil shear strength values are used in analyses.

- Results of the analyses:
a) With 4:1 side slopes and sand/gravel fill ($\gamma = 21.7$ kN/m^3), FS $\approx$ 0.72.
b) Since FS was substantially less than 1 for 4H:1V slopes, flatter slopes were evaluated, even though additional right-of-way would be required. With 8:1 side slopes and sand/gravel fill ($\gamma = 21.7$ kN/m^3), a FS $\approx$ 0.87 was computed.
c) Light-weight fill ($\gamma = 15.7$ kN/m^3) was also considered, with it, the FS varied between $\approx$0.90 to 1.15

- Transportation Department required safety factors are:
  Fs_{min} > 1.5 for long-term conditions
  FS_{allow} ≈ 1.3 for short-term conditions

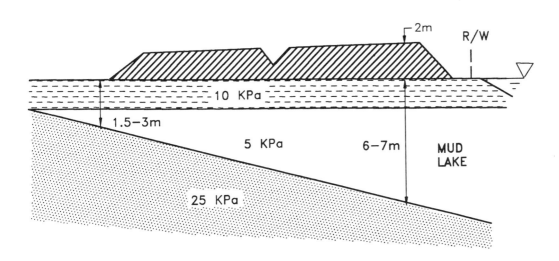

Project Cross Section

REQUIRED

Design geotextile reinforcement to provide a stable embankment.

DEFINE

A. Geotextile function(s)

B. Geotextile properties required

C. Geotextile specification

SOLUTION

A. Geotextile function(s):
Primary   -   reinforcement (for short-term conditions)
Secondary -   separation and filtration

B. Geotextile properties required:
tensile characteristics
interface shear strength
survivability
apparent opening size (AOS)

DESIGN

Design embankment with geotextile reinforcement to meet short-term stability requirements.

STEP 1. DEFINE DIMENSIONS AND LOADING CONDITIONS

See project cross section figure.

STEP 2. SUBSURFACE CONDITIONS AND PROPERTIES

Undrained shear strength provided in given data. Design for end-of-construction. Long-term design with drained shear strength parameters not covered within this example.

STEP 3. EMBANKMENT FILL PROPERTIES

sand and gravel, with
$\gamma_m = 21.7 \text{ kN/m}^3$ $\phi' = 35°$

STEP 4. ESTABLISH DESIGN REQUIREMENTS

Transportation Department required safety factors are:
$FS_{min} > 1.5$ for *long-term* conditions
$FS_{min} \approx 1.3$ for *short-term* conditions

settlement -

Primary consolidation must be completed prior to paving roadway. A total fill height of 2 m is anticipated to reach design elevation. This height includes the additional fill material thickness to compensate for anticipated settlements.

STEP 5. CHECK OVERALL BEARING CAPACITY

Recommended minimum safety factor (section 7.3-2) is 2.

A. Overall bearing capacity of soil, ignoring footing size is

$$q_{ult} = c\, N_c$$
$$q_{ult} = 5\text{ kPa} \times 5.14 = 25.7\text{ kPa}$$

Considering depth of embedment (*i.e.*, shearing will have to occur through the embankment for a bearing capacity failure) the bearing capacity is more accurately computed (see Meyerhof) as follows.

$$N_c = 4.14 + 0.5\, B/D \quad \text{where,} \quad B = \text{the base width of the embankment}$$
$$(\sim 31\text{ m}), \text{ and}$$
$$D = \text{the average depth of soft soil } (\sim 4.5\text{ m})$$
$$N_c = 4.14 + 0.5\,(31\text{ m} / 4.5\text{ m}) = 7.6$$
$$q_{ult} = 5\text{ kPa} \times 7.6 = 38\text{ kPa}$$

maximum load, $P_{max} = \gamma_m\, H$

w/o a geotextile -
$$P_{max} = 21.7\text{ kN/m}^3 \times 2\text{ m} = 43.4\text{ kPa}$$
$$\text{implies FS} = 38 \div 43.4 = 0.88 \qquad \therefore \text{ NO GOOD}$$

with a geotextile, and assuming that the geotextile will result in an even distribution of the embankment load over the width of the geotextile (*i.e.*, account for the slopes at the embankment edges),

$$P_{avg} = A_E\, \gamma_m\, / B \qquad \text{where,} \quad A = \text{cross section area of embankment,}$$
$$B = \text{base width of the embankment}$$
$$P_{avg} = \{[½\,(31\text{ m} + 15\text{ m})\, 2\text{ m}]\, 21.7\text{ kN/m}^3\} / 31\text{ m}$$
$$P_{avg} = 32.2\text{ kPa} < q_{ult} \text{ worst case} \qquad \text{Safety Factor Marginal}$$

Add berms to increase bearing capacity. Berms, 3 m wide, can be added within the existing right-of-way, increasing the base width to 37 m. With this increase in width,

$$N_c = 4.14 + 0.5\,(37\text{ m} / 4.5\text{ m}) = 8.3$$
$$q_{ult} = 5\text{ kPa} \times 8.3 = 41.5\text{ kPa}$$
and,
$$P_{avg} = 32.2\text{ kPa}\,(31 / 37) = 27.0\text{ kPa}$$
$$\text{FS} = 41.5\text{ kPa} / 27.0\text{ kPa} = 1.54 \qquad \text{Safety Factor O.K.}$$

B. Lateral squeeze

From FHWA Foundation Manual (Cheney and Chassie, 1993) -

If $\gamma_{fill} \times H_{fill} > 3c$, then lateral squeeze of the foundation soil can occur. Since $P_{max} = 43.4$ kPa is much greater than 3c, even considering the crust layer (c = 10 kPa), a rigorous lateral squeeze analysis was performed using the method by Jürgeson (1934). In this method, the lateral stress beneath the toe of the embankment is determined through charts or finite element analysis and compared to the shear strength of the soil. This method indicated a safety factor of approximately 1 for the 31 m base width. Adding the berm and extending the reinforcement to the toe of the berm decreases the potential for lateral squeeze as the lateral stress is reduced at the toe of the berm. The berms increased $FS_{SQUEEZZE}$ to greater than 1.5.

Also, comparing the reinforced design with Figure 7-5 indicates that the reinforced structure should be stable.

STEP 6. PERFORM ROTATIONAL SHEAR STABILITY ANALYSIS

Recommended minimum safety factor at end of construction (section 7.3-2) is 1.3.

The critical unreinforced failure surface is found through rotational stability methods. For this project, STABL4M was used and the critical, unreinforced surface FS = 0.72. As the soil supporting the embankment was highly compressible peat, the reinforcement was assumed to rotate such that $\beta = \theta$ (Figure 7-3 and Eq. 7-4b). Thus,

$$FS_{req} = \frac{M_R + T_g R}{M_D} \geq 1.3$$

$$T_r = \frac{1.3 M_D - M_R}{R}$$

therefore, $\qquad\qquad\qquad\qquad\qquad\qquad T_g \approx 263 \text{ kN/m}$

Feasible - yes. Geosynthetics are available which exceed this strength requirement, especially if multiple layers are used. For this project, an installation damage factor of approximately equal to 0, and 2 layers were used:

Bottom: 90 kN/m
Top: 180 kN/m

The use of 2 layers allowed the lower cost bottom material to be used over the full embankment plus berm width, while the higher strength and more expensive geotextile was only placed under the embankment section where it was required.

STEP 7. CHECK LATERAL SPREADING (SLIDING) STABILITY

Recommended minimum safety factor (section 7.3-2) is 1.5.

A. from Figure 7-4b:

$T = FS \times P_A = FS \times 0.5 \, K_a \, \gamma_m \, H^2$

$T = 1.5 \, (0.5) \, [\tan^2 (45 - 35/2)] \, (21.7 \text{ kN/m}^3) \, (2 \text{ m})^2$

$T = 17.6 \text{ kN/m}$

Use FS = 3 for creep and installation damage
therefore, $T_{ls} = 53$ kN/m

$T_{ls} < T_g$, therefore $T_{design} = T_g = 263$ kN/m

B. check sliding:

FS > 6, OK

$$FS = \frac{b \, \tan \, \phi_{sg}}{K_a \, H}$$

$$FS = \frac{8m \times \tan 23}{0.27 \times 2m}$$

STEP 8. ESTABLISH TOLERABLE DEFORMATION (LIMIT STRAIN) REQUIREMENTS

For cohesionless sand and gravel over deformable peat use $\varepsilon = 10\%$

STEP 9. EVALUATE GEOSYNTHETIC STRENGTH REQUIRED IN LONGITUDINAL DIRECTION

From Step 7,
use $T_L = T_{ls} = 53$ kN/m for reinforcement and seams in the cross machine (X-MD) direction

STEP 10. ESTABLISH GEOSYNTHETIC PROPERTIES

A. Design strength and elongation based upon ASTM D 4595
Ultimate tensile strength

$T_{d1} = T_{ult} \geq 90$ kN/m in MD - Layer 1
$T_{d2} = T_{ult} \geq 180$ kN/m in MD - Layer 2
$T_{ult} \geq 53$ kN/m in X-MD - both layers

Reinforcement Modulus, J

$$J = T_{ls} / 0.10 = 530 \text{ kN/m for limit strain of } 10\%$$
$$J \geq 530 \text{ kN/m - MD and X-MD, both directions}$$

B. seam strength

$$T_{seam} \geq 53 \text{ kN/m with controlled fill placement}$$

C. soil-geosynthetic adhesion

from testing, per ASTM D 5321, $\phi_{sg} \geq 23°$

D. geotextile stiffness based upon site conditions and experience

E. survivability and constructability requirements

Assume: 1. medium ground pressure equipment
               2. 300 mm first lift
               3. uncleared subgrade

Use a High Survivability geotextile with elongation < 50% (from Tables 7-2 and 7-3):

Geotextile separator shall meet or exceed the minimum average roll values of:

Property	Test Method	Value
Grab Strength	ASTM D 4632	1400 N
Puncture Resistance	ASTM D 4833	500 N
Tear Resistance	ASTM D 4533	500 N

Drainage and filtration requirements -
    Use Table 5-4
    Need grain size distribution of subgrade soils
    Determine: maximum AOS for retention
                    minimum $k_g > k_s$
                    minimum AOS for clogging resistance

Complete Steps 11 through 15 to finish design.

STEP 11.   PERFORM SETTLEMENT ANALYSIS

STEP 12.   ESTABLISH CONSTRUCTION SEQUENCE REQUIREMENTS

STEP 13.   ESTABLISH CONSTRUCTION OBSERVATION REQUIREMENTS

STEP 14.   HOLD PRECONSTRUCTION MEETING

STEP 15.   OBSERVE CONSTRUCTION

7.6   SPECIFICATIONS

Because the reinforcement requirements for soft-ground embankment construction will be project and site specific, standard specifications, which include suggested geosynthetic properties, are not appropriate, and special provisions or a separate project specification must be used. The following example includes most of the items that should be considered in a reinforced embankment project.

HIGH STRENGTH GEOTEXTILE FOR EMBANKMENT REINFORCEMENT
(from Washington Department of Transportation, November 1994)

Description

This work shall consist of furnishing and placing construction geotextile in accordance with the details shown in the plans.

Materials

Geotextile and Thread for Sewing

The material shall be a woven geotextile consisting only of long chain polymeric filaments or yarns formed into a stable network such that the filaments or yarns retain their position relative to each other during handling, placement, and design service life. At least 85 percent by weight of the of the material shall be polyolefins or polyesters. The material shall be free from defects or tears. The geotextile shall be free of any treatment or coating which might adversely alter its hydraulic or physical properties after installation. The geotextile shall conform to the properties as indicated in Table 1.

Thread used shall be high strength polypropylene, polyester, or Kevlar thread. Nylon threads will not be allowed.

Geotextile Properties

Table 1.

Properties for high strength geotextile for embankment reinforcement.

Property	Test Method[1]	Geotextile Property Requirements[2]
AOS	ASTM D4751	.84 mm max. (#20 sieve)
Water Permittivity	ASTM D4491	0.02/sec. min.
Tensile Strength, min. in machine direction	ASTM D4595	(to be based on project specific design)
Tensile Strength, min. in x-machine direction	ASTM D4595	(to be based on project specific design)
Secant Modulus at 5% strain	ASTM D4595	(to be based on project specific design)
Seam Breaking Strength	ASTM D4884	(to be based on project specific design)
Puncture Resistance	ASTM D4833	330 N min.
Tear Strength, min. in machine and x-machine direction	ASTM D4533	330 N min.
Ultraviolet (UV) Radiation Stability	ASTM D4355	70% Strength Retained min., after 500 hrs in weatherometer

[1] The test procedures are essentially in conformance with the most recently approved ASTM geotextile test procedures, except geotextile sampling and specimen conditioning, which are in accordance with WSDOT Test Methods 914 and 915, respectively. Copies of these test methods are available at the Headquarters Materials Laboratory in Tumwater, Washington.

[2] All geotextile properties listed above are minimum average roll values (i.e., the test result for any sampled roll in a lot shall meet or exceed the values listed).

Geotextile Approval
Source Approval

The Contractor shall submit to the Engineer the following information regarding each geotextile proposed for use:

> Manufacturer's name and current address,
> Full Product name,
> Geotextile structure, including fiber/yarn type, and
> Geotextile polymer type(s).

If the geotextile source has not been previously evaluated, a sample of each proposed geotextile shall be submitted to the Headquarters Materials Laboratory in Tumwater for evaluation. After the sample and required information for each geotextile type have arrived at the Headquarters Materials Laboratory in Tumwater, a maximum of 14 calendar days will be required for this testing. Source approval will be based on conformance to the applicable values from Table 1. Source approval shall not be the basis of acceptance of specific lots of material unless the lot sampled can be clearly identified, and the number of samples tested and approved meet the requirements of WSDOT Test Method 914.

Geotextile Samples for Source Approval

Each sample shall have minimum dimensions of 1.5 meters by the full roll width of the geotextile. A minimum of 6 square meters of geotextile shall be submitted to the Engineer for testing. The geotextile machine direction shall be marked clearly on each sample submitted for testing. The machine direction is defined as the direction perpendicular to the axis of the geotextile roll.

The geotextile samples shall be cut from the geotextile roll with scissors, sharp knife, or other suitable method which produces a smooth geotextile edge and does not cause geotextile ripping or tearing. The samples shall not be taken from the outer wrap of the geotextile nor the inner wrap of the core.

Acceptance Samples

Samples will be randomly taken by the Engineer at the jobsite to confirm that the geotextile meets the property values specified.

Approval will be based on testing of samples from each lot. A "lot" shall be defined for the purposes of this specification as all geotextile rolls within the consignment (i.e., all rolls sent to the project site) which were produced by the same manufacturer during a continuous period of production at the same manufacturing plant and have the same product name. After the samples and manufacturer's certificate of compliance have arrived at the Headquarters Materials Laboratory in Tumwater, a maximum of 14 calendar days will be required for this testing. If the results of the testing show that a geotextile lot, as defined, does not meet the properties required in Table 1, the roll or rolls which were sampled will be rejected. Two additional rolls for each roll tested which failed from the lot previously tested will then be selected at random by the Engineer for sampling and retesting. If the retesting shows that any of the additional rolls tested do not meet the required properties, the entire lot will be rejected. If the test results from all the rolls retested meet the required properties, the entire lot minus the roll(s) which failed will be accepted. All geotextile which has defects, deterioration, or damage, as determined by the Engineer, will also be rejected. All rejected geotextile shall be replaced at no expense to the Contracting Agency.

Certificate of Compliance

The Contractor shall provide a manufacturer's certificate of compliance to the Engineer which includes the following information about each geotextile roll to be used:

> Manufacturer's name and current address,
> Full product name,

Geotextile structure, including fiber/yarn type,
Geotextile polymer type(s),
Geotextile roll number, and
Certified test results.

Approval of Seams

If the geotextile seams are to be sewn in the field, the Contractor shall provide a section of sewn seam before the geotextile is installed which can be sampled by the Engineer.

The seam sewn for sampling shall be sewn using the same equipment and procedures as will be used to sew the production seams. The seam sewn for sampling must be at least 2 meters in length. If the seams are sewn in the factory, the Engineer will obtain samples of the factory seam at random from any of the rolls to be used. The seam assembly description shall be submitted by the Contractor to the Engineer and will be included with the seam sample obtained for testing. This description shall include the seam type, stitch type, sewing thread type(s), and stitch density.

Construction Requirements

Geotextile Roll Identification, Storage, and Handling

Geotextile roll identification, storage, and handling of the geotextile shall be in conformance to ASTM D 4873. During periods of shipment and storage, the geotextile shall be kept dry at all times and shall be stored off the ground. Under no circumstances, either during shipment or storage, shall the materials be exposed to sunlight, or other form of light which contains ultraviolet rays, for more than five calendar days.

Preparation and Placement of the Geotextile Reinforcement

The area to be covered by the geotextile shall be graded to a smooth, uniform condition free from ruts, potholes, and protruding objects such as rocks or sticks. The Contractor may construct a working platform, up to 0.6 meters in thickness, in lieu of grading the existing ground surface. A working platform is required where stumps or other protruding objects which cannot be removed without excessively disturbing the subgrade are present. All stumps shall be cut flush with the ground surface and covered with at least 150 mm of fill before placement of the first geotextile layer. The geotextile shall be spread immediately ahead of the covering operation. The geotextile shall be laid with the machine direction perpendicular or parallel to centerline as shown in Plans. Perpendicular and parallel directions shall alternate. All seams shall be sewn. Seams to connect the geotextile strips end to end will not be allowed, as shown in the Plans. The geotextile shall not be left exposed to sunlight during installation for a total of more than 10 calendar days. The geotextile shall be laid smooth without excessive wrinkles. Under no circumstances shall the geotextile be dragged through mud or over sharp objects which could damage the geotextile. The cover material shall be placed on the geotextile in such a manner that a minimum of 200 mm of material will be between the equipment tires or tracks and the geotextile at all times. Construction vehicles shall be limited in size and weight such that rutting in the initial lift above the geotextile is not greater than 75 mm deep, to prevent overstressing the geotextile. Turning of vehicles on the first lift above the geotextile will not be permitted. Compaction of the first lift above the geotextile shall be limited to routing of placement and spreading equipment only. No vibratory compaction will be allowed on the first lift.

Small soil piles or the manufacturer's recommended method shall be used as needed to hold the geotextile in place until the specified cover material is placed.

Should the geotextile be torn or punctured or the sewn joints disturbed, as evidenced by visible geotextile damage, subgrade pumping, intrusion, or roadbed distortion, the backfill around the damaged or displaced area shall be removed and the damaged area repaired or replaced by the Contractor at no expense to the Contracting Agency. The repair shall consist of a patch of the same type of geotextile placed over the damaged area. The patch shall be sewn at all edges.

If geotextile seams are to be sewn in the field or at the factory, the seams shall consist of two parallel rows of stitching, or shall consist of a J-seam, Type Ssn-1, using a single row of stitching. The two rows of stitching shall be 25 mm apart with a tolerance of plus or minus 13 mm and shall not cross, except for restitching. The stitching shall be a lock-type stitch. The minimum seam allowance, i.e., the minimum distance from the geotextile edge to the stitch line nearest to that edge, shall be 40 mm if a flat or prayer seam, Type SSa-2, is used. The minimum seam allowance for all other seam types shall be 25 mm. The seam, stitch type, and the equipment used to perform the stitching shall be as recommended by the manufacturer of the geotextile and as approved by the Engineer.

The seams shall be sewn in such a manner that the seam can be inspected readily by the Engineer or his representative. The seam strength will be tested and shall meet the requirements stated in this Specification.

Embankment construction shall be kept symmetrical at all times to prevent localized bearing capacity failures beneath the embankment or lateral tipping or sliding of the embankment. Any fill placed directly on the geotextile shall be spread immediately. Stockpiling of fill on the geotextile will not be allowed.

The embankment shall be compacted using Method B of Section 2-03.3(14)C. Vibratory or sheepsfoot rollers shall not be used to compact the fill until at least 0.5 meters of fill is covering the bottom geotextile layer and until at least 0.3 meters of fill is covering each subsequent geotextile layer above the bottom layer.

The geotextile shall be pretensioned during installation using either Method 1 or Method 2 as described herein. The method selected will depend on whether or not a mudwave forms during placement of the first one or two lifts. If a mudwave forms as fill is pushed onto the first layer of geotextile, Method 1 shall be used. Method 1 shall continue to be used until the mudwave ceases to form as fill is placed and spread. Once mudwave formation ceases, Method 2 shall be used until the uppermost geotextile layer is covered with a minimum of 0.3 meters of fill. These special construction methods are not needed for fill construction above this level. If a mudwave does not form as fill is pushed onto the first layer of geotextile, then Method 2 shall be used initially and until the uppermost geotextile layer is covered with at least 0.3 meters of fill.

Method 1

After the working platform, if needed, has been constructed, the first layer of geotextile shall be laid in continuous transverse strips and the joints sewn together. The geotextile shall be stretched manually to ensure that no wrinkles are present in the geotextile. The fill shall be end-dumped and spread from the edge of the geotextile. The fill shall first be placed along the outside edges of the geotextile to form access roads. These access roads will serve three purposes: to lock the edges of the geotextile in place, to contain the mudwave, and to provide access as needed to place fill in the center of the embankment. These access roads shall be approximately 5 meters wide. The access roads at the edges of the geotextile shall have a minimum height of 0.6 meters when completed. Once the access roads are approximately 15 meters in length, fill shall be kept ahead of the filling operation, and the access roads shall be kept approximately 15 meters ahead of this filling operation as shown in the Plans. Keeping the mudwave ahead of this filling operation and keeping the edges of the geotextile from moving by use of the access roads will effectively pre-tension the geotextile. The geotextile shall be laid out no more than 6 meters ahead of the end of the access roads at any time to prevent overstressing of the geotextile seams.

Method 2

After the working platform, if needed, has been constructed, the first layer of geotextile shall be laid and sewn as in Method 1. The first lift of material shall be spread from the edge of the geotextile, keeping the center of the advancing fill lift ahead of the outside edges of the lift as shown in the Plans. The geotextile shall be manually pulled taut prior to fill placement. Embankment construction shall continue in this manner for subsequent lifts until the uppermost geotextile layer is completely covered with 0.3 meters of compacted fill.

Measurement

High strength geotextile for embankment reinforcement will be measured by the square meter for the ground surface area actually covered.

Payment

The unit contract price per square meter for "High Strength Geotextile For Embankment Reinforcement," shall be full pay to complete the work as specified.

7.7   COST CONSIDERATIONS

The cost analysis for a geosynthetic reinforced embankment includes:
1. Geosynthetic cost: including purchase price, factory prefabrication, and shipping.
2. Site preparation: including clearing and grubbing, and working table preparation.
3. Geosynthetic placement: related to field workability (see Christopher and Holtz, 1989),
    a) with no working table, or
    b) with a working table.

4. Fill material: including purchasing, hauling, dumping, compaction, allowance for additional fill due to embankment subsidence. (NOTE: Use free-draining granular fill for the lifts adjacent to geosynthetic to provide good adherence and drainage.)

7.8   CONSTRUCTION PROCEDURES

The construction procedures for reinforced embankments on soft foundations are extremely important. Improper fill placement procedures can lead to geosynthetic damage, nonuniform settlements, and even embankment failure. By the use of low ground pressure equipment, a properly selected geosynthetic, and proper procedures for placement of the fill, these problems can essentially be eliminated. Essential construction details are outlined below. The Washington State DOT Special Provision in Section 7.6 provides additional details.

A. Prepare subgrade:
1. Cut trees and stumps flush with ground surface.
2. Do not remove or disturb root or meadow mat.
3. Leave small vegetative cover, such as grass and reeds, in place.
4. For undulating sites or areas where there are many stumps and fallen trees, consider a working table for placement of the reinforcement. In this case, a lower strength sacrificial geosynthetic designed only for constructability can be used to construct and support the working table.

B. Geosynthetic placement procedures:
1. Orient the geosynthetic with the machine direction perpendicular to the embankment alignment. No seams should be allowed parallel to the alignment. Therefore,
    - The geosynthetic rolls should be shipped in unseamed machine direction lengths equal to one or more multiples of the embankment design base width.
    - The geosynthetic should be manufactured with the largest machine width possible.
    - These widths should be factory-sewn to provide the largest width compatible with shipping and field handling.

2. Unroll the geosynthetic as smoothly as possible transverse to the alignment. (Do not drag it.)

3. Geotextiles should be sewn as required with all seams up and every stitch inspected. Geogrids should be positively joined by clamps, cables, pipes, etc.

4. The geosynthetic should be manually pulled taut to remove wrinkles. Weights (sand bags, tires, etc.) or pins may be required to prevent lifting by wind.

5. Before covering, the Engineer should examine the geosynthetic for holes, rips, tears, etc. Defects, if any, should be repaired by:
   - Large defects, should be replaced by cutting along the panel seam and sewing in a new panel.
   - Smaller defects, can be cut out and a new panel resewn into that section, if possible.
   - Defects less than 150 mm, can be overlapped a minimum of 1 m or more in all directions from the defective area. (Additional overlap may be required, depending on the geosynthetic-to-geosynthetic friction angle).

   NOTE: If a weak link exists in the geosynthetic, either through a defective seam or tear, the system will tell the engineer about it in a dramatic way -- spectacular failure! (Holtz, 1990)

C. Fill placement, spreading, and compaction procedures:

1. Construction sequence for extremely soft foundations (when a mudwave forms) is shown in Figure 7-6.

   a. End-dump fill along edges of geosynthetic to form toe berms or access roads.
      - Use trucks and equipment compatible with constructability design assumptions (Table 7-1).
      - End-dump on the previously placed fill; do not dump directly on the geosynthetic.
      - Limit height of dumped piles, *e.g.*, to less than 1 m above the geosynthetic layer, to avoid a local bearing failure. Spread piles immediately to avoid local depressions.
      - Use lightweight dozers and/or front-end loaders to spread the fill.
      - Toe berms should extend one to two panel widths ahead of the remainder of the embankment fill placement.

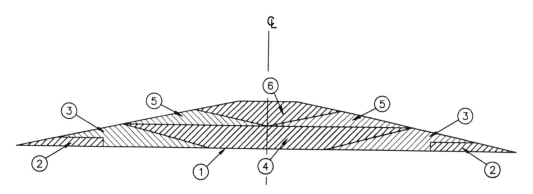

SEQUENCE OF CONSTRUCTION

1. LAY GEOSYNTHETIC IN CONTINUOUS TRAVERSE STRIPS, SEW STRIPS TOGETHER
2. END DUMP ACCESS ROADS
3. CONSTRUCT OUTSIDE SECTIONS TO ANCHOR GEOSYNTHETIC
4. CONSTRUCT INTERIOR SECTIONS TO "SET" GEOSYNTHETIC
5. CONSTRUCT INTERIOR SECTIONS TO TENSION GEOSYNTHETIC
6. CONSTRUCT FINAL CENTER SECTION

Figure 7-6   *Construction sequence for geosynthetic reinforced embankments for extremely weak foundations (from Haliburton, Douglas and Fowler, 1977).*

b. After constructing the toe berms, spread fill in the area between the toe berms.

   •   Placement should be parallel to the alignment and symmetrical from the toe berm inward toward the center to maintain a U-shaped leading edge (concave outward) to contain the mudwave (Figure 7-7).

c. Traffic on the first lift should be parallel to the embankment alignment; no turning of construction equipment should be allowed.

   •   Construction vehicles should be limited in size and weight to limit initial lift rutting to 75 mm. If rut depths exceed 75 mm, decrease the construction vehicle size and/or weight.

d. The first lift should be compacted only by tracking in place with dozers or end-loaders.

e. Once the embankment is at least 600 mm above the original ground, subsequent lifts can be compacted with a smooth drum vibratory roller or other suitable compactor.   If localized liquefied conditions occur, the vibrator should be turned off and the weight of the drum alone should be used for compaction. Other types of compaction equipment also can be used for nongranular fill.

2. After placement, the geosynthetic should be covered within 48 hours.

For less severe foundation conditions (*i.e.*, when no mudwave forms):

a. Place the geosynthetic with no wrinkles or folds; if necessary, manually pull it taut prior to fill placement.

b. Place fill symmetrically from the center outward in an inverted U (convex outward) construction process, as shown in Figure 7-8. Use fill placement to maintain tension in the geosynthetic.

c. Minimize pile heights to avoid localized depressions.

d. Limit construction vehicle size and weight so initial lift rutting is no greater than 75 mm.

e. Smooth-drum or rubber-tired rollers may be considered for compaction of first lift; however, do not overcompact. If weaving or localized quick conditions are observed, the first lift should be compacted by tracking with construction equipment.

D. Construction monitoring:

1. Monitoring should include piezometers to indicate the magnitude of excess pore pressure developed during construction. If excessive pore pressures are observed, construction should be halted until the pressures drop to a predetermined safe value.

2. Settlement plates should be installed at the geosynthetic level to monitor settlement during construction and to adjust fill requirements appropriately.

Inclinometers should be considered at the embankment toes to monitor lateral displacement.

Photographs of reinforced embankment construction are shown in Figure 7-9.

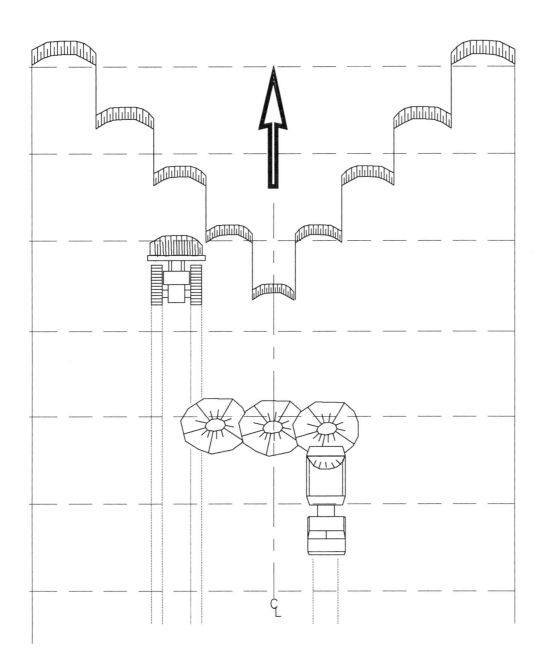

Figure 7-7   Placement of fill between toe berms on extremely soft foundations (CBR < 1) with a mud wave anticipated.

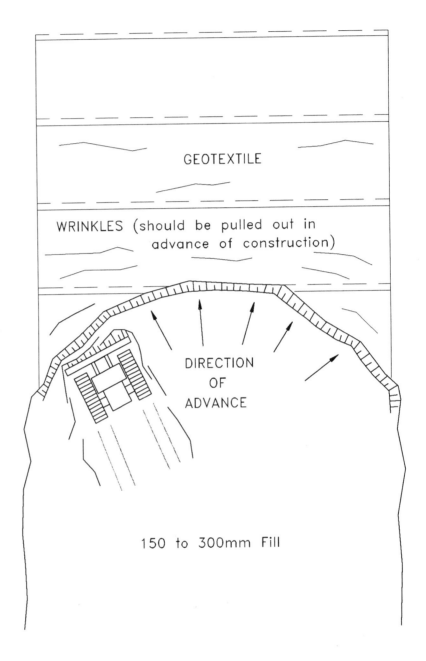

Figure 7-8   *Fill placement to tension geotextile on moderate ground conditions; moderate subgrade (CBR > 1); no mud wave.*

a.

b.

c.

d.

Figure 7-9   Reinforced embankment construction; a) geosynthetic placement; b) geosynthetic placement; c) fill dumping; and d) fill spreading.

7.9   INSPECTION

Since implemented construction procedures are crucial to the success of reinforced embankments on very soft foundations, competent and professional construction inspection is absolutely essential. Field personnel must be properly trained to observe every phase of the construction and to ensure that (1) the specified material is delivered to the project, (2) the geosynthetic is not damaged during construction, and (3) the specified sequence of construction operations are explicitly followed. Field personnel should review the checklist in Section 1.7.

7.10   REINFORCEMENT OF EMBANKMENTS COVERING LARGE AREAS

Special considerations are required for constructing large reinforced areas, such as parking lots, toll plazas, storage yards for maintenance materials and equipment, and construction pads. Loads are more biaxial than conventional highway embankments, and design strengths and strain considerations must be the same in all directions. Analytical techniques for geosynthetic reinforcement requirements are the same as those discussed in Section 7.3. Because geosynthetic strength requirements will be the same in both directions, including across the seams, special seaming techniques must often be considered to meet required strength requirements. Ends of rolls may also require butt seaming. In this case, rolls of different lengths should be used to stagger the butt seams. Two layers of fabric should be considered, with the bottom layer seams laid in one direction, and the top layer seams laid perpendicular to the bottom layer. The layers should be separated by a minimum lift thickness, usually 300 mm, soil layer.

For extremely soft subgrades, the construction sequence must be well planned to accommodate the formation and movement of mudwaves. Uncontained mudwaves moving outside of the construction can create stability problems at the edges of the embankment. It may be desirable to construct the fill in parallel embankment sections, then connect the embankments to cover the entire area. Another method staggers the embankment load by constructing a wide, low embankment with a higher embankment in the center. The outside low embankments are constructed first and act as berms for the center construction. Next, an adjacent low embankment is constructed from the outside into the existing embankment; then the central high embankment is spread over the internal adjacent low embankment. Other construction schemes can be considered depending on the specific design requirements. In all cases, a perimeter berm system is necessary to contain the mudwave.

7.11   REFERENCES

AASHTO (1990), Standard Specifications for Geotextiles - M 288, Standard Specifications for Transportation Materials and Methods of Sampling and Testing, American Association of State Transportation and Highway Officials, Washington, D.C., pp 689-692.

Bell, J.R. (1980), Design Criteria for Selected Geotextile Installations, Proceedings of the 1st Canadian Symposium on Geotextiles, pp. 35-37.

Bonaparte, R. and Christopher, B.R. (1987), Design and Construction of Reinforced Embankments Over Weak Foundations, Proceedings of the Symposium on Reinforced Layered Systems, Transportation Research Record 1153, Transportation Research Board, Washington, D.C., pp. 26-39.

Bonaparte, R., Holtz, R.D. and Giroud, J.P. (1987), Soil Reinforcement Design Using Geotextiles and Geogrids, Geotextile Testing and The Design Engineer, Special Technical Publication 952, American Society for Testing and Materials, Philadelphia, PA, pp. 69-118.

Cheney, R.S. and Chassie, R.G. (1993), Soils and Foundations Workshop Manual, FHWA Report No. HI-88-099, Federal Highway Administration, Washington, D.C., July, 395 p.

Christopher, B.R. and Holtz, R.D. (1989), Geotextile Design and Construction Guidelines, Federal Highway Administration, National Highway Institute, Report No. FHWA-HI-90-001, 297 p.

Christopher, B.R., Holtz, R.D. and Bell, W.D. (1986), New Tests for Determining the In-Soil Stress-Strain Properties of Geotextiles, Proceedings of the Third International Conference on Geotextiles, Vol. II, Vienna, Austria, pp. 683-688.

Christopher, B.R. and Holtz, R.D. (1985), Geotextile Engineering Manual, Report No. FHWA-TS-86/203, Federal Highway Administration, Washington, D.C., 1044 p.

Fowler, J. (1981), Design, Construction and Analysis of Fabric-Reinforced Embankment Test Section at Pinto Pass, Mobile, Alabama, Technical Report EL-81-7, USAE Waterways Experiment Station, 238 p.

Fowler, J. and Haliburton, T.A. (1980), Design and Construction of Fabric Reinforced Embankments, The Use of Geotextiles for Soil Improvement, Preprint 80-177, ASCE Convention, pp. 89-118.

Haliburton T.A., Lawmaster, J.D. and McGuffey, V.E. (1982), Use of Engineering Fabrics in Transportation Related Applications Final Report Under Contract No. DTFH61-80-C-0094.

Haliburton, T.A., Anglin, C.C. and Lawmaster, J.D. (1978a), Selection of Geotechnical Fabrics for Embankment Reinforcement, Report to U.S. Army Engineer District, Mobile, Oklahoma State University, Stillwater, 138 p.

Haliburton, T.A., Anglin, C.C. and Lawmaster, J.D. (1978b), Testing of Geotechnical Fabric for Use as Reinforcement, Geotechnical Testing Journal, American Society for Testing and Materials, Vol. 1, No. 4, pp. 203-212.

Haliburton, T.A., Douglas, P.A. and Fowler, J. (1977), Feasibility of Pinto Island as a Long-Term Dredged Material Disposal Site, Miscellaneous Paper, D-77-3, U.S. Army Waterways Experiment Station.

Hird, C.C. and Jewell, R.A. (1990), Theory of Reinforced Embankments, Reinforced Embankments - Theory and Practice, Shercliff, D.A., Ed., Thomas Telford Ltd., London, UK, pp. 117-142.

Holtz, R.D. (1990), Design and Construction of Geosynthetically Reinforced Embankments on Very Soft Soils, State-of-the-Art Paper, Session 5, Performance of Reinforced Soil Structure, Proceedings of the International Reinforced Soil Conference, Glasgow, British Geotechnical Society, pp. 391-402.

Holtz, R.D. (1989), Treatment of Problem Foundations for Highway Embankments, Synthesis of Highway Practice 147, National Cooperative Highway Research Program, Transportation Research Board, Washington, D.C., 72 p.

Holtz, R.D. (1977), Laboratory Studies of Reinforced Earth Using a Woven Polyester Fabric, Proceedings of the International Conference on the Use of Fabrics in Geotechnics, Paris, Vol. 3, pp. 149-154.

Humphrey, D.N. and Rowe, R.K. (1991), Design of Reinforced Embankments - Recent Developments in the State of the Art, Geotechnical Engineering Congress 1991, McLean, F., Campbell, D.A. and Harris, D.W., Eds., ASCE Geotechnical Special Publication No. 27, Vol. 2, June, pp. 1006-1020.

Humphrey, D.N. and Holtz, R.D. (1989), Effects of a Surface Crust on Reinforced Embankment Design, Proceedings of Geosynthetics '89, Industrial Fabrics Association International, St. Paul, MN, Vol. 1, pp. 136-147.

Humphrey, D.N. (1987), Discussion of Current Design Methods by R.M. Koerner, B-L Hwu and M.H. Wayne, Geotextiles and Geomembranes, Vol. 6, No. 1, pp. 89-92.

Humphrey, D.N. and Holtz, R.D. (1987), Use of Reinforcement for Embankment Widening, Proceedings of Geosynthetics '87, Industrial Fabrics Association International, St. Paul, MN, Vol. 1, pp. 278-288.

Humphrey, D.N. and Holtz, R.D. (1986), Reinforced Embankments - A Review of Case Histories, Geotextiles and Geomembranes, Vol. 4, No. 2, pp.129-144.

Jewell, R.A. (1988), The Mechanics of Reinforced Embankments on Soft Soils, Geotextiles and Geomembranes, Vol. 7, No. 4, pp.237-273.

Jürgenson, L. (1934), The Shearing Resistance of Soils, Journal of the Boston Society of Civil Engineers. Also in Contribution to Soil Mechanics, 1925-1940, BSCE, pp. 134-217.

Koerner, R.M., Editor (1990), The Seaming of Geosynthetics, Special Issue, Geotextiles and Geomembranes, Vol. 9, Nos. 4-6, pp. 281-564.

Ladd, C.C. (1991), Stability Evaluation During Staged Construction, 22nd Terzaghi Lecture, Journal of Geotechnical Engineering, American Society of Civil Engineers, Vol. 117, No. 4, pp. 537-615.

Leshchinsky, D. (1987), Short-Term Stability of Reinforced Embankment over Clayey Foundation, Soils and Foundations, The Japanese Society of Soil Mechanics and Foundation Engineering, Vol. 27, No. 3, pp. 43-57.

McGown, A., Andrawes, K.Z., Yeo, K.C. and DuBois, D.D., (1982), Load-Extension Testing of Geotextiles Confined in Soil, Proceedings of the Second International Conference on Geotextiles, Las Vegas, Vol. 3, pp. 793-798.

Perloff, W.H. and Baron, W. (1976), Soil Mechanics: Principles and Applications, Ronald, 745 p.

Rowe, R.K. and Mylleville, B.L.J. (1990), Implications fo Adopting an Allowable Geosynthetic Strain in Estimating Stability, Proceedings of the 4th International Conference on Geotextiles, Geomembranes, and Related Products, The Hague, Vol. 1, pp. 131-136.

Rowe, R.K. and Mylleville, B.L.J. (1989), Consideration of Strain in the Design of Reinforced Embankments, Proceedings of Geosynthetics '89, Industrial Fabrics Association International, St. Paul, MN, Vol. 1, pp. 124-135.

Rowe, R.K. and Soderman, K.L. (1987a), Reinforcement of Embankments on Soils Whose Strength Increases With Depth, Proceedings of Geosynthetics '87, Industrial Fabrics Association International, St. Paul, MN, Vol. 1, pp. 266-277.

Rowe, R.K. and Soderman, K.L. (1987b), Stabilization of Very Soft Soils Using High Strength Geosynthetics: The Role of Finite Element Analyses, Geotextiles and Geomembranes, Vol. 6, No. 1, pp. 53-80.

Silvestri, V. (1983), The Bearing Capacity of Dykes and Fills Founded on Soft Soils of Limited Thickness, Canadian Geotechnical Journal, Vol. 20, No. 3, pp. 428-436.

Tan, S.L. (1990), Stress-Deflection Characteristics of Soft Soils Overlain with Geosynthetics, MSCE Thesis, University of Washington, 146 p.

Terzaghi, K. and Peck, R.B. (1967), Soil Mechanics in Engineering Practice, 2nd Edition, John Wiley & Sons, New York, 729 p.

U.S. Department of the Navy (1982), Soil Mechanics, Design Manual 7.1, Naval Facilities Engineering Command, Alexandria, VA.

U.S. Department of the Navy (1982), Foundations and Earth Structures, Design Manual 7.2, Naval Facilities Engineering Command, Alexandria, VA.

Vesic, A.A. (1975), Bearing Capacity of Shallow Foundations, Chapter 3 in Foundation Engineering Handbook, Winterkorn and Fang, Editors, Van Nostrand Reinhold, pp. 121-147.

Wager, O. (1981), Building of a Site Road over a Bog at Kilanda, Alvsborg County, Sweden in Preparation for Erection of Three 400kV Power Lines, Report to the Swedish State Power Board, AB Fodervävnader, Borås, Sweden, 16 p.

8.0 REINFORCED SLOPES

8.1 BACKGROUND

Even if foundation conditions are satisfactory, slopes may be unstable at the desired slope angle. For new construction, the cost of fill, right-of-way, and other considerations may make a steeper slope desirable. Existing slopes, natural or manmade, may also be unstable, as is painfully obvious when they fail. As shown in Figure 8-1, multiple layers of geogrids or geotextiles may be placed in an earthfill slope during construction or reconstruction to reinforce the soil and provide increased slope stability. Reinforced slopes with face inclinations of 70° to 90° are usually designed as retaining walls. These are addressed in Chapter 9.

In this chapter, analysis of the reinforcement and construction details required to provide a safe slope will be reviewed. The design method included in this chapter was first developed in the early 1980s for landslide repair in northern California. This approach has been validated by the thousands of reinforced soil slopes constructed over the last decade and through results of an extensive FHWA research program on reinforced soil structures (Christopher, et al.,1990). Contracting options and guideline specifications are from (Berg, 1993).

8.2 APPLICATIONS

Geosynthetics are primarily used as slope reinforcement for construction of slopes to angles steeper than those constructed with the fill material being used, as illustrated in Figure 8-1a. Geosynthetics used in this manner can provide significant project economy by:

- creating usable landspace at the crest or toe of the reinforced slope;
- reducing the volume of fill required;
- allowing the use of less-than-high-quality fill; and
- eliminating the expense of facing elements required by reinforced walls.

Applications which highlight some of these advantages, illustrated in Figure 8-2, include:

- construction of new embankments;
- construction of alternatives to retaining walls;
- widening of existing embankments; and
- repair of failed slopes.

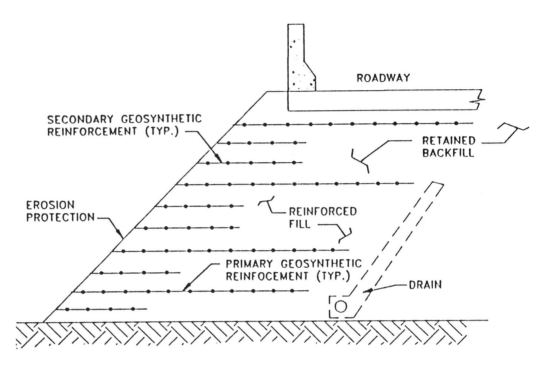

ROADWAY

SECONDARY GEOSYNTHETIC
REINFORCEMENT (TYP.)

RETAINED
BACKFILL

EROSION
PROTECTION

REINFORCED
FILL

PRIMARY GEOSYNTHETIC
REINFOCEMENT (TYP.)

DRAIN

a.

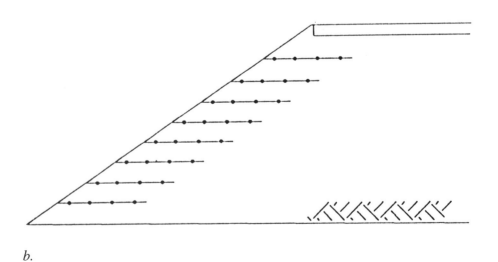

b.

Figure 8-1   Use of geosynthetics in engineered slopes: (a) to increase
stability of a slope; and (b) to provide improved compaction and
surficial stability at edge of slopes (after Berg, et al., 1990).

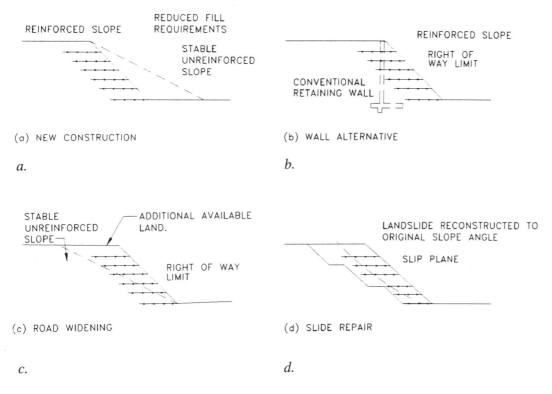

Figure 8-2   *Applications of reinforced slopes: (a) construction of new embankments; (b) alternative to retaining walls; (c) widening existing embankments; and (d) repair of landslides (after Tensar, 1987).*

The design of reinforcement for safe, steep slopes requires rigorous analysis. The design of reinforcement for these applications is critical because reinforcement failure results in slope failure. To date, several thousand reinforced slope structures have been successfully constructed at various slope face angles. The tallest structure constructed in the U.S. to date, is a 1H:1V reinforced slope 33.5 m high (Bonaparte, et al., 1989).

A second purpose of geosynthetics placed at the edges of a compacted fill slope is to provide lateral resistance during compaction (Iwasaki and Watanabe, 1978) and surficial stability (Thielen and Collin, 1993). The increased lateral resistance allows for increased compacted soil density over that normally achieved and provides increased lateral confinement for soil at the face. Even modest amounts of reinforcement in compacted slopes have been found to prevent sloughing and reduce slope erosion. Edge reinforcement also allows compaction equipment to operate safely near the edge of the slope. Further compaction improvements have been found in cohesive soils using geosynthetics with in-plane drainage capabilities (*e.g.*, nonwoven geotextiles) which allow for rapid pore pressure dissipation in the compacted soil (Zornberg and Mitchell, 1992).

For the compaction improvement application, the design is simple. Place a geogrid or geotextile that will survive construction at every lift or every other lift in a continuous plane along the edge of the slope. Only narrow strips, about 1 to 2 m in width, are required. No reinforcement design is required if the overall slope is found to be safe without reinforcement. Where the slope angle approaches the angle of repose of the soil, a face stability analysis should be performed using the method presented in Section 8.3. Where reinforcement is required by analysis, the narrow strip geosynthetic may be considered as a secondary reinforcement used to improve compaction and to stabilize the slope face between primary reinforcing layers.

Other applications of reinforced soil slopes include:
- upstream/downstream face stability and increased height of dams;
- construction of permanent levees and temporary flood control structures;
- steepening abutments and decreasing bridge spans;
- temporary road widening for detours; and
- embankment construction with wet, fine-grained soils.

8.3 DESIGN GUIDELINES FOR REINFORCED SLOPES

8.3-1 Design Concepts

The overall design requirements for reinforced slopes are similar to those for unreinforced slopes: the factor of safety must be adequate for both the short-term and long-term conditions and for all possible modes of failure.

Permanent, critical reinforced structures should be designed using comprehensive slope stability analyses. A structure may be considered permanent if its design life is greater than 1 to 3 years. A reinforcement application is considered critical if there is mobilized tension in the reinforcement for the life of the structure, if failure of the reinforcement results in failure of the structure, or if the consequences of failure include personal injury or significant property damage (Bonaparte and Berg, 1987). A reinforced slope is typically not considered critical if the safety factor against instability of the same unreinforced slope is greater than 1.1, and the reinforcement is used to increase the safety factor.

Failure modes of reinforced slopes (Berg, et al., 1989) include:
1. internal, where the failure plane passes through the reinforcing elements;
2. external, where the failure surface passes behind and underneath the reinforced mass; and
3. compound, where the failure surface passes behind and through the reinforced soil mass.

In many cases, the stability safety factor will be approximately equal in two or all three modes.

Reinforced slopes are currently analyzed using modified versions of the classical limit equilibrium slope stability methods. A circular or wedge-type potential failure surface is assumed, and the relationship between driving and resisting forces or moments determines the slope's factor of safety. Based on their tensile capacity and orientation, reinforcement layers intersecting the potential failure surface are assumed to increase the resisting moment or force. The tensile capacity of a reinforcement layer is the minimum of its allowable pullout resistance behind, or in front of, the potential failure surface and/or its long-term design tensile strength, whichever is smaller. A wide variety of potential failure surfaces must be considered, including deep-seated surfaces through or behind the reinforced zone. The slope stability factor of safety is taken from the critical surface requiring the maximum reinforcement. Detailed design of reinforced slopes is performed by determining the factor of safety with sequentially modified reinforcement layouts until the target factor of safety is achieved.

Ideally, reinforced slope design is accomplished using a conventional slope stability computer program modified to account for the stabilizing effect of reinforcement. Such programs should account for reinforcement strength and pullout capacity, compute reinforced and unreinforced safety factors automatically, and have a searching routine to help locate critical surfaces.

Several reinforced slope programs are commercially available, though some are limited to specific soil and reinforcement conditions. These programs generally do not design the reinforcement but allow for an evaluation of a given reinforcement layout. An iterative approach then follows to optimize either the reinforcement or the layout. Some of these programs are limited to simple soil profiles and, in some cases, reinforcement layouts. Vendor-supplied programs are, in many cases, reinforcement specific. These programs could be used to provide a preliminary evaluation or to check a detailed analysis.

A generic program, Reinforced Soil Slopes (RSS), for both reinforcement design and evaluation of almost any condition, is currently being developed by FHWA. The program is based on the design method presented in Section 8.3-2, Design of Reinforced Slopes, Steps 5 and 6; and in Christopher, et al. (1990).

8.3-2 Design of Reinforced Slopes

The steps for design of a reinforced soil slope are:

STEP 1. Establish the geometric, loading, and performance requirements for design.

STEP 2. Determine the subsurface stratigraphy and the engineering properties of the natural soils.

STEP 3. Determine the engineering properties of the available fill soils.

STEP 4. Establish design parameters for the reinforcement (design reinforcement strength, durability criteria, soil-reinforcement interaction).

STEP 5. Determine the factor of safety of the unreinforced slope.

STEP 6. Design reinforcement to provide stable slope.
Method A - Direct reinforcement design
Method B - Trial reinforcement layout analysis

STEP 7. Check external stability.

STEP 8. Evaluate requirements for subsurface and surface water control.

Details required for each step, along with equations for analysis, are presented in section 8.3-3. The procedure in section 8.3-3 assumes that the slope will be constructed on a stable foundation (*i.e.*, a circular or wedge-shaped failure surface through the foundation is not critical and local bearing support is clearly adequate). It does not include recommendations for deep-seated failure analysis. The user is referred to Chapter 7 for use of reinforcement in embankments over weak foundation soils.

For slide repair applications, it is also very important that solutions address the cause of original failure. Make sure that the new reinforced soil slope will not have the same problems. If water table or erratic water flows exist, particular attention must be paid to drainage. In natural soils, it is also necessary to identify any weak seams that could affect stability.

8.3-3   Reinforced Slope Design Guidelines

The following provides design procedure details for reinforced soil slopes.

STEP 1.   Establish the geometric, load, and performance requirements for design (Figure 8-3).

Geometric and load requirements:
a.   Slope height, H.
b.   Slope angle, β.
c.   External (surcharge) loads:

- Surcharge load, q
- Temporary live load, Δq
- Design Seismic acceleration, $\propto_g$

Performance requirements:
a.   External stability and settlement

- Horizontal sliding of the RSS mass along its base, $FS_{min} = 1.5$
- External, deep-seated, $FS_{min} = 1.3$
- Dynamic loading: $FS \geq 1.1$.
- Settlement--maximum, based on project requirements.

b.   Compound failure modes (for planes passing behind and then through the reinforced mass)

- Compound failure surfaces, $FS_{min} = 1.3$

c.   Internal stability

- Internal failure surfaces, $FS_{min} = 1.3$

STEP 2.   Determine the engineering properties of the natural soils in the slope.
a.   Determine foundation soil profile below the slope's base and along the alignment (every 30 to 60 m, depending on the homogeneity of the subsurface profile) deep enough to evaluate a potential deep-seated failure (recommended exploration depth is twice the height of the slope or to refusal).
b.   Determine the foundation soil strength parameters (c_u, ϕ_u or c' and ϕ'), unit weight (wet and dry), and consolidation parameters (C_c, C_r, c_v and σ'_p).
c.   Locate the groundwater table, d_w, (especially important if water will exit slope).

d. For slope and landslide repairs, identify the cause of instability and locate the previous failure surface.

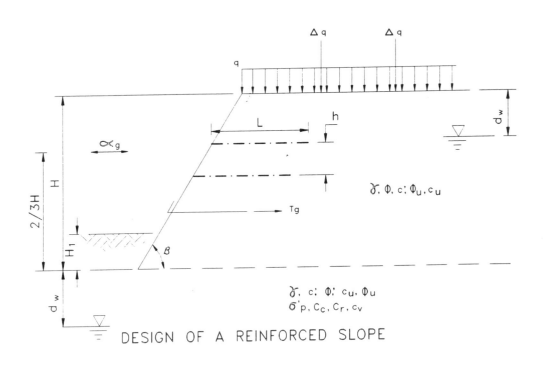

DESIGN OF A REINFORCED SLOPE

Figure 8-3   Requirements for design of a reinforced slope.

STEP 3.   Determine properties of reinforced fill and, if different, the fill behind the reinforced zone.

See recommendations in Section 8.4-1.

a. Gradation and plasticity index
b. Compaction characteristics and placement requirements
c. Shear strength parameters, c_u, ϕ_u or c', ϕ'
d. Chemical composition of soil that may affect durability of reinforcement

STEP 4.   Establish design parameters for the reinforcement.

See recommendations in section 8.4-1 and Appendix F

a. Design tensile strength (T_d): $T_d \leq T_a$ @ 10% total strain

b. Allowable geosynthetic strength, T_a = ultimate strength (T_{ULT}) ÷ reduction factors for creep, installation damage, and durability (RF)

c. Pullout Resistance: FS = 1.5 with a 1 m minimum for granular soils. Use FS = 2 for cohesive soils.

STEP 5.   Check unreinforced stability.

Perform a stability analysis using conventional stability methods to determine safety factors and driving moments for potential failure surfaces. Use both circular arc and sliding wedge methods, and consider failure through the toe, through the face (at several elevations), and deep seated below the toe. Failure surface exit points should be defined within each of the potential failure zones. A number of stability analysis computer programs are available for rapid evaluation (*e.g.*, the STABL family of programs developed at Purdue University including the current version, STABL5M, and the program XSTABL developed at the University of Idaho). In all cases, you should perform a few calculations by hand to verify reasonableness of the computer program.

To determine the size of the critical zone to be reinforced, examine the full range of potential failure surfaces with safety factors less than or equal to the slope's target safety. Plot all of these surfaces on the slope's cross-section. Surfaces that just meet the target factor of safety roughly envelope the limits of the critical zone to be reinforced.

Critical failure surfaces extending below the toe of the slope indicate deep foundation and edge bearing capacity problems that must be addressed prior to design completion. For such cases, a more-extensive foundation analysis is warranted. Geosynthetics may be used to reinforce the base of the embankment and to construct toe berms for improved embankment stability, as reviewed in Chapter 7. Other foundation improvement measures should be considered.

STEP 6.   Design reinforcement to provide for a stable slope.

Several approaches are available for the design of slope reinforcement, many of which are contained in Christopher and Holtz (1985, Chapter 5). Two methods are presented in this section. The first method uses a direct design approach to obtain the reinforcing requirements. The second method, analyzes trial reinforcement layouts. The computer program RSS developed by the FHWA incorporates both approaches.

Method A - Direct design approach.

The first method, presented in Figure 8-4 for a rotational slip surface, uses any conventional slope stability computer program, and the steps necessary to manually calculate the reinforcement requirements. This design approach can accommodate fairly complex conditions depending on the analytical method used (*e.g.*, Bishop, Janbu, etc.).

The assumed orientation of the reinforcement tensile force influences the calculated slope safety factor. In a conservative approach, the deformability of the reinforcements is not taken into account; therefore, the tensile forces per unit width of reinforcement, T_r, are always assumed to be horizontal to the reinforcements, as illustrated in Figure 8-4. However, close to failure, the reinforcements may elongate along the failure surface, and an inclination from the horizontal can be considered. Tensile force direction is therefore dependent on the extensibility of the reinforcements used, and for extensible geosynthetic reinforcement, a T inclination tangent to the sliding surface is recommended.

Figure 8-5 provides a method for quickly checking the computer-generated results. The charts are based upon simplified analysis methods of two-part and one-part wedge-type failure surfaces and are limited by the assumptions noted on the figure. Note that Figure 8-5 is not intended to be a single design tool. Other design charts are also available from Jewell et al. (1984); Jewell (1989); Werner and Resl (1986); Ruegger (1986); and Leshchinsky and Boedeker (1989). Several computer programs are also available for analyzing a slope with given reinforcement and can also be used as a check.

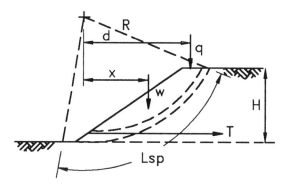

Step A   Locate most critical surface through the toe of the slope (*e.g.*, use XSTABL) and find M_R available.

Factor of unreinforced safety:

$$FS = \frac{Resisting\ Moment\ (M_R)}{Driving\ Moment\ (M_D)} = \frac{\tau_f\ L_{sp}\ R}{(Wx + qd)}$$

where: W = weight of earth segment
L_{sp} = length of slip plane
q = surcharge
τ_f = shear strength of soil

Step B   Once the critical surface and factor of safety have been found, if the program does not calculate M_R, it can be easily calculated from M_D, where:

M_R = $FS_u (Wx + qd)$
FS_u = calculated from computer program for several slip circles

Step C   To determine strength of reinforcement, T:

$$T_{total} = \frac{FS_R\ M_D - M_R}{R} = \frac{FS_R\ M_D - FS_u\ M_R}{R} = \frac{(FS_R - FS_u)\ M_D}{R}$$

where: FS_R = Factor of safety required (*e.g.*, $FS_R = 1.5$)
R = Radius of slip surface

Step D   Strength of geosynthetic required $T_d = T_{total}$ ÷ spacing requirements.

Step E   Repeat for failure surface above each layer to ensure distribution is adequate.

Step F   For large H (e.g., H > 10 m), increase T in lower half and use Step C to decrease reinforcement in upper half to decrease overall reinforcement requirements.

Figure 8-4   *Rotational shear approach to determine required strength of reinforcement.*

Judgment and experience in selecting of the most appropriate design is required. The following design steps are necessary:

a. Calculate the total reinforcement tension, T_s, required to obtain the required factor of safety for each potential failure circle inside the critical zone (Step 5) that extends through or below the toe of the slope. Use the following equation:

$$T_S = (FS_R - FS_u) \frac{M_D}{D}$$

where:

T_s = sum of required tensile force per unit width of reinforcement (considering rupture and pullout) in all reinforcement layers intersecting the failure surface;

M_D = driving moment about the center of the failure circle;

D = the moment arm of T_s about the center of failure circle, $D =$ radius of circle R for extensible geosynthetic reinforcement (*i.e.*, assumed to act tangentially to the circle).

FS_R = target minimum slope safety factor; and

FS_U = unreinforced slope safety factor.

The largest T_s calculated establishes the required design tension, T_{max}.

b. Determine the required design tension, T_{max}, using the charts in Figure 8-5. Compare results with Step 6a. If substantially different, recheck Steps 5 and 6a.

c. Determine the distribution of reinforcement:

For low slopes (H ≤ 6 m) assume a uniform distribution of reinforcement and use T_{max} to determine spacing or reinforcement requirements in Step 6d.

For high slopes (H > 6 m), divide the slope into two (top and bottom) or three (top, middle, and bottom) reinforcement zones of equal height and use a factored T_{max} in each zone for spacing or reinforcement requirements in Step 6d. The total required tension in each zone is found from the following equations:

For two zones:

$$T_{Bottom} = \tfrac{3}{4} T_{max}$$
$$T_{Top} = \tfrac{1}{4} T_{max}$$

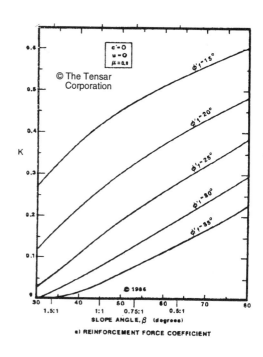

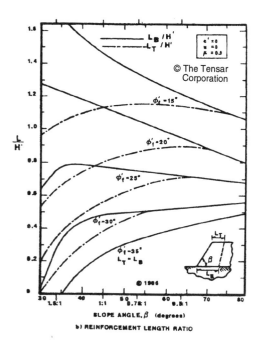

CHART PROCEDURE:

$$\phi_f \;=\; \tan^{-1} \left(\frac{\tan \phi_r}{FS_R} \right)$$

1) Determine force coefficient K from figure above , where:

   where: ϕ_r = friction angle of reinforced fill

$$T_{\max} \;=\; 0.5\, K\, \gamma_r\, (\,H'\,)^2$$

2) Determine:

   where: H' = H + q/γ_r

   q = a uniform load

3) Determine the required reinforcement length at the top L_T and bottom L_B of the slope from the figure above.

LIMITING ASSUMPTIONS

- Extensible reinforcement.
- Slopes constructed with uniform, cohesionless soil (c = 0).
- No pore pressures within slope.
- Competent, level foundation soils.
- No seismic forces.
- Uniform surcharge not greater than 0.2 γ_r H.
- Relatively high soil/reinforcement interface friction angle, ϕ_{sg} = 0.9 ϕ_r (may not be appropriate for some geotextiles).

Figure 8-5   *Sliding wedge approach to determine the coefficient of earth pressure K (after Schmertmann, et al., 1987).*

For three zones:

$$T_{Bottom} = \tfrac{1}{2} T_{max}$$
$$T_{Middle} = \tfrac{1}{3} T_{max}$$
$$T_{Top} = \tfrac{1}{6} T_{max}$$

d. Determine reinforcement vertical spacing S_v.

For each zone, calculate the design tension, T_d, requirements for each reinforcing layer based on an assumed S_v. If the allowable reinforcement strength is known, calculate the minimum vertical spacing and number of reinforcing layers, N, required for each zone based on:

$$T_d = T_a R_c = \frac{T_{zone} S_v}{H_{zone}} = \frac{T_{zone}}{N}$$

where:

R_c = percent coverage of reinforcement, in plan view ($R_c = 1$ for continuous sheets);

S_v = vertical spacing of reinforcement; should be a multiple of compaction layer thickness for ease of construction;

T_{zone} = maximum reinforcement tension required for each zone; T_{zone} equals T_{max} for low slopes (H $\leq$ 6 m);

H_{zone} = height of zone, and is equal to T_{top}, T_{middle}, and T_{Bottom} for high slopes (H > 6 m).

Short (1 to 2 m) lengths of intermediate reinforcement layers can be used to maintain a maximum vertical spacing of 600 mm or less for face stability and compaction quality (Figure 8-6). For slopes less than 1H:1V, closer spaced reinforcements (*i.e.*, every lift or every other lift, but no greater than 400 mm) generally preclude having to wrap the face. Wrapped faces are usually required for steeper slopes to prevent face sloughing. Alternative vertical spacings can be used to prevent face sloughing, but in these cases a face stability analysis should be performed using the method presented in Section 8.3 or as recommended by Thielen and Collin (1993). Intermediate reinforcement should be placed in continuous layers and need not be as strong as the primary reinforcement, but in any case, all reinforcements should be strong enough to survive installation (*e.g.*, see Chapter 5, Tables 5-1 and 5-2). For planar reinforcements, if ϕ_{sg} is less than ϕ, then ϕ_{sg} should be used in the analysis of the failure surface in the reinforced soil zone.

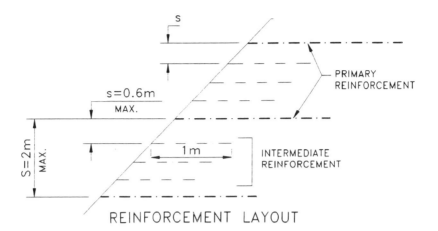

REINFORCEMENT LAYOUT

Figure 8-6   *Spacing and embedding requirements for slope reinforcement showing: (a) primary and intermediate reinforcement layout; and (b) uniform reinforcement layout.*

e. For critical or complex structures, and when checking a complex design, Step 6a should be repeated for a potential failure directly above each layer of primary reinforcement to ensure adequate distribution.

f. Determine the reinforcement lengths required.

The embedment length, L_e, of each reinforcement layer beyond the most critical sliding surface found in Step 6a (*i.e.*, circle found for T_{max}) must be sufficient to provide adequate pullout resistance. For the method illustrated in Figure 8-4, use:

$$L_e = \frac{T_a \; FS}{F^* \; \alpha \; \sigma'_v \; 2}$$

where:

F^*, α, and σ'_v are defined in Section 8.4 and Appendix F.

Minimum value of L_e is 1 m. For cohesive soils, check L_e for both short- and long-term pullout conditions. For long-term design, use ϕ'_r with $c_r = 0$. For short-term evaluation, conservatively use ϕ_r and $c_r = 0$ from consolidated-undrained tests, or run pullout tests.

On a slope cross section containing the rough limits of the critical zone determined in Step 5, plot the reinforcement lengths obtained from the pullout evaluation. The length of the lower layers must extend to or beyond the limits of the critical zone. The length required for sliding stability at the base will generally control the length of the lower reinforcement levels. Upper levels of reinforcement may not have to extend to the critical zone limits if sufficient reinforcement exists in the lower levels to provide FS_R for all circles within the critical zone (*e.g.*, see Step 6g). Make sure that the sum of the reinforcement passing though each failure surface is greater than T_s, from Step 6a, required for that surface. To account for pullout resistance, only consider reinforcement that extends 1 meter beyond the surface. If the available reinforcement is not sufficient, increase the length of reinforcement not passing through the surface or increase the strength of lower-level reinforcement.

Simplify the layout by lengthening some reinforcement layers to create two or three sections of equal reinforcement length. Reinforcement layers generally do not need to extend to the limits of the critical zone, except for the lowest levels of each reinforcement section.

Check the obtained length using Chart B in Figure 8-5. L_e is already included in the total length, L_t and L_B from Chart B.

g. Check design lengths of complex designs. When checking a design that has differing reinforcement length zones, lower zones may be over-reinforced to provide reduced lengths of upper reinforcement levels. In evaluating the length requirements for such cases, the reinforcement pullout stability must be carefully checked in each zone for the critical surfaces exiting at each length zone base.

STEP 6.   Method B - Trial reinforcement analysis.

Another way to design reinforcement for a stable slope is to develop a trial layout of reinforcement and analyze the reinforced slope with a computer program, such as the new FHWA program RSS. Layout includes number, length, design strength, and vertical distribution of the geosynthetic reinforcement. The charts presented in Figure 8-5 provide a method for generating a preliminary layout. Note that these charts were developed with the specific assumptions noted on the figure.

Analyze the reinforced slope with trial geosynthetic reinforcement layouts. The most economical reinforcement layout will be one which results in approximately equal, but greater than the minimum required, stability safety factor for internal, external, and compound failure planes. A contour plot of lowest safety factor values about the trial failure circle centroids is recommended to map and locate the minimum safety factors for the three modes of failure.

External stability analysis in Step 7 will then include an evaluation of local bearing capacity, foundation settlement, and dynamic stability.

STEP 7.   Check external stability.

The external stability of a reinforced soil mass depends on the soil mass's ability to act as a stable block and withstand all external loads without failure. Failure possibilities include sliding, deep-seated overall instability, local bearing capacity failure at the toe (lateral squeeze-type failure), as well as compound failures initiating internally and externally through the short- and long-term conditions.

a.   Sliding resistance.

The reinforced mass must be wide enough at any level to resist sliding along the reinforcement. A wedge-type failure surface defined by the reinforcement limits (the length of the reinforcement from the toe) identified in Step 5 can be checked to ensure it is sufficient to resist sliding from the following relationships:

$$\text{Resisting Force} = FS \times \text{Sliding Force}$$
$$(W + P_A \sin \phi) \tan \phi_{sg} = FS \, P_A \cos \phi$$

with:

W	=	$1/2 \, L^2 \, \gamma \tan \beta$ for L < H
W	=	$[LH - H^2/(2\tan\beta)] \, \gamma$ for L > H

where:

L	=	length of bottom reinforcing layer in each zone where there is a reinforcement length change;
H	=	height of slope;
FS	=	factor of safety for sliding (>1.5);
P_A	=	active earth pressure;
ϕ_{sg}	=	angle of shearing friction between soil and geosynthetic; and
β	=	slope angle.

b.   Deep-seated global stability.

As a check, potential deep-seated failure surfaces behind the reinforced soil mass should be reevaluated. The analysis performed in Step 5 should provide this information. However, as a check, classical rotational slope stability methods such as simplified Bishop (1955), Morgenstern and Price (1965), Spencer (1981), or others, may be used. Appropriate computer programs may also be used.

c. Local bearing failure at the toe (lateral squeeze).

Consideration must be given to the bearing capacity at the toe of the slope. High lateral stresses in a confined soft stratum beneath the embankment could lead to a lateral squeeze-type failure. This must be analyzed if the slope is on a soft -- not firm foundation. Refer to Chapter 7 for design equations and references.

d. Foundation settlement.

The magnitude of foundation settlement should be determined using ordinary geotechnical engineering procedures. If the calculated settlement exceeds project requirements, then foundation soils must be improved.

e. Dynamic stability (Berg, 1993).

If the slope is located in an area subject to potential seismic activity, then some type of dynamic analysis is warranted. Usually a simple pseudo-static type analysis is carried out using a seismic coefficient obtained from local or national codes. For critical projects in areas of potentially high seismic risk, a complete dynamic analysis should be performed.

STEP 8. Evaluate requirements for subsurface and surface water control.

a. Subsurface water control.

Uncontrolled subsurface water seepage can decrease slope stability and ultimately result in slope failure. Hydrostatic forces on the rear of the reinforced mass and uncontrolled seepage into the reinforced mass will decrease stability. Seepage through the mass can reduce pullout capacity of the geosynthetic and create erosion at the face.

Consider the water source and the permeability of the natural and fill soils through which water must flow when designing subsurface water drainage features. Flow rate, filtration, placement, and outlet details should be addressed. Drains are typically placed at the rear of the reinforced mass. Lateral spacing of outlets is dictated by site geometry, expected flow, and existing agency standards. Outlet design should address long-term performance and maintenance requirements. Geocomposite drainage systems or conventional granular blanket and trench drains could be used. The design of geocomposite drainage materials is addressed in Chapter 2.

Geotextile reinforcements (primary and intermediate layers) must be more permeable that the reinforced fill material to prevent a hydraulic build up above the geotextile layers during precipitation.

Where drainage is critical for maintaining slope stability, special emphasis is recommended on the design and construction of subsurface drainage features. Redundancy in the drainage system is also recommended in these cases.

b. Surface water runoff.

Slope stability can be threatened by erosion due to surface water runoff. Erosion rills and gullies can lead to surface sloughing and possibly deep-seated failure surfaces. Erosion control and revegetation measures must, therefore, be an integral part of all reinforced slope system designs and specifications.

Surface water runoff should be collected above the reinforced slope and channeled or piped below the base of the slope. Standard agency drainage details should be utilized.

If not otherwise protected, reinforced slopes should be vegetated after construction to prevent or minimize erosion due to rainfall and runoff on the face. Vegetation requirements will vary by geographic and climatic conditions and are therefore project-specific. Geosynthetic reinforced slopes are inherently difficult sites to establish and maintain vegetative cover due to these steep slopes. The steepness of the slope limits the amount of water absorbed by the soil before runoff occurs. Once vegetation is established on the face, it must be maintained to ensure long-term survival.

A synthetic (permanent) erosion control mat that is stabilized against ultraviolet light and is inert to naturally occurring soil-born chemicals and bacteria may be required with seeding. The erosion control mat serves three functions: 1) to protect the bare soil face against erosion until vegetation is established, 2) to reduce runoff velocity for increased water absorption by the soil, thereby promoting long-term survival of the vegetative cover, and 3) to reinforce the root system of the vegetative cover. Maintenance of vegetation will still be required.

A permanent synthetic mat may not be required in applications characterized by flatter slopes (less than 1:1), low height slopes, and/or moderate runoff. In these cases, a temporary (degradable) erosion blanket may be specified to protect the slope face and promote growth until vegetative cover is firmly established. Refer to Chapter 4 for design of erosion mats and blankets.

Erosion control mats and blankets vary widely in type, cost, and -- more importantly -- applicability to project conditions. **Slope protection should not be left to the contractor's or vendor's discretion.** Guideline material specifications are provided in Section 8.8.

8.4   MATERIAL PROPERTIES

8.4-1   Reinforced Slope Systems

Reinforced soil systems consist of planar reinforcements arranged in horizontal planes in the fill soil to resist outward movement of the reinforced soil mass. Facing treatments ranging from vegetation to flexible armor systems are applied to prevent raveling and sloughing of the face. These systems are generic in nature and can incorporate any of a variety of reinforcements and facing systems. This section provides the material properties required for design.

8.4-2   Soils

Any soil meeting the requirements for embankment construction can be used in a reinforced slope system. From a reinforcement point of view alone, even lower-quality soil than that conventionally used in unreinforced slope construction could be used; however, a higher-quality material offers less durability concerns and is easier to handle, place, and compact, which tends to speed up construction. Therefore, the following guidelines are provided as recommended backfill requirements for reinforced engineered slopes.

Gradation (Christopher et al., 1990):   Recommended backfill requirements for reinforced engineered slopes are:

Sieve Size	Percent Passing
100 mm	100 - 75
4.75 mm	100 - 20
0.425 mm	0 - 60

0.075 mm   0 - 50

Plasticity Index (PI) $\leq$ 20 (AASHTO T-90)

Soundness: Magnesium sulfate soundness loss less than 30% after 4 cycles.

The maximum size should be limited to 19 mm unless tests have been, or will be, performed to evaluate potential strength reduction due to installation damage (see Appendix F). In any case, geosynthetic strength reduction factors for site damage should be checked in relation to particle size and angularity of the largest particles.

Definition of total and effective stress shear strength properties becomes more important as the percentage passing the 0.075 mm sieve increases. Likewise, drainage and filtration design are more critical. Fill materials outside of these gradation and plasticity index requirements have been used successfully (Christopher et al., 1990; Hayden et al., 1991); however, long-term (> 5 years) performance field data is not available. Performance monitoring is recommended if fill soils fall outside of the requirements listed above.

Chemical Composition (Berg, 1993): The chemical composition of the fill and retained soils should be assessed for effect on reinforcement durability (pH, chlorides, oxidation agents, etc.). Some of the soil environments posing potential concern when using geosynthetics are listed in Appendix F. Soils with pH $\geq$ 12 or with pH $\leq$ 3 should not be used in reinforced slopes and a pH range $\geq$ 3 to $\leq$ 9 is recommended. Specific supporting test data should be required if pH > 9.

Compaction (Christopher et al., 1990): Soil fill shall be compacted to 95% of optimum dry density (γ_d) and + or - 2 % of the optimum moisture content, w_{opt}, according to the standard Proctor test. Cohesive soils should be compacted in 150 to 200 mm compacted lifts, and granular soils in 200 to 300 mm compacted lifts.

Shear Strength (Berg, 1993): Peak shear strength parameters determined using direct shear or consolidated-drained (CD) triaxial tests should be used in the analysis (Christopher et al., 1990). Effective stress strength parameters should be used for granular soils with less than 15% passing the 0.075 mm sieve.

For all other soils, peak effective stress and total stress strength parameters should be determined. These parameters should be used in the analyses to check stability for the immediately-after-construction and long-term cases. Use CD direct shear tests (sheared slowly enough for adequate sample drainage), or consolidated-undrained (CU) triaxial tests with pore water pressures measured for determination of effective stress parameters. Use CU direct shear or triaxial tests for determination of total stress parameters.

Shear strength testing is recommended. However, use of assumed shear values based on local experience may be acceptable for some projects. Verification of site soil type(s) should be completed following excavation or identification of borrow pit, as applicable.

Unit Weights: Dry unit weight for compaction control, moist unit weight for analyses, and saturated unit weight for analyses (where applicable) should be determined for the fill soil.

8.4-3 Geosynthetic Reinforcement

Geosynthetic design strength must be determined by testing and analysis methods that account for long-term interaction (*e.g.*, grid/soil stress transfer) and durability of all of the geosynthetic components. Geogrids transfer stress to the soil through passive soil resistance on the grid's transverse members and through friction between the soil and the geogrid's horizontal surfaces (Mitchell and Villet, 1987). Geotextiles transfer stress to the soil through friction.

An inherent advantage of geosynthetics is their longevity in fairly aggressive soil conditions. The anticipated half-life of some geosynthetics in normal soil environments is in excess of 1000 years. However, as with steel reinforcements, strength characteristics must be adjusted to account for potential degradation in the specific environmental conditions, even in relatively neutral soils.

Allowable Tensile Strength: Allowable tensile strength (T_a) of the geosynthetic shall be determined using a partial factor of safety approach (Bonaparte and Berg, 1987). Reduction factors are used to account for installation damage, chemical and biological conditions and to control potential creep deformation of the polymer. Where applicable, a reduction is also applied for seams and connections. The total reduction factor is based upon the mathematical product of these factors. The allowable tensile strength, T_a, (after Berg, 1993) is thus can be obtained from:

$$T_a = \frac{T_{uit}}{RF}$$

with:

$$RF = RF_{CR} \times RF_{ID} \times RF_{CD} \times RF_{BD} \times RF_{INT}$$

where:

T_a = allowable geosynthetic tensile strength,(kN/m), for use in stability analyses;

T_{ult} = ultimate geosynthetic tensile strength,(kN/m);

RF_{CR} = partial factor of creep deformation, ratio of T_{ult} to creep-limiting strength, (dimensionless);

RF_{ID} = partial factor of safety for installation damage, (dimensionless);

RF_{CD} = partial factor of safety for chemical degradation, (dimensionless);

RF_{BD} = partial factor of safety for biological degradation, used in environments where biological degradation potential may exist, (dimensionless); and

RF_{JNT} = partial factor of safety for joints (seams and connections), (dimensionless).

The procedure presented in Appendix F (Berg, 1993) is derived from the Task Force 27 guidelines for geosynthetic reinforced soil retaining walls (1990), the Geosynthetic Research Institute's Methods GG4a and GG4b - Standard Practice for Determination of the Long Term Design Strength of Geogrids (1990, 1991), and the Geosynthetic Research Institute's Method GT7 - Standard Practice for Determination of the Long Term Design Strength of Geotextiles (1992).

RF values for durable geosynthetics in non-aggressive, granular soil environments range from 2.5 to 7. Appendix F suggests that a default value of RF = 7 may be used with geosynthetic products that have not been fully tested or evaluated, for design of routine, non critical structures **which meet the soil, geosynthetic and structural limitations listed in the appendix**. However, as indicated by the range of RF values, there is a potential to significantly reduce the reinforcing requirements and the corresponding cost of the structure by obtaining a reduced RF from test data.

Soil-Reinforcement Interaction: Two types of soil-reinforcement interaction coefficients or interface shear strengths must be determined for design: pullout coefficient, and interface friction coefficient (Task Force 27 Report, 1990). Pullout coefficients are used in stability analyses to compute mobilized tensile force **at the front** and **tail** of each reinforcement layer. Interface friction coefficients are used to check factors of safety against outward sliding of the entire reinforced mass.

Detailed procedures for quantifying interface friction and pullout interaction properties are presented in Appendix F. The ultimate pullout resistance, P_r, of the reinforcement per unit width of reinforcement is given by:

$$P_r = 2F^* \cdot \alpha \cdot \sigma'_v \cdot L_e$$

where:

$2 \cdot L_e$ = the total surface area per unit width of the reinforcement in the resistance zone behind the failure surface

L_e = the embedment or adherence length in the resisting zone behind the failure surface

F^* = the pullout resistance (or friction-bearing-interaction) factor

α = a scale effect correction factor

σ'_v = effective vertical stress at the soil-reinforcement interfaces

For preliminary design and in absence of specific geosynthetic test data, F^* may be conservatively taken as:

F^* = 2/3 tan ϕ for geotextiles, and

F^* = 0.8 tan ϕ for geogrids.

8.5   PRELIMINARY DESIGN AND COST EXAMPLE

EVALUATION AND COST ESTIMATE EXAMPLE

A 1 kilometer long, 5 m high, 2.5H:1V side slope road embankment in a suburban area is to be widened by one lane. At least a 6 m width extension is required to allow for the additional lane plus shoulder improvements. Several options are being considered.

1. Simply extend the slope of the embankment.

2. Construct a 2.5 m high concrete, cantilever retaining wall at the toe of the slope, extend the alignment, and slope down at 2.5H:1V to the wall. (Of course a geosynthetically reinforced retaining wall should also be considered, but that's covered in the next chapter.)

3. Construct a 1H:1V reinforced soil slope up from the toe of the existing slope, which will add 7.5 m to the alignment, enough for future widening, if required.

Guardrails are required for all options and is not included in the cost comparisons.

Option 1

The first alternative will require 30 m^3 fill per meter of embankment length. The fill is locally available with some hauling required and has an estimated in-place cost of $8/$m^3$ (about $4.00 per 1000 kg). The cost of the 6 m right of way is $15/$m^2$, for a cost of $90 per meter of embankment length. Finally, hydroseeding and mulching will cost approximately $0.75 per meter of face, or approximately $10 per meter of embankment. Thus the total cost of embankment will be $340 for the full height per meter length of embankment or $68/m² of vertical face. There will also be a project delay while the additional right of way is obtained.

Option 2

Based on previous projects in this area, the concrete retaining wall is estimated to cost $400/m² of vertical wall face including backfill. Thus, the 2.5 m high wall will cost $1,000 per meter length of embankment. This leads to a cost of $200/m² of vertical embankment per meter length of structure. In addition 18 m³ of fill will be required to construct the sloped portion, adding $144 per meter of embankment, or $28.80/m² of vertical face, to the cost. Hydroseeding and mulching of the slope will add about a $1/m² of vertical face to the cost. Thus, the total cost of this option is estimated at $230/m² of vertical embankment face. This option will require an additional 2 weeks of construction time to allow the concrete to cure. On some projects, additional costs can be incurred due to the delay plus additional traffic control and highway personnel required for inspection during removal of the forms. Since this project was part of a larger project, such delays were not considered.

Option 3

This option will require a preliminary design to determine the quantity of reinforcement.

STEP 1. Slope description
      a.   H = 5 m
      b. $\beta = 45°$
      c.   q = 10 kPa (for pavement section) + 2% road grade

Performance requirements

      a.   External Stability:
            Sliding Stability: $FS_{min} = 1.5$
            Overall slope stability and deep seated: $FS_{min} = 1.3$
            Dynamic loading: no requirement
            Settlement: analysis required
      b.   Compound Failure: $FS_{min} = 1.3$
      c.   Internal Stability: $FS_{min} = 1.3$

STEP 2. Engineering properties of foundation soils.
      a.   Review of soil borings from the original embankment construction indicates foundation soils consisting of stiff to very stiff, low-plasticity silty clay with interbedded seams of sand and gravel. The soils tend to increase in density and strength with depth.
      b. $\gamma_d = 19$ kN/m³, $\omega_{opt} = 15\%$, UU = 100 kPa, $\phi' = 28°$, c' = 0
      c.   At the time of the borings, $d_w = 2$ m below the original ground surface.
      d.   Not applicable

STEP 3. Properties of reinforced and embankment fill

(The existing embankment fill is a clayey sand and gravel). For preliminary evaluation, the properties of the embankment fill are assumed for the reinforced section as follows:

a.
Sieve Size	Percent Passing
100 mm	100%
20 mm	99%
4.75 mm	63%
0.425 mm	45%
0.075 mm	25%

PI (of fines) = 10

Gravel is competent

pH = 7.5

b. $\gamma_r = 21$ kN/m^3, $\omega_{opt} = 15\%$

c. $\phi' = 33°$, c' = 0

d. Soil is relatively inert

STEP 4. Design parameters for reinforcement

For preliminary analysis use default values.

a. $T_d = T_a$ @ 10% strain

b. $T_a = T_{ult}/R_f$

c. $FS_{po} = 1.5$

STEP 5. Check unreinforced stability

Using STABL5M the minimum unreinforced factor of safety was 0.68 with the critical zone defined by the target factor of safety FS_R as shown in Design Figure A.

STEP 6. Calculate T_s for the FS_R

Option A. From the computer runs, obtain FS_U, M_D and R for each failure surface within the critical zone, and calculate T_s as follows. (NOTE: With minor code modification, this could easily be done as part of the computer analysis.)

a.

$$T_s = (1.3 - FS_u) \frac{M_D}{R}$$

Evaluating all of the surfaces in the critical zone indicates maximum $T_s = 49.7$ kN/m for $FS_U = 0.89$ as shown in Design Figure B.

b. Checking T_s by using Figure 8-5:

$$\phi_f = \tan^{-1}\left(\frac{\tan\phi_r}{FS_R}\right) = \tan^{-1}\left(\frac{\tan 33^0}{1.3}\right) = 26.5^o$$

From Figure 8-5a, K ≈ 0.14

and,

$$H' = H + q/\gamma_r + 0.25 \text{ m (for 2% road grade)}$$
$$= 5 \text{ m} + (10 \text{ kN/m}^2 \div 21 \text{ kN/m}^3) + 0.25 \text{ m} = 5.75 \text{ m}$$

then,

$$T_R = 0.5 \text{ K}\gamma_r \text{ H'}$$
$$= 0.5(0.14)(21\text{kN/m}^3)(5,75\text{m})^2$$
$$= 48.6 \text{ kN/m}$$

The evaluation using Figure 8-5 appears to be in good agreement with the computer analysis.

c. Determine the distribution of reinforcement.

    Since H < 6m, use a uniform spacing. Due to the cohesive nature of the backfill, maximum compaction lifts of 200 mm are recommended.

d. To avoid wrapping the face, use S_v = 400 mm reinforcement spacing; therefore, N = 5m/0.4m = 12.5. Use 12 layers with the bottom layer placed after the first lift of embankment fill.

$$T_d = \frac{T_{max}}{N} = \frac{49.7 \; kN\,/\,m}{12} = 4.14 \; kN\,/\,m$$

    (NOTE: Other reinforcement options such as using short secondary reinforcements at every lift with spacing and strength increased for primary reinforcements could be considered, and should be evaluated, for selecting the most cost-effective option for final design.)

e. Not required.

f. For preliminary analysis, the critical zone found in the computer analysis (Design Figure A) can be used to define the reinforcement limits. This is especially true for this problem, since the factor of safety for sliding ($FS_{sliding} \geq$ 1.5) is greater than the internal stability requirement ($FS_{internal} \geq 1.3$); thus, the sliding failure surface well encompasses the most critical reinforcement surface.

    As measured at the bottom and top of the sliding surface in Design Figure A, the required lengths of reinforcement are: L_{bottom} = 6.4 m
    L_{top} = 3.8 m

    Check the length requirement using Figure 8-5. For L_B

$$\phi_f = \tan^{-1}\left(\frac{\tan 28^o}{1.5}\right) = 19.5^o$$

From Figure 8-5: L_B/H' = 1.17
thus L_B = 5.75 m (1.17) = 6.7 m

For L_T

$$\phi_f = \tan^{-1}\left(\frac{\tan 33^o}{1.5}\right) = 23.4^o$$

From Figure 8-5: L_T/H' = 0.63
thus L_T = 5.75 m (0.63) = 3.6 m

Using Figure 8-5, the evaluation again appears to be in good agreement with the computer analysis.

g.   Not applicable

Option B.   Since this is a preliminary analysis and a fairly simple problem, Figure 8-5 or any number of proprietary computer programs, can be used to rapidly evaluate T_s and T_d.

In summary, 12 layers of reinforcement are required with a design strength, T_d, of 4.14 kN/m and an average length of 5 m over the full height of embankment. This would result in a total of 60 m^2 reinforcement per meter length of embankment or 12 m^2 per vertical meter of height. Adding 10% to 15% for overlaps and overages results in an anticipated reinforcement volume of 13.5 m^2 per vertical embankment face. Based on the cost information in Appendix E, reinforcement with an allowable strength $T_a \geq 4.14$ kN/m would cost approximately \$1.00 to \$1.50/m^2. Assuming \$0.50 m^2 for handling and placement, the in-place cost of reinforcement would be approximately \$25/m^2 of vertical embankment face. Approximately 18.8 m^3 of additional backfill would be required for this option, adding \$30/m^2 to the cost of this option. In addition, overexcavation and backfill of existing embankment material will be required to allow for reinforcement placement. Assuming \$2/m^3 for overexcavation and replacement will add approximately \$4/m^2 of vertical face. Erosion protection for the face will also add a cost of \$5/m^2 of vertical face. Thus, the total estimated cost for this option totals approximately \$64/m^2 of vertical embankment face.

Option 3 provides a slightly lower cost than Option 1 plus it does not require additional right-of-way.

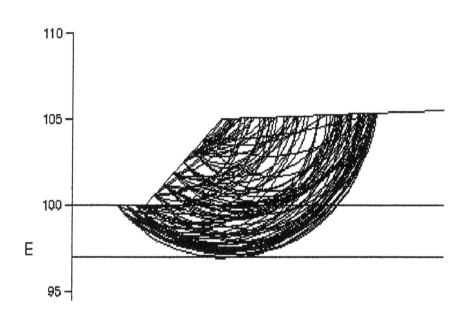

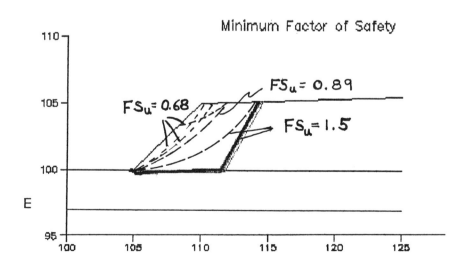

Design Figure A   *Unreinforced stability analysis.*

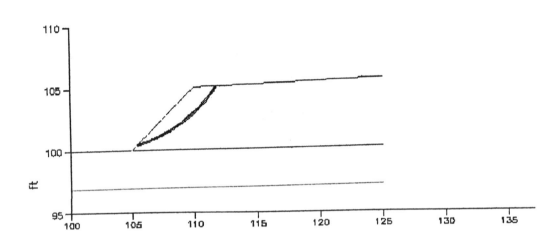

Strength Design — Reinforcement for Critical Surface

Factor of Safety for Circle with Tmax : 0.888
Total Required Reinforcement : 49.7 kN/m

Design Figure B   Surface requiring maximum reinforcement (i.e., most critical reinforced surface).

8.6   COST CONSIDERATIONS

As with any other reinforcement application, an appropriate benefit-to-cost ratio analysis should be completed to determine if the steeper slope with reinforcement is justified economically over the flatter slope with increased right-of-way and materials costs, etc. In some cases, however, the height of the embankment will be controlled by grade requirements, and the slope might as well be as steep as possible. With respect to economy, the factors to consider are:

- cut or fill earthwork quantities;
- size of slope area;
- average height of slope area;
- angle of slope;
- cost of nonselect versus select backfills;
- erosion protection requirements;
- cost and availability of right-of-way needed;
- complicated horizontal and vertical alignment changes;
- safety equipment (guard rails, fences, etc.);
- need for temporary excavation support systems;
- maintenance of traffic during construction; and
- aesthetics.

Figure 8-7 provides a rapid first order assessment of cost for comparing a flatter unreinforced slope with a steeper reinforced slope.

8.7   IMPLEMENTATION

The recent availability of many new geosynthetic reinforcement materials -- as well as drainage and erosion control products -- requires Engineers to consider many alternatives before preparing contract bid documents so that proven, cost-effective materials can be chosen. Reinforced soil slopes may be contracted using two different approaches. Slope structures can be contracted on the basis of (Berg, 1993):

- In-house (owner's engineers or design consultants) design with geosynthetic reinforcement, drainage details, erosion measures, and construction execution specified.
- System or end-result approach using approved systems, similar to reinforced soil walls, with lines and grades noted on the drawings.

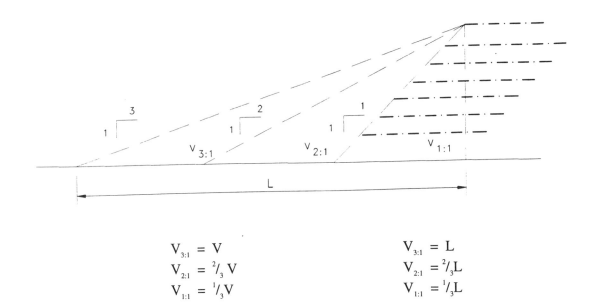

$$V_{3:1} = V \qquad\qquad V_{3:1} = L$$
$$V_{2:1} = {}^{2}/_{3}\,V \qquad\qquad V_{2:1} = {}^{2}/_{3}L$$
$$V_{1:1} = {}^{1}/_{3}V \qquad\qquad V_{1:1} = {}^{1}/_{3}L$$

COST:

3:1 = V_{SOIL} + L_{LAND} + Guardrail (?) + Hydroseeding(?)

2:1 = ${}^{2}/_{3}V_{SOIL}$ + ${}^{2}/_{3}L_{LAND}$ + Guardrail + Erosion Control + High Maintenance

1:1 = ${}^{1}/_{3}V_{SOIL}$ + ${}^{1}/_{3}L_{LAND}$ + Reinforcement + Guardrail + Erosion Control

Figure 8-7   Cost evaluation of reinforced soil slopes.

For either approach, the following assumptions should be considered:

- Geosynthetic reinforced slope systems can successfully compete with select embankment fill requirements, other landslide stabilization techniques, and unreinforced embankment slopes in urban areas.

- Though they may incorporate proprietary materials, reinforced slope systems are non-proprietary and may be bid competitively with geosynthetic reinforcement material alternatives. Geosynthetic reinforcement design parameters must be based upon documentation that is provided by the manufacturer, submitted and approved by the engineer, or based upon default partial safety values as described in Section 8.3 and Appendix F.

- Designers contemplating the use of reinforced slope systems should offer the same degree of involvement to all suppliers who can accomplish the project objectives.

It is highly recommended that each Transportation Agency develop documented procedures for review and approval of geosynthetic soil reinforcement and associated materials. Guidelines for preparation of geosynthetic review and approval documents are included in (Berg, 1993).

8.8   SPECIFICATIONS

Guideline specifications are included from Guidelines for Design, Specification, and Contracting of Geosynthetic Mechanically Stabilized Earth Slopes on Firm Foundations (Berg, 1993).

1.   DESCRIPTION:

Work shall consist of furnishing geosynthetic soil reinforcement for use in construction of reinforced soil slopes.

2.   GEOSYNTHETIC REINFORCEMENT MATERIAL:

2.1 The specific geosynthetic reinforcement material and supplier shall be preapproved by the Agency as outlined in the Agency's reinforced slope policy.

2.2 The geosynthetic reinforcement shall be a regular network of integrally connected polymer tensile elements with aperture geometry sufficient to permit significant mechanical interlock with the surrounding soil or rock. The geosynthetic reinforcement structure shall be dimensionally stable and able to retain its geometry under construction stresses and shall have high resistance to damage during construction, to ultraviolet degradation, and to all forms of chemical and biological degradation encountered in the soil being reinforced.

2.3 The geosynthetics shall have an Allowable Strength (T_a) and Pullout Resistance, for the soil type(s) indicated, as listed in Table 8.1 for geotextiles and/or 8.2 for geogrids.

2.4 The permeability of a geotextile reinforcement shall be greater than the permeability of the fill soil.

2.5 Certification: The contractor shall submit a manufacturer's certification that the geosynthetics supplied meet the respective index criteria set when geosynthetic was approved by the Agency, measured in full accordance with all test methods and standards specified.   In case of dispute over validity of values, the Engineer can require the Contractor to supply test data from an Agency-approved laboratory to support the certified values submitted.

2.6 Quality Assurance/Index Properties: Testing procedures for measuring design properties require elaborate equipment, tedious set up procedures and long durations for testing. These tests are inappropriate for quality assurance (QA) testing of geosynthetic reinforcements received on site. In lieu of these tests for design properties, a series of index criteria may be established for QA testing. These index criteria include mechanical and geometric properties that directly impact the design strength and soil interaction behavior of geosynthetics. It is likely that each family of products will have varying index properties and QC/QA test methods. QA testing should measure the respective index criteria set when geosynthetic was approved by the Agency. Minimum average roll values, per ASTM D 4759, shall be used for conformance.

3.   CONSTRUCTION:

3.1 Delivery, Storage, and Handling: Follow requirements set forth under materials specifications for geosynthetic reinforcement, drainage composite, and geosynthetic erosion mat.

3.2 On-Site Representative: Geosynthetic reinforcement material suppliers shall provide a qualified and experienced representative on site, for a minimum of three days, to assist the Contractor and Agency inspectors at the start of construction. If there is more than one slope on a project then this criteria will apply to construction of the initial slope only. The representative shall also be available on an as-needed basis, as requested by the Agency Engineer, during construction of the remaining slope(s).

3.3 Site Excavation: All areas immediately beneath the installation area for the geosynthetic reinforcement shall be properly prepared as detailed on the plans, specified elsewhere within the specifications, or directed by the Engineer. Subgrade surface shall be level, free from deleterious materials, loose, or otherwise unsuitable soils. Prior to placement of geosynthetic reinforcement, subgrade shall be proofrolled to provide a uniform and firm surface. Any soft areas, as determined by the Owner's Engineer, shall be excavated and replaced with suitable compacted soils. Foundation surface shall be inspected and approved by the Owner's Geotechnical Engineer prior to fill placement. Benching the backcut into competent soil is recommended to improve stability.

3.4 Geosynthetic Placement: The geosynthetic reinforcement shall be installed in accordance with the manufacturer's recommendations. The geosynthetic reinforcement shall be placed within the layers of the compacted soil as shown on the plans or as directed.

The geosynthetic reinforcement shall be placed in continuous, longitudinal strips in the direction of main reinforcement. However, if the Contractor is unable to complete a required length with a single continuous length of geogrid, a joint may be made with the Engineer's approval. Only one joint per length of geogrid shall be allowed. This joint shall be made for the full width of the strip by using a similar material with similar strength. Joints in geogrid reinforcement shall be pulled and held taut during fill placement. Joints shall not be used with geotextiles.

Adjacent strips, in the case of 100% coverage in plan view, need not be overlapped. The minimum horizontal coverage is 50%, with horizontal spacings between reinforcement no greater than 40 inches. Horizontal coverage of less than 100% shall not be allowed unless specifically detailed in the construction drawings.
Adjacent rolls of geosynthetic reinforcement shall be overlapped or mechanically connected where exposed in a wrap-around face system, as applicable.

Place only that amount of geosynthetic reinforcement required for immediately pending work to prevent undue damage. After a layer of geosynthetic reinforcement has been placed, the next succeeding layer of soil shall be placed and compacted as appropriate. After the specified soil layer has been placed, the next geosynthetic reinforcement layer shall be installed. The process shall be repeated for each subsequent layer of geosynthetic reinforcement and soil.

Geosynthetic reinforcement shall be placed to lay flat and pulled tight prior to backfilling. After a layer of geosynthetic reinforcement has been placed, suitable means, such as pins or small piles of soil, shall be used to hold the geosynthetic reinforcement in position until the subsequent soil layer can be placed. Under no circumstances shall a track-type vehicle be allowed on the geosynthetic reinforcement before at least 150 mm of soil has been placed.

During construction, the surface of the fill should be kept approximately horizontal. Geosynthetic reinforcement shall be placed directly on the compacted horizontal fill surface. Geosynthetic reinforcements are to be placed within 75 mm of the design elevations and extend the length as shown on the elevation view, unless otherwise directed by the Owner's Engineer. Correct orientation of the geosynthetic reinforcement shall be verified by the Contractor.

3.5 Fill Placement: Fill shall be compacted as specified by project specifications or to at least 95 percent of the maximum density determined in accordance with AASHTO T-99, whichever is greater.

Density testing shall be made every 1,000 cubic meters of soil placement or as otherwise specified by the Owner's Engineer or contract documents.

Backfill shall be placed, spread, and compacted in such a manner to minimize the development of wrinkles and/or displacement of the geosynthetic reinforcement.

Fill shall be placed in 300 mm maximum lift thickness where heavy compaction equipment is to be used, and 150 mm maximum uncompacted lift thickness where hand-operated equipment is used.

Backfill shall be graded away from the slope crest and rolled at the end of each workday to prevent ponding of water on the surface of the reinforced soil mass.

Tracked construction equipment shall not be operated directly upon the geosynthetic reinforcement. A minimum fill thickness of 150 mm is required prior to operation of tracked vehicles over the geosynthetic reinforcement. Turning of tracked vehicles should be kept to a minimum to prevent tracks from displacing the fill and the geosynthetic reinforcement.

If recommended by the geogrid manufacturer and approved by the Engineer, rubber-tired equipment may pass over the geogrid reinforcement at slow speeds, less than 10 mph. Sudden braking and sharp turning shall be avoided.

3.6 Erosion Control Material Installation: See Erosion Control Material Specification for installation notes.

3.7 Geosynthetic Drainage Composite: See Geocomposite Drainage Composite Material Specification for installation notes.

3.8 Final Slope Geometry Verification: Contractor shall confirm that as-built slope geometries conform to approximate geometries shown on construction drawings.

4.   METHOD OF MEASUREMENT:

Measurement of geosynthetic reinforcement is on a square-meter basis and is computed on the total area of geosynthetic reinforcement shown on the construction drawings, exclusive of the area of geosynthetics used in any overlaps. Overlaps are an incidental item.

5.   BASIS OF PAYMENT:

5.1 The accepted quantities of geosynthetic reinforcement by type will be paid for per square-meter in-place.

5.2 Payment will be made under:

Pay Item	Pay Unit
Geogrid Soil Reinforcement - Type I	square meter
Geogrid Soil Reinforcement - Type II	square meter
Geogrid Soil Reinforcement - Type III	square meter
or	
Geotextile Soil Reinforcement - Type I	square meter
Geotextile Soil Reinforcement - Type II	square meter
Geotextile Soil Reinforcement - Type III	square meter
Material Supplier Representative (exceeding 3 days)	person-day

Table 8-1

Allowable Geotextile Strength and Pullout Resistance with Various Soil Types[1] for Geosynthetic Reinforcement in Earth Slopes

(Geotextile Pullout Resistance and Allowable Strengths vary with reinforced backfill used due to soil anchorage and site damage factors. Guidelines are provided below.)

Soil Type	Minimum Pullout $F^* \alpha C$ Resistance[2,3]	Allowable Strength, T_a (kN/m)[2,4]		
		Geotextile Type I	Geotextile Type II	Geotextile Type III
Gravels, sandy gravels, and gravel-sand-silt mixtures (GW & GM)[5]				
Well graded sands, gravelly sands, and sand-silt mixtures (SW & SM)[5]				
Silts, very fine sands, clayey sands and clayey silts (SC & ML)[5]				
Gravelly clays, sandy clays, silty clays, and lean clays (CL)[5]				

NOTES:
1. Design values shall be verified via testing in accordance with procedures referenced in Appendix F.
2. Values for pullout and allowable strength should be inserted from prequalified geosynthetic suppliers.
3. See Appendix F.
4. All partial factors of safety for reduction of design strength are included in listed values. No additional factors of safety are required to further reduce these design strengths.
5. Unified Soil Classification.

Table 8-2

Allowable Geogrid Strength and Pullout Resistance with Various Soil Types[1] for Geosynthetic Reinforcement Earth Slopes

(Geogrid Pullout Resistance and Allowable Strengths vary with reinforced backfill used due to soil anchorage and site damage factors. Guidelines are provided below.)

Soil Type	Minimum Pullout Resistance[2,3] $F^* \alpha C$	Allowable Strength, T_a (kN/m)[2,4]		
		Geogrid Type I	Geogrid Type II	Geogrid Type III
Gravels, sandy gravels, and gravel-sand-silt mixtures (GW & GM)[5]				
Well graded sands, gravelly sands, and sand-silt mixtures (SW & SM)[5]				
Silts, very fine sands, clayey sands and clayey silts (SC & ML)[5]				
Gravelly clays, sandy clays, silty clays, and lean clays (CL)[5]				

NOTES:
1. Design values shall be verified via testing in accordance with procedures referenced in Appendix F.
2. Values for pullout and allowable strength should be inserted from prequalified geogrid suppliers.
3. See Appendix F.
4. All partial factors of safety for reduction of design strength are included in listed values. No additional factors of safety are required to further reduce these design strengths.
5. Unified Soil Classification.

8.9   INSTALLATION PROCEDURES

Reinforcement layers are easily incorporated between the compacted lifts of fill. Therefore, construction of reinforced slopes is very similar to normal embankment construction. The following is the usual construction sequence:

A. Site preparation.
- Clear and grub site.
- Remove all slide debris (if a slope reinstatement project).
- Prepare level subgrade for placement of first level of reinforcing.
- Proofroll subgrade at the base of the slope with roller or rubber-tired vehicle.

B. Place the first reinforcing layer.
- Reinforcement should be placed with the principal strength direction perpendicular to the face of slope.
- Secure reinforcement with retaining pins to prevent movement during fill placement.
- A minimum overlap of 150 mm is recommended along the edges perpendicular to the slope for wrapped-face structures. Alternatively, with geogrid reinforcement, the edges may be clipped or tied together. When geosynthetics are not required for face support, no overlap is required and edges should be butted.

C. Place backfill on reinforcement.
- Place fill to required lift thickness on the reinforcement using a dozer or front-end loader operating on previously placed fill or natural ground.
- Maintain a minimum of 150 mm between reinforcement and tracks or wheels of construction equipment. This requirement may be waived for rubber-tired equipment provided that field trials, including geosynthetic strength tests, have demonstrated that anticipated traffic conditions will not damage the specific geosynthetic reinforcement.
- Compact with a vibratory roller or plate-type compactor for granular materials, or a rubber-tired vehicle for cohesive materials.
- When placing and compacting the backfill material, avoid any deformation or movement of the reinforcement.
- Use lightweight compaction equipment near the slope face to help maintain face alignment.

D. Compaction control.
- Provide close control on the water content and backfill density. It should be compacted at least 95% of the standard Proctor maximum density within 2% of optimum moisture.
- If the backfill is a coarse aggregate, then a relative density or a method type compaction specification should be used.

E. Face construction.

As indicated in the design section (8.3-3), a face wrap generally is not required for slopes up to 1H:1V, if the reinforcement is maintained at close spacing (*i.e.*, every lift or every other lift, but no greater than 400 mm). In this case the reinforcement can be simply extended to the face. For this option, a facing treatment should be applied to prevent erosion during and after construction. If slope facing is required to prevent sloughing (*i.e.*, slope angle β is greater than ϕ_{soil}) or erosion, sufficient reinforcement lengths could be provided for a wrapped-face structure. The following procedures are recommended for wrapping the face.
- Turn up reinforcement at the face of the slope and return the reinforcement a minimum of 1 to 1.2 m into the embankment below the next reinforcement layer (see Figure 8-8).

- For steep slopes, form work may be required to support the face during construction, especially if lift thicknesses of (0.5 to 0.6 m or greater) are used.
- For geogrids, a fine mesh screen or geotextile may be required at the face to retain backfill materials.

F. Continue with additional reinforcing materials and backfill.

> NOTE: If drainage layers are required, they should be constructed directly behind or on the sides of the reinforced section.

Several construction photos from reinforced slope projects are shown in Figure 8-9.

8.10   FIELD INSPECTION

As with all geosynthetic construction, and especially with critical structures such as reinforced slopes, competent and professional field inspection is absolutely essential for successful construction. Field personnel must be properly trained to observe every phase of the construction. They must make sure that the specified material is delivered to the project, that the geosynthetic is not damaged during construction, and that the specified sequence of construction operations are explicitly followed. Field personnel should review the checklist items in Section 1.7. Other important details include construction of the slope face and application of the facing treatment to minimize geosynthetic exposure to ultraviolet light.

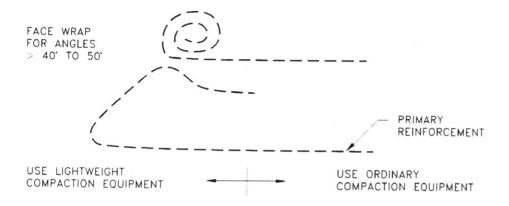

FACE WRAP
FOR ANGLES
> 40° TO 50°

PRIMARY
REINFORCEMENT

USE LIGHTWEIGHT
COMPACTION EQUIPMENT

USE ORDINARY
COMPACTION EQUIPMENT

(a) LIFT 1 PLUS REINFORCING FOR LIFT 2

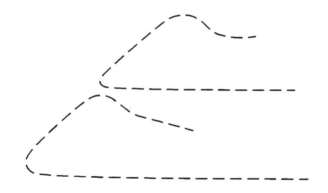

(b) LIFT 2 WITH FACE WRAPPED

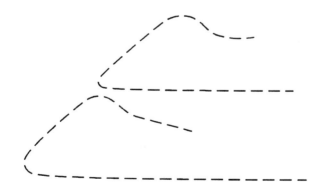

(c) LIFT 2 COMPLETED

Figure 8-8   Construction of reinforced slopes.

a.

b.

c.

Figure 8-9   *Reinforced slope construction: a) geogrid and fill placement; b) soil fill erosion control mat facing; and c) finished, vegetated 1:1 slope.*

8.11 REFERENCES

References quoted within this section are listed below. The Berg (1993) reference is a recent, comprehensive guideline specifically directed at reinforced slopes in transportation applications. It is a key reference for design, specification, and contracting. Two FHWA publications on reinforced slopes are nearing finalization; they are the RSS computer program and users manual mentioned earlier and the Demo 82 manual on reinforced soil systems.

AASHTO (1990), Design Guidelines for Use of Extensible Reinforcements (Geosynthetic) for Mechanically Stabilized Earth Walls in Permanent Applications, Task Force 27 Report - In Situ Soil Improvement Techniques, American Association of State Transportation and Highway Officials, Washington, D.C.

ASTM (1994), Annual Books of ASTM Standards, American Society for Testing and Materials, Philadelphia, PA:
> Volume 4.08 (I), Soil and Rock
> Volume 4.08 (II), Soil and Rock; Geosynthetics

Berg, R.R. (1993), Guidelines for Design, Specification, & Contracting of Geosynthetic Mechanically Stabilized Earth Slopes on Firm Foundations, Report No. FHWA-SA-93-025, Federal Highway Administration, Washington, D.C., 87 p.

Berg, R.R., Anderson, R.P., Race, R.J., and Chouery-Curtis, V.E. (1990), Reinforced Soil Highway Slopes, Transportation Research Record No. 1288, Geotechnical Engineering, Transportation Research Board, Washington, D.C., pp. 99-108.

Berg, R.R., Chouery-Curtis, V.E. and Watson, C.H. (1989), Critical Failure Planes in Analysis of Reinforced Slopes, Proceedings of Geosynthetics '89, Volume 1, San Diego, CA, Industrial Fabrics Association International, February, pp. 269-278.

Bishop, A.W. (1955), The use of the Slip Circle in the Stability Analysis of Slopes, Geotechnique, Vol. 5, No. 1.

Bonaparte, R., Schmertmann, G.R., Chu, D. and Chouery-Curtis, V.E. (1989), Reinforced Soil Buttress to Stabilize a High Natural Slope, Proceedings of the 12th International Conference on Soil Mechanics and Foundation Engineering, Vol. 2, Rio de Janeiro, Brazil, August, pp. 1227-1230.

Bonaparte, R. and Berg, R.R. (1987), Long-Term Allowable Tension for Geosynthetic Reinforcement, Proceedings of Geosynthetics '87 Conference, Volume 1, New Orleans, LA, Industrial Fabrics Association International, pp. 181-192.

Christopher, B.R. and Holtz, R.D. (1985), Geotextile Engineering Manual, Report No. FHWA-TS-86/203, Federal Highway Administration, Washington, D.C., March, 1044 p.

Christopher, B.R., Gill, S.A., Giroud, J.P., Juran, I. Scholsser, F., Mitchell, J.K. and Dunnicliff, J. (1990), Reinforced Soil Structures, Volume I. Design and Construction Guidelines, Federal Highway Administration, Washington, D.C., Report No. FHWA-RD--89-043, November, 287 p.

GRI Test Standards (1994), Geosynthetic Research Institute, Drexel University, Philadelphia, PA.

Hayden, R.F., Schmertmann, G.R., Qedan, B.Q., and McGuire, M.S. (1991), High Clay Embankment Over Cannon Creek Constructed With Geogrid Reinforcement, Proceedings of Geosynthetics '91, Volume 2, Atlanta, GA, Industrial Fabrics Association International, pp. 799-822.

Iwasaki, K. and Watanabi, S. (1978), Reinforcement of Highway Embankments in Japan, Proceedings of the Symposium on Earth Reinforcement, American Society of Civil Engineers, Pittsburgh, PA, pp. 473-500.

Jewell, R.A. (1989), Revised Design Charts for Steep Reinforced Slopes, Reinforced Embankments: Theory and Practice in the British Isles, Shercliff, Ed., Thomas Telford Ltd., London, U.K., pp. 1-30.

Jewell, R.A., Paine, N. and Woods, R.I. (1984), Design Methods for Steep Reinforced Embankments, Proceedings of the Symposium on Polymer Grid Reinforcement, Institute of Civil Engineering, London, U.K., pp. 18-30.

Leshchinsky, D. and Boedeker, R.H. (1989), Geosynthetic Reinforced Soil Structures, Journal of Geotechnical Engineering, ASCE, Vol. 115, No. 10, pp. 1459-1478.

Mitchell, J.K. and Villet, W.C.B. (1987), Reinforcement of Earth Slopes and Embankments, NCHRP Report No. 290, Transportation Research Board, Washington, D.C., 323 p.

Morgenstern, N. and Price, V.E. (1965), The Analysis of the Stability of General Slip Surfaces, Geotechnique, Vol. 15, No. 1, pp. 79-93.

Ruegger, R. (1986), Geotextile Reinforced Soil Structures on which Vegetation can be Established, Proceedings of the 3rd International Conference on Geotextiles, Vienna, Austria, Vol. II, pp. 453-458.

Schmertmann, G.R., Chouery-Curtis, V.E., Johnson, R.D. and Bonaparte, R. (1987), Design Charts for Geogrid-Reinforced Soil Slopes, Proceedings of Geosynthetics '87, New Orleans, LA, Vol. 1, Industrial Fabrics Association International, pp. 108-120.

Spencer, E. (1981), Slip Circles and Critical Shear Planes, Journal of the Geotechnical Engineering Division, American Society of Civil Engineers, Vol. 107, No. GT7, pp. 929-942.

The Tensar Corporation (1987), Slope Reinforcement, Brochure, 10 p.

Thielen, D.L. and Collin, J.G. (1993), Geogrid Reinforcement for Surficial Stability of Slopes, Proceedings of Geosynthetics '93, Vancouver, B.C., Vol. 1, Industrial Fabrics Association International, pp. 229-244.

Werner, G. and Resl, S. (1986), Stability Mechanisms in Geotextile Reinforced Earth-Structures, Proceedings of the 3rd International Conference on Geotextiles, Vienna, Austria, Vol. II, pp. 465-470.

Zornberg, J.G. and J.K. Mitchell (1992), Poorly Draining Backfills for Reinforced Soil Structures - A State of the Art Review, Geotechnical Research Report No. UCB/GT/92-10, Department of Civil Engineering, University of California, Berkeley, 101 p.

9.0 REINFORCED SOIL RETAINING WALLS AND ABUTMENTS

9.1 BACKGROUND

Retaining walls in transportation engineering are quite common. They are required where a slope is uneconomical or not technically feasible. When selecting a retaining wall type, reinforced soil retaining walls should always be considered. Reinforced soil walls basically consist of some type of reinforcing element in the soil fill to help resist lateral earth pressures. When compared with conventional retaining wall systems, there are often significant advantages to using walls with reinforced backfills. They are very cost effective, especially for higher walls in fill sections. Furthermore, these systems are more flexible than conventional earth retaining walls such as reinforced concrete cantilever or gravity walls. Therefore, they are very suitable for sites with poor foundations and for seismically active areas.

The modern invention of reinforcing the soil backfill behind retaining walls was developed by Vidal in France in the mid-1960s. The Vidal system, called Reinforced Earth™, uses metal strips for reinforcement, as shown schematically in Figure 9-1. The design and construction of Reinforced Earth™ walls is quite well established, and thousands have been successfully built worldwide in the last 25 years. During this time, other similar reinforcing systems, both proprietary and nonproprietary, utilizing different types of metallic reinforcement have been developed (*e.g.*, VSL, Hilfiker, etc.; see Mitchell and Villet, 1987, and Christopher et al., 1990).

The use of geogrids or geotextiles rather than metallic strips (ties), shown conceptually in Figure 9-2, is a further development of the Reinforced Earth™ concept. Geosynthetics offer a viable and often very economical alternative to metallic reinforcement for both permanent and temporary walls, especially under certain environmental conditions. Although the first geotextile wall was built in France in 1971, the first in the United States was in 1974 (Bell et al., 1975). The maximum heights of geosynthetic reinforced walls constructed to date are about 15 m, whereas Reinforced Earth walls have exceeded 30 m in height. A significant benefit of using geosynthetics is the wide variety of wall facings available, resulting in greater aesthetic and economic options. Metallic reinforcement is typically used with articulated precast concrete panels. Alternate facing systems for geosynthetic reinforced walls are described in Section 9.5.

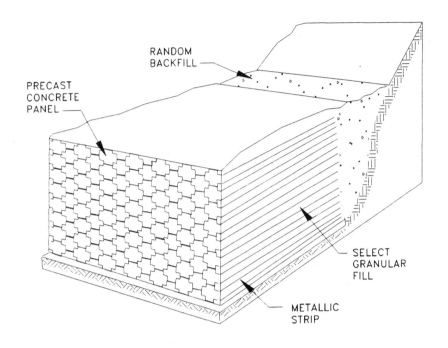

Figure 9-1   Component parts of a Reinforced Earth wall

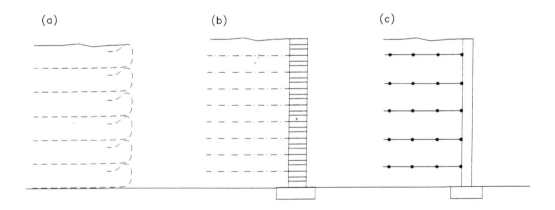

Figure 9-2   Reinforced retaining wall systems using geosynthetics: (a) with wrap-around geosynthetic facing, (b) with segmental or modular concrete block, and (c) with full-height (propped) precast panels.

9.2   APPLICATIONS

Flexible reinforced soil structures including those reinforced with geosynthetics should be considered as cost-effective alternates for all applications where conventional gravity, cast-in-place concrete cantilever, bin-type, or metallic reinforced soil retaining walls are specified.   This includes bridge abutments as well as locations where conventional earthen embankments cannot be constructed due to right-of-way restrictions (another alternative is a steep reinforced slope; see Chapter 8). Conceptually, geosynthetic reinforced walls can be used for any fill wall and low- to moderate-height cut-wall situations. Similar to other reinforced wall types, the relatively wide base width required for geosynthetic walls typically precludes their use in tall cut situations.   Figure 9-3 shows several completed geosynthetic reinforced retaining walls.

Geosynthetic reinforced walls are generally less expensive than conventional earth retaining systems. Using geogrids or geotextiles as reinforcement has been found to be 30 to 50% less expensive than other reinforced soil construction with concrete facing panels, especially for small- to medium-sized projects (Allen and Holtz, 1991).   They may be most cost-effective in temporary or detour construction, and in low-volume road construction (*e.g.*, national forests and parks).

Due to their greater flexibility, reinforced walls offer significant technical advantages over conventional gravity or reinforced concrete cantilever walls at sites with poor foundations and/or slope conditions.   These sites commonly require costly additional construction procedures, such as deep foundations, excavation and replacement, or other foundation soil improvement techniques.

The level of confidence needed for the design of a geosynthetic wall depends on the criticality of the project (Carroll and Richardson, 1986). The criticality depends on the design life, maximum wall height, the soil environment, risk of loss of life, and impact to the public and to other structures if failure occurs. Assessment of criticality is rather subjective, and sound engineering judgement is required. See Allen and Holtz (1991) and Simac et al. (1993) for approaches. The Engineer or regulatory authority should determine the critical nature of a given application.

Geosynthetic reinforced walls may utilize geogrids or geotextiles as soil reinforcing elements, although the most common material used in walls today is geogrid reinforcement. This trend is driven both by geosynthetic manufacturers and suppliers of packaged wall systems. Enhanced aesthetics of the completed wall is obviously controlled by the facing used, and as discussed in Section 9.6, the facing can dictate the type of geosynthetic reinforcement.

a.

b.

c.

d.

Figure 9-3   *Examples of geosynthetic reinforced walls: a. full-height panels, geogrid-reinforced walls, Arizona; b. full-height panels, geogrid-reinforced walls Utah; c. modular concrete units, geogrid-reinforced walls, Texas; and d. temporary geotextile wrap around wall, Washington.*

9.3 DESIGN GUIDELINES FOR REINFORCED WALLS

9.3-1 Approaches and Models

A number of approaches to geotextile and geogrid reinforced retaining wall design have been proposed, and these are summarized by Christopher and Holtz (1985), Mitchell and Villet (1987), Christopher, et al. (1990), and Claybourn and Wu (1993). The most commonly used method is classical Rankine earth pressure theory combined with tensile-resistant tie-backs, in which the reinforcement extends beyond an assumed Rankine failure plane. Figure 9-4 shows a segmental retaining wall (SRW), or modular concrete block (MCB), faced system and the model typically analyzed. Because this design approach was first proposed by Steward, Williamson, and Mohney (1977) of the U.S. Forest Service, it is often referred to as the Forest Service or tie-back wedge method. An alternative approach is the structural stiffness method of calculating internal lateral earth pressures, presented in Christopher et al. (1990).

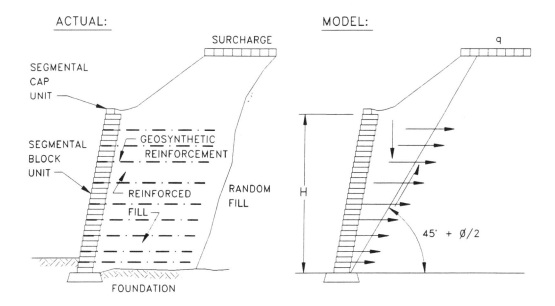

Figure 9-4   Actual geosynthetic reinforced soil wall in contrast to the design model.

Both options use an assumed Rankine failure surface. The tie-back wedge procedure considers an active Rankine state of stress (K_a) for all geosynthetics. Computation of the normal pressure is based on overburden load, surcharge loads, and vertical pressure exerted by the retained backfill imposed on the reinforcing layer, and an equivalent Meyerhof-type distribution is assumed. The structural stiffness method uses a lateral earth pressure coefficient that is a function of the reinforcement stiffness. Computation of lateral pressure is based on overburden and surcharge loads. For geotextiles, a minimum coefficient equal to the active Rankine state (K_a) is used from top to bottom. For stiffer geogrids a coefficient equal to 1.5 active Rankine (1.5 K_a) at the top of wall to active Rankine (K_a) at a depth of 6 m and below is recommended.

A third method is that presented in the National Concrete Masonry Association's (NCMA) Design Manual for Segmental Retaining Walls (Simac et al., 1993). A tie-back wedge approach is used, but with lateral earth pressures and design failure planes computed using a Coulomb earth pressure theory. This is a comprehensive manual (336 pages) on design of walls, and a computer program following this procedure is also available from NCMA.

The basic approach for internal stability is a limiting equilibrium analysis, with consideration of possible failure modes of the reinforced soil mass as given in Table 9-1. These failure modes are analogous to those of metallic reinforced walls.

Table 9-1
Failure Modes and Required Properties for Reinforced Walls

FAILURE MODE FOR GEOSYNTHETIC REINFORCEMENT	FAILURE MODE FOR METALLIC REINFORCEMENT	PROPERTY REQUIRED
Geogrid or geotextile rupture	Strips or meshes break	Tensile strength
Geogrid or geotextile pullout	Strips or meshes pullout	Soil-reinforcement interaction (passive resistance, frictional resistance)
Excessive creep of geogrid or geotextile	N/A	Creep resistance
Connection Failure	Strips or meshes break, or strips or meshes pullout	Tensile failure or pullout

As with conventional retaining structures, overall (external) stability and wall settlement must also be satisfactory. In fact, external stability considerations (*i.e.*, sliding) generally control the length of the reinforcement required.

9.3-2 Design Steps

The following is a step-by-step procedure for the design of geosynthetic reinforced walls.

STEP 1.   Establish design limits, scope of project, and external loads (Figure 9-5).

    A. Wall height, H

    B. Wall length, L

    C. Face batter angle

    D. External loads:
    1. Temporary concentrated live loads, Δq
    2. Uniform surcharge loads, q
    3. Seismic loads, $\propto_g$

    E. Type of facing and connections:
    1. Segmental or modular concrete units, timbers, incremental height precast panels, etc.
    2. Full-height concrete panels
    3. Wrapped

    F. Spacing requirements, s_i, (if any) based on facing connections, stability during construction, lift thickness, and placement considerations (*i.e.*, maximum s = 0.5 m for geotextile- and geogrid-wrapped faced walls), and reinforcement strength.

    G. Environmental conditions such as frost action, scour, shrinkage and swelling, drainage, seepage, rainfall runoff, chemical nature of backfill and seepage water (*e.g.*, pH range, hydrolysis potential, chlorides, sulfates, chemical solvents, diesel fuel, other hydrocarbons, etc.), etc.

    H. Design and service life periods

STEP 2.   Determine engineering properties of foundation soil (Figure 9-5).

    A. Determine the soil profile below the base of the wall; exploration depth should be at least twice the height of the wall or to refusal. Borings should be spaced at least every 30 to 50 m along the alignment of the wall.

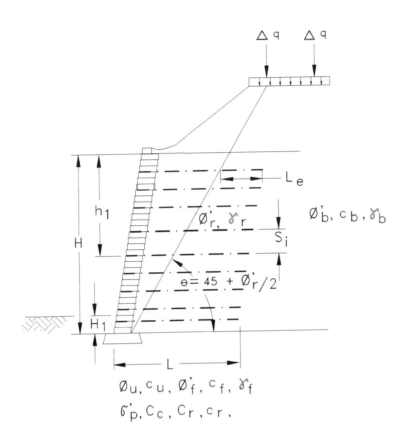

Figure 9-5   *Geometric and loading characteristics of geosynthetic reinforced walls.*

B. Determine the foundation soil strength parameters (c_u, ϕ_u, c', and ϕ'), unit weight (γ), and consolidation parameters (C_c, C_r, c_v and σ_p') for each foundation stratum.

C. Establish location of groundwater table. Check need for drainage behind and beneath the wall.

STEP 3.   Determine properties of both the reinforced fill and retained backfill soils (see Section 9.6 for recommended soil fill requirements).

A. Water content, gradation and plasticity. Note that soils with appreciable fines (silts and clays) are *not* recommended for geosynthetic reinforced walls.

B. Compaction characteristics (maximum dry unit weight, γ_d, and optimum water content, w_{opt}, or relative density).

C. Angle of internal friction, ϕ_r'.

D. pH, chlorides, oxidation agents, etc. (For a discussion of chemical and biological characteristics of the backfill that could affect geosynthetic durability, see Section 9.6.)

STEP 4. Establish design factors of safety (the values below are recommended minimums; local codes may require greater values).

A. External stability:
1. Sliding: FS ≥ 1.5
2. Bearing capacity: FS ≥ 2
3. Overturning: FS ≥ 2
4. Deep-seated (overall) stability: FS ≥ 1.5
5. Settlement: Maximum allowable total and differential based on performance requirements of the project.
6. Dynamic loading: FS ≥ 1.1 or greater, depending on local codes.

B. Internal stability:
1. Determine the allowable long-term tensile strength, T_a, of the reinforcement; see Appendix F. Remember to consider connection strength between the reinforcement and facing element, which may limit the reinforcement's design tensile strength.
2. Determine the long-term design strength, T_d, of the reinforcement, where:

$$T_d = T_a / FS$$

A minimum FS against internal stability failure of 1.5 is normally used. (See Appendix F for additional discussion).

3. Pullout resistance: FS ≥ 1.5; minimum embedment length is approximately 1 m. Use FS ≥ 1.1 for seismic pullout.

STEP 5. Determine preliminary wall dimensions.

A. For analyzing a first trial section, assume a reinforced section length of L= 0.7H.

B. Determine wall embedment depth.

1. Minimum embedment depth H_1, at the front of the wall (Figure 9-5):

Slope in Front of Wall	Minimum H_1
horizontal (walls)	H/20
horizontal (abutments)	H/10
3H:1V	H/10
2H:1V	H/7
3H:2V	H/5

2. Consider possible frost action, shrinkage, and swelling potential of foundation soils, global stability, and seismic activity. In any case, the minimum H_1 is 0.45 m.

STEP 6. Develop the internal and external lateral earth pressure diagrams for the reinforced section. Use either the tie-back wedge method or the structural stiffness approach, as presented in Christopher et al. (1990).

A. Consider the reinforced soil fill and design approach selected.

B. Consider the retained backfill soil properties.

C. Consider dead load and live load surcharges.

D. Combine earth, surcharge, and live load pressure diagrams into a composite diagram for design.

STEP 7. Check external wall stability. (See Section 9.3-3 for additional comments.)

A. Sliding resistance (Figure 9-6a or 9-6b). Check with and without surcharge.

B. Bearing capacity of the foundation (Figure 9-7a or 9-7b).

C. Overturning of the wall (Figure 9-8a or 9-8b).

D. Deep-seated (overall) stability (Figure 9-9).

E. Seismic analysis. See Section 9.3-4 and Figure 9-10.

STEP 8. Estimate settlement of the reinforced section using conventional settlement analyses. (See Section 9.3-3.)

STEP 9.   Calculate the maximum horizontal stress at each level of reinforcement.

    A. Determine, at each reinforcement level, the vertical stress distribution due to reinforced fill weight and the uniform surcharge.

    B. Determine, at each reinforcement level, the additional vertical stress due to any concentrated surcharges.

    C. Calculate the horizontal stresses, σ_h, using the lateral earth pressure diagram from Step 6.

STEP 10.   Check internal stability and determine reinforcement requirements. (See Section 9.3-3).

Use the lateral earth pressure diagrams developed in Step 6 for the reinforced section.

    A. Determine the long-term design strength of the reinforcement, T_d, and vertical spacings, s_i (Figure 9-5) of reinforcing layers to resist the internal lateral pressures. The long-term design strength is equal to the long-term allowable strength, defined in Appendix F, divided by the FS against internal failure selected in Step 4.

    B. Ensure that the connection strength at SRW and timber facings does not control the magnitude of reinforcement strength, T_d, that can be mobilized. Also, check the local stability of SRWs, timber, or concrete panels that are used for the wall facing. If a wrap-around face is used, determine overlap length, L_o, for the folded portion of the geosynthetic at the face (Figure 9-11).

    C. Check length of the reinforcement, L_e, required to develop pullout resistance beyond the Rankine failure wedge (Figure 9-11). A minimum $L_e = 1$ m is recommended.

STEP 11.   Evaluate requirements for backfill drainage and surface runoff control.

STEP 12.   Evaluate anticipated lateral displacement (see Section 9.4) for establishing initial construction face batter.

STEP 13.   Prepare plans and specifications.

To calculate $FS_{SLIDING}$:

$$V_q = \gamma_s h_s L \qquad\qquad \mu = \text{Minimum of } \tan \phi_r, \tan \phi_f, \text{ or } \tan \phi_{sg}$$

$$W = \gamma_r H L \qquad\qquad K_{a,b} = \tan^2 (45 - \phi_b/2)$$

$$P_b = 0.5 K_{a,b} \gamma_b H^2 \qquad\qquad P_q = K_{a,b} \gamma_s h_s H$$

c = cohesion of foundation soil or adhesion between soil and reinforcement

P_Q = as determined from Boussinesq equation -- if sustained loading

$$FS_{SLIDING} = \frac{\sum Horizontal\ Resisting\ Forces}{\sum Horizontal\ Sliding\ Forces} = \frac{(V_q + W)\mu + cL}{P_b + P_q + P_Q}$$

$$= \frac{[(\gamma_s h_s + \gamma_r H)\mu + c]L}{K_{a,b} H (0.5 \gamma_b H + \gamma_s h_s) + P_Q} \geq 1.5$$

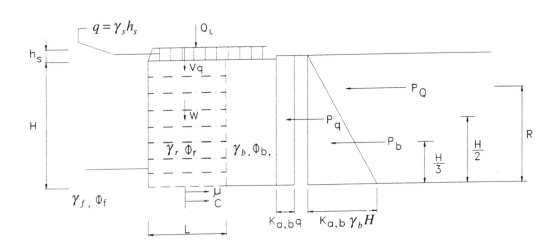

Figure 9-6a   *External sliding stability of a geosynthetic reinforced wall with a uniform surcharge load (Christopher and Holtz, 1989).*

To Calculate $FS_{SLIDING}$

$V_q = (\gamma_s h_2 L) / 2$ μ =   Minimum of $\tan \phi_r$, $\tan \phi_r$, or $\tan \phi_{sg}$

$W = \gamma_r h_1 L$ P_b = $(0.5 K_{a,b} \gamma_b H^2) \cos \beta$

$$K_{a,b} = \cos \beta \left(\frac{\cos \beta - \sqrt{\cos^2 \beta - \cos^2 \phi_b}}{\cos \beta + \sqrt{\cos^2 \beta - \cos^2 \phi_b}} \right)$$

c   =   cohesion of foundation soil or adhesion between soil and reinforcement

$$FS_{SLIDING} = \frac{\Sigma\ Horizontal\ Resisting\ Forces}{\Sigma\ Horizontal\ Sliding\ Forces} = \frac{(V_q + W)\mu + cL}{P_b}$$

$$= \frac{[(\gamma_s h_2 + 2\gamma_r h_1)\mu + 2c]L}{K_{a,b}\ \gamma_b\ H^2\ \cos \beta} \geq 1.5$$

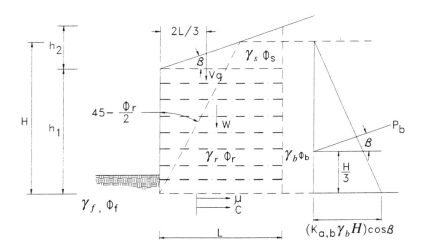

Figure 9-6b   *External sliding stability of a geosynthetic reinforced wall with an inclined surcharge load (Christopher and Holtz 1989).*

To calculate the bearing capacity:

$$V_q = \gamma_s\, h_s\, L \qquad\qquad K_{a,b} = \tan^2(45 - \phi_b/2)$$
$$W = \gamma_r\, H\, L \qquad\qquad q_a = \text{allowable bearing capacity of the soil}$$
$$P_b = 0.5\, K_{a,b}\, \gamma_b\, H^2 \qquad P_q = K_{a,b}\, \gamma_s\, h_s\, H$$
$$P_Q \text{ and } R = \text{as determined from Boussinesq equation, if sustained loading}$$

1) The eccentricity, e, of the resultant loads:

$$e = \frac{\sum Driving\ Moments}{\sum Vertical\ Forces} = \frac{P_b\,(H/3) + P_q\,(H/2) + P_Q\,R}{W + V_q}$$

$$= \frac{K_{a,b}\,H^2\,(\gamma_b\,H + 3\,\gamma_s\,h_s) + 6\,P_Q\,R}{6\,L\,(\gamma_r\,H + \gamma_s\,h_s)} \le L/6$$

2) The magnitude of the maximum vertical stress, $\sigma_{v\,max}$:

$$\sigma_{v\,max} = \frac{V_q + W}{L - 2e} = \frac{\gamma_r\,H + \gamma_s\,h_s}{1 - 2e/L} \le q_a$$

where:
$$q_a = q_{ult}/2$$

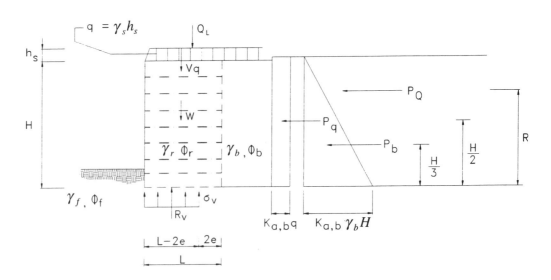

Figure 9-7a   Bearing capacity for external stability of a geosynthetic reinforced soil wall with uniform surcharge load (Christopher and Holtz, 1989).

To calculate the bearing capacity:

$$V_q = (\gamma_s\, h_2\, L)\,/\,2 \qquad\qquad P_b = (0.5\, K_{a,b}\, \gamma_b\, H^2)\cos\beta$$

$$W = \gamma_r\, h_1\, L \qquad\qquad\qquad q_a = \text{allowable bearing capacity of the soil}$$

$$K_{a,b} = \cos\beta\left(\frac{\cos\beta - \sqrt{\cos^2\beta - \cos^2\phi_b}}{\cos\beta + \sqrt{\cos^2\beta - \cos^2\phi_b}}\right)$$

1) The eccentricity, e, of the resultant loads:

$$e = \frac{\Sigma\,Driving\ Moments}{\Sigma\,Vertical\ Forces} = \frac{P_b\,(H/3)}{W + V_q} = \frac{H^3 K_{a,b}\gamma_b\cos\beta}{(3\gamma_s h_2 + 6\gamma_r h_1)L} \le L/6$$

2) The magnitude of the maximum vertical stress, $\sigma_{v\ max}$:

$$\sigma_{v\ max} = \frac{V_q + W}{L - 2e} = \frac{(\gamma_s\, h_2)/2 + \gamma_r\, h_1}{1 - 2e/L} \le q_a$$

where:

$$q_a = q_{ult}/2$$

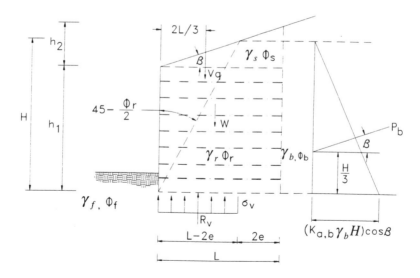

Figure 9-7b   *Bearing capacity for external stability of a geosynthetic reinforced soil wall with an inclined surcharge load (Christopher and Holtz, 1989).*

To Calculate $FS_{OVERTURNING}$

V_q = $\gamma_s h_s L$ $\qquad\qquad$ μ = $\tan\phi_r$ or $\tan\phi_f$

W = $\gamma_r H L$ $\qquad\qquad$ $K_{a,b}$ = $\tan^2(45 - \phi_b/2)$

P_b = $0.5 K_{a,b} \gamma_b H^2$ $\qquad\qquad$ P_q = $K_{a,b} \gamma_s h_s H$

c = cohesion of foundation soil or adhesion between soil and reinforcement

P_Q = as determined from Boussinesq equation - if sustained loading

$$FS_{OVERTURNING} = \frac{\sum Moments\ Resisting}{\sum Moments\ Overturning} = \frac{(V_q + W)(L/2)}{P_b(H/3) + P_q(H/2) + P_Q R}$$

$$= \frac{3 L^2 (\gamma_s h_s + \gamma_r H)}{H^2 K_{a,b}(\gamma_b H_b + 3\gamma_s h_s) + 6 P_Q R} \geq 2.0$$

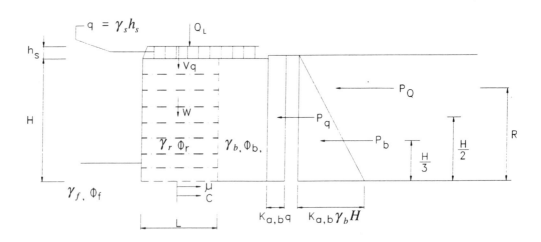

Figure 9-8a *External overturning stability of a geosynthetic reinforced soil wall with uniform surcharge load (Christopher and Holtz, 1989).*

To Calculate $FS_{OVERTURNING}$

$$V_q = (\gamma_s\, h_2\, L)\, /\, 2 \qquad\qquad \mu \;=\; \tan\phi_r \text{ or } \tan\phi_f$$

$$W = \gamma_r\, h_1\, L \qquad\qquad\qquad P_b \;=\; (0.5\, K_{a,b}\, \gamma_b\, H^2)\, \cos\beta$$

$$K_{a,b} = \cos\beta\left(\frac{\cos\beta - \sqrt{\cos^2\beta - \cos^2\phi_b}}{\cos\beta + \sqrt{\cos^2\beta - \cos^2\phi_b}}\right)$$

c = cohesion of foundation soil or adhesion between soil and reinforcement

$$FS_{OVERTURNING} \;=\; \frac{\Sigma\ Moments\ Resisting}{\Sigma\ Moments\ Overturning} \;=\; \frac{V_q\,(\,2L/3\,) + W\,(L/2\,)}{P_b\,(\,H/3\,)}$$

$$=\; \frac{(\,2\,\gamma_s\, h_2 + 3\,\gamma_r\, h_1\,)\, L^2}{K_{a,b}\ \gamma_b\ H^3\, \cos\beta} \;\geq\; 2.0$$

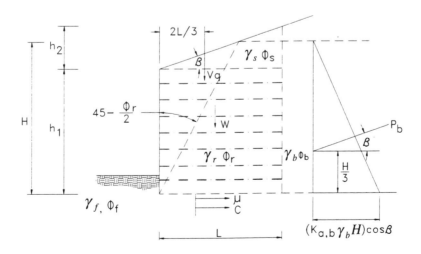

Figure 9-8b   *External overturning stability of a geosynthetic reinforced soil wall with an inclined surcharge load (Christopher and Holtz, 1989).*

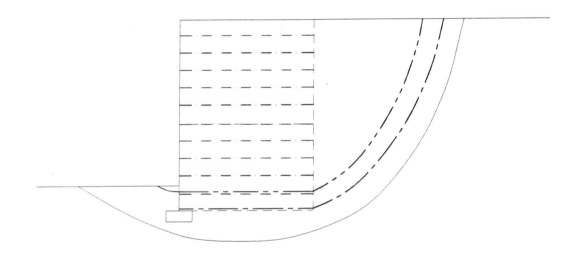

Figure 9-9   *Potential slip surfaces for geosynthetic reinforced soil walls (Christopher and Holtz, 1989).*

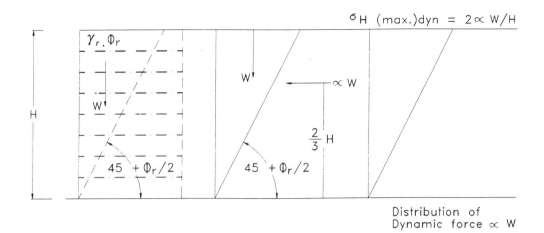

$\alpha \quad = \quad$ 0.1g to 0.2g

W   =   weight of soil mass in the active zone

$$W \;=\; \frac{\gamma_r \, H^2}{2 \tan \left(45 + \dfrac{\phi_r}{2} \right)}$$

Total dynamic force $= \alpha\, W$

$$\sigma_{H(\,max\,)\,dyn} \;=\; 2\,\alpha\,W\,/\,H$$

Figure 9-10   Horizontal stress increase due to seismic loading (Christopher and Holtz, 1989).

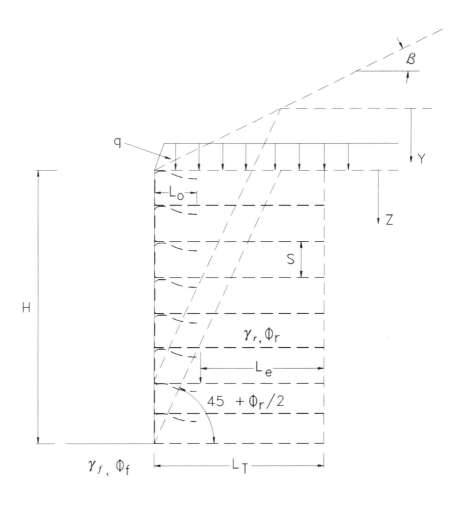

To calculate embedment length to resist pullout:

$$L_o = \frac{T_d\,(FS)}{2\,\gamma_r\,z\,F^*} > 0.9\,m$$

$$L_o = \frac{\sigma_h\,(FS)}{2\,\gamma_r\,z\,F^*} > 0.9\,m$$

where:

F^* = pullout resistance of shearing friction between soil and geosynthetic

σ_h = total horizontal stress at considered depth, as determined by Figure 9-7a or 9-7b

$L_T \geq L_{SLIDING}$

Figure 9-11   Embedment length for reinforcement required to resist pullout.

9.3-3 Comments on the Design Procedure

STEPS 1 and 2 need no elaboration.

STEP 3.   Determine reinforced fill and retained backfill properties.

Requirements for reinforced fill are presented in Section 9.6-1 and Christopher et al. (1990). Retained backfill soils should meet usual embankment fill soil requirements. The engineering properties of these two separate fill materials has a significant influence on the design of the reinforced soil volume.

The moist unit weights, γ_m, of the reinforced fill and retained backfill soils can be determined from the standard Proctor test or alternatively, from a vibratory-type relative density test. The angles of internal friction, ϕ', should be consistent with the respective design value of unit weight. Conservative estimates can be made for granular materials, or alternatively for major projects, this soil property can be determined by drained direct shear (ASTM D 3080) or triaxial tests.

Conventional compaction control density measurements should be performed for fills where a majority of the material passes a 2.0 mm sieve. For coarse, gravelly backfills, use either relative density for compaction control or a method-type compaction specification for fill placement. The latter is appropriate if the fill contains more than 30% of 19 mm or larger materials.

STEP 4.   Establish design factors of safety.

Review local codes and experience for safe and economical design values.

STEP 5.   Determine preliminary wall dimensions, including wall embedment depth.

As the design process is trial and error, it is necessary to initially analyze a set of assumed trial wall dimensions. The recommended minimum value of $L \approx 0.7H$ is a good place to start. Surcharges and sloping fills will likely increase the reinforcement length requirements. For low (H < 3 m) walls, you may wish to use a minimum of L = 2 m. In any case, be sure the overall stability (Step 7) is satisfactory.

Unless the foundation is on rock, a minimum embedment depth is required to provide adequate bearing capacity and to provide for environmental considerations such as frost action, shrinkage and swelling clays, or earthquakes. The recommendations given earlier under Step 5 are conservative.

Embedment of the wall also helps resist the lateral earth pressure exerted by the reinforced fill through passive resistance at the toe. This resistance is neglected for design purposes because it may not always be there. Construction sequence, possible scour, or future excavation at the front of the wall may eliminate it.

STEP 6.   Develop lateral earth pressure diagrams for both the reinforced fill and retained backfill.

A. Consider the reinforced fill:

Using the reinforced fill properties as determined in Step 3, calculate the lateral earth pressure coefficient and develop the internal stability lateral earth pressure diagram for the design height of the wall.

The lateral earth pressure coefficient, K_a, may be computed with the Rankine earth pressure theory. For horizontal backfill and near-vertical (*e.g.*, $\leq 10°$) face batter:

$$K_a = \tan^2 (45° - \phi/2) \tag{9-1}$$

For backfill at an angle β,

$$K_a = \cos \beta \, \frac{\cos \beta - \sqrt{\cos^2 \beta - \cos^2 \phi}}{\cos \beta + \sqrt{\cos^2 \beta - \cos^2 \phi}}$$

$$\tag{9-2}$$

Determine the appropriate lateral earth pressure distribution diagram for the design height of the retaining wall, using the selected design method (*i.e.*, tie-back wedge or stiffness approach).

In conventional retaining wall design, active earth pressure conditions (earth pressure coefficient = K_a) are normally assumed. There may be some situations, however, where the wall is prevented from moving (examples include abutments of rigid frame bridges; walls on bedrock), and at rest earth pressure conditions (K_o), or even greater pressures due to compaction, are appropriate.

K_o may be estimated from the Jaky (1948) relationship:

$$K_o = 1 - \sin \phi' \qquad\qquad [9\text{-}3]$$

B. Consider the retained backfill:

Using the retained backfill properties as determined in Step 3, calculate the lateral earth pressure coefficient and develop the external stability lateral earth pressure diagram for the wall. This pressure acts along the height, measured from bottom of wall to top of finished grade, at a vertical line at the back of the reinforced soil mass.

The lateral earth pressure coefficient, K_a, may be computed with the Rankine earth pressure theory. For horizontal backfill use Equation 9-1 and for sloping fills use Equation 9-2. Be sure to use the appropriate ϕ' in these equations.

C. Consider any dead load and live load surcharges:

Various approaches for considering the lateral earth pressures due to distributed surcharges, concentrated surcharges, and live loads are discussed by Christopher and Holtz (1985) and Christopher et al., (1990). Terzaghi and Peck (1967), Wu (1975), Perloff and Baron (1976), the U.S. Forest Service (Steward et al., 1977), and the U.S. Navy DM-7 (1982) provide suitable methods.

D. Develop the composite pressure diagram:

The earth pressure and surcharge pressure diagrams are combined to develop a composite earth pressure diagram which is used for design. See the standard references for procedures on locating the resultant forces.

STEP 7. Check external wall stability.

As with conventional retaining wall design, the overall stability of a geosynthetic wall must be satisfactory. External stability is evaluated by assuming that the reinforced soil mass acts as a rigid body, although in reality the wall system is really quite flexible. It must resist the earth pressure imposed by the backfill which is retained by the reinforced mass and any surcharge loads. Potential external modes of failure to be considered are:
*   sliding of the wall;
*   overturning of the wall;
*   bearing capacity of the wall foundation; and

- stability of the slope created by the wall.

These failure modes and methods of design against them are discussed by Christopher and Holtz (1985, 1989), Mitchell and Villett (1987) and Christopher et al. (1990).

The potential for sliding along the base is checked by equating the external horizontal forces with the shear stress at the base of the wall (Figure 9-6a and b). Sliding must be evaluated with respect to the minimum frictional resistance provided by either the reinforced soil, ϕ_r, the foundation soil, ϕ_f, or the interaction between the geosynthetic and the reinforced soil, ϕ_{sg}. Often, external stability, particularly sliding, controls the length of reinforcement required. Reinforcement layers at the base of the wall may be considerably longer than required by internal earth pressure considerations alone. For construction simplicity, reinforcement layers of the same length are generally used throughout the entire height of the wall. The factor of safety against sliding should be at least 1.5.

Design for bearing capacity follows the same procedures as those outlined for an ordinary shallow foundation. The entire reinforced soil mass is assumed to act as a footing. Because there is a horizontal earth pressure component in addition to the vertical gravitational component, the resultant is inclined and should pass through the middle third of the foundation to insure there is no uplift (tension) in the base of the wall.

Appropriate bearing capacity factors or allowable bearing pressures must be used as in conventional geotechnical practice. The ultimate bearing capacity, q_{ult}, is determined using classical soil mechanics:

$$q_{ult} = c_r N_c + 0.5 \gamma_f L' N_\gamma \qquad [9\text{-}4]$$

where:
$$L' = L - 2e$$

The total bearing capacity per unit length of wall parallel to the wall face is

$$q_{ult} = c_r N_c (1 - 2e/L) + 0.5 \, \gamma_f \, N_\gamma (1 - 2e/L)^2 \qquad [9\text{-}5]$$

N_c and N_γ are dimensionless bearing capacity coefficients that can be readily obtained from foundation engineering textbooks.

Due to the flexibility of geosynthetic reinforced walls, the factor of safety for bearing capacity is lower than normally used for stiffer reinforced concrete cantilever or gravity structures. The factor of safety must be at least two with respect to the ultimate bearing capacity.

Loads tending to cause overturning are developed from the resultants of the horizontal earth pressure and surcharge diagrams for the retained backfill portion of the wall. Overturning is checked by summing moments of external forces at the toe of the wall (Figures 9-8a and b). The factor of safety must be greater than two.

Other foundation design considerations include environmental factors such as frost action, drainage, shrinkage or swelling of the foundation soils, and potential seismic activity at the site. Each of these items must be checked to ensure adequate wall performance is maintained throughout the wall's design and service life.

Overall stability (Figure 9-9) of the slope in which the wall sits typically requires a factor of safety of at least 1.5 for long-term conditions. Note that the reinforced mass should not be considered as a solid block (that failure planes cannot penetrate) for overall slope stability analyses. Slope stability analysis methods that model the reinforced fill and reinforcement as discrete elements should be used, as presented in Chapter 8.

In seismically active areas, the reinforced wall and facing system, if any, must be stable during earthquakes. Seismic stability is discussed in Section 9.3-4.

STEP 8.   Estimate settlement of the reinforced section.

Potential settlement of the wall structure should be assessed and conventional settlement analyses for shallow foundations carried out to ensure that immediate, consolidation, and secondary settlements of the wall are less than the performance requirements of the project. Both total and differential settlements along the wall length should be considered. For specific procedures, consult standard textbooks on foundation engineering.

STEP 9.   Calculate the maximum horizontal stress at each level of reinforcement.

Calculate, at each reinforcement level, the horizontal stress, T_h, along the potential active earth pressure failure surface, as shown inclined at the angle θ in Figure 9-5. From Rankine earth pressure theory, θ is inclined at $45° + \phi_r'/2$, where ϕ_r' is the internal friction angle appropriate for the reinforced soil section. Use the moist unit weight of the reinforced backfill plus, if present, uniform and concentrated surcharge loads.

Use K_a and the lateral earth pressure diagram from the selected design procedure, as discussed in Step 6.

STEP 10.   Check internal stability and determine reinforcement requirements.

Use the lateral earth pressure diagrams developed in Step 6 for the reinforced section.

A. Determine vertical spacings of the geosynthetic reinforcing layers and the strength of the reinforcement, T_i, required at each level to resist the internal lateral pressures.

The required tensile strength, T_i, of the geosynthetic is controlled by the vertical spacing of the layers of the reinforcing, and it is obtained from:

$$T_i = s_i \sigma_h \qquad\qquad [9\text{-}6]$$

where:

s_i = ½ (distance to reinforcing layer above + distance to reinforcing layer below)

σ_h = horizontal earth pressure at middle of the layer

Vertical spacings should be based on multiples of the compacted fill lift thickness. From Equation 9-6, it is obvious that greater vertical spacing between the horizontal layers is possible if stronger geosynthetics are used. (NOTE: Vertical spacing may be governed by the connection strength between the reinforcement and facing.) This may reduce the cost of the reinforcement, as well as increase the fill placement rate to some extent. On the other hand, large spacings may require temporary external face support during construction. Typical reinforcement spacing for geosynthetic reinforced walls varies between 200 mm to 1.2 m for geogrids and rigid facings, and between 200 to 400 mm for geotextile wrapped walls. For spacings greater than 600 mm, unless the wall has a rigid face, intermediate layers will be required to prevent excessive bulging of the face between the layers.

For design, a spacing based on these considerations is assumed, and the required tensile strength is calculated from Equation 9-6. Since the largest horizontal stress will be at or near the bottom of the wall, this equation will yield the maximum required strength of the geosynthetic. Although conservative, usually for construction simplicity, layer spacings are kept constant and the same strength material is used throughout the reinforced section. For walls higher than 5 or 6 m, it is possible that using two or three different strength geosynthetics and/or some variation in layer spacing will result in a more economical design; but it will be more complex to build and it will require careful inspection to avoid construction mistakes.

B. Determine the length, L_e, of geosynthetic reinforcement required to develop pullout resistance beyond the Rankine failure wedge.

This design step is necessary to calculate embedment length, L_e, behind the assumed failure plane (Figure 9-11). The angle of the assumed failure plane is taken to be the Rankine failure angle, or $45° + \phi/2$. Also, this plane is usually assumed to initiate from the toe of the wall and proceed upward at that angle. This assumption results in conservative embedment lengths. The formula for the embedment length, L_e, is:

$$L_e = \frac{T_i}{2\,\gamma_r\,z\,\alpha\,F^*}\,(FS) \qquad\qquad [9\text{-}7]$$

where:

T_i	=	computed tensile load in the geosynthetic;
z	=	depth of the layer being designed;
γ_r	=	unit weight of backfill (reinforced section);
F^*	=	coefficient of pullout interaction between soil and geosynthetic;
α	=	a scale effect correction factor; and
FS	=	factor of safety against pullout failure.

The factor of safety for embedment should be 1.5, with a minimum embedment length of approximately 1 m.

For wrap-around walls, the overlap length, L_o, must be long enough to transfer stresses from the lower portion to the longer layer above it. The equation for geosynthetic overlap length, L_o, is:

$$L_o = \frac{\sigma_h}{2\,\gamma_r\,z\,F^*}\,(FS) \qquad\qquad [9\text{-}8]$$

Again, a minimum value of approximately 1 m is recommended for L_o to insure adequate anchorage of reinforcement layers.

STEP 11. Evaluate requirements for backfill drainage and surface runoff control

Just as with conventional earth retaining structures, drainage of the backfill as a geosynthetic reinforced wall is very important for the stability of the structure as well as the wall's appearance. Providing for good backfill drainage is the reason why it was recommended in Step 3 that soils with appreciable fines not be used in the backfill. If the backfill becomes saturated because of surface infiltration or groundwater seepage, both the overall wall stability and the stability of the reinforced backfill with be reduced. If water cannot easily drain out of the backfill, seepage pressures acting on the back of an impermeable permanent facing can potentially damage the facing or at best cause unsightly staining. The suggestions for retaining wall drainage by Cedergren (1989) and Terzaghi and Peck (1967) are also appropriate for most reinforced structures.

STEP 12. Evaluate anticipated lateral displacement.

See Section 9.4.

STEP 13. Prepare plans and specifications.

Specifications are discussed in Section 9.9.

9.3-4 Seismic Design (Allen and Holtz, 1991)

In seismically active areas, an analysis of the geosynthetic MSE wall stability under seismic conditions must be performed. For temporary structures, a formal analysis is probably not necessary. For permanent structures, seismic analyses can range from a simple pseudo-static analysis to a complete dynamic soil-structure interaction analysis such as might be performed on earth dams and other critical structures.

Seismic analysis procedures for walls with metallic reinforcement and concrete facings are well established; see Vrymoed (1990) for a review of these procedures. The generally conservative pseudo-static Mononabe-Okabe analysis is recommended for geosynthetic walls (Mitchell and Villet, 1987; Christopher et al., 1990). This analysis as shown in Figure 9-10, correctly includes the horizontal inertial forces for internal seismic resistance, as well as the pseudo-static thrust imposed by the retained fill on the reinforced section.

Because of their inherent flexibility, properly designed and constructed geosynthetic walls are probably better able to resist seismic loadings, but high walls in earthquake-prone regions should be checked. The facing connections must also resist the inertial force of the wall facia which can occur during the design seismic event. Stress build-up behind the face, resulting from strain incompatibility between a relatively stiff facing system and the extensible geosynthetic reinforcement must also be resisted by facing connections. Additional research is needed to evaluate the effect of seismic forces on geosynthetic walls with stiff facings.

9.4   LATERAL DISPLACEMENT

Lateral displacement of the wall face occurs primarily during construction, although some movement also can occur due to post construction surcharge loads. Post-construction deformations can also occur due to settlement of the structure. As noted by Christopher et al. (1990), there is no standard method for evaluating the overall lateral displacement of reinforced soil walls.

The major factors influencing lateral displacements during construction include compaction intensity, reinforcement to soil stiffness ratio (*i.e.*, the modulus and the area of reinforcement as compared to the modulus and area of the soil), reinforcement length, slack in reinforcement connections at the wall face, and deformability of the facing system. An empirical relationship for estimating relative lateral displacements during construction of walls with granular backfills is presented in Figure 9-12 (Christopher et al., 1990). The relationship was developed from finite element analyses, small-scale model tests, and very limited field evidence from 6 m high test walls. Note that as L/H decreases, the lateral deformation increases. The procedure predicted wall face movements of a 12.6 m high geotextile wall that were slightly greater than observed (Holtz et al., 1991).

For critical structures such as bridge abutments that require precise tolerances, the lateral displacement of the wall must be estimated more accurately, taking into account any tilting due to the thrust at the back of the wall. The finite element method is recommended in this case.

Two major factors influencing lateral displacements -- compaction intensity and slack in the reinforcement at the wall face -- are contractor controlled. Therefore, geosynthetic wall specifications should state acceptable horizontal and vertical erected face tolerances.

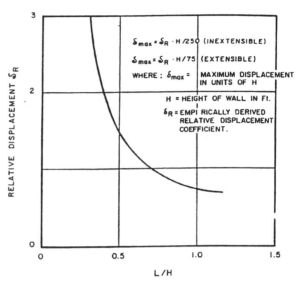

Figure 9-12   Empirical relation for estimating lateral movement during construction of geosynthetic-reinforced walls ≤ 6 m high (after Christopher et al., 1990).

9.5   WALL FACE DESIGN (after Allen and Holtz, 1991)

A significant advantage of geosynthetic reinforced walls over other earth retaining structures is the variety of facings that can be used and the resulting aesthetic options that can be provided. They can be installed (1) as the wall is constructed or (2) after the wall is built. Facings installed as the wall is constructed include segmental and full height precast concrete panels, interlocking precast concrete blocks, welded wire panels, gabion baskets, treated timber facings, and geosynthetic face wraps. In these cases, the geosynthetic reinforcement is attached directly to the facing element. Shotcrete, cast-in-place concrete facia, as well as precast concrete or timber panels, can be attached to steel bars placed or driven between the layers of the geosynthetic wrapped wall face at the end of wall construction and after wall movements are complete. Sketches of some of these systems are shown in Figure 9-13.

Segmental Retaining Wall (SRW) units are the most common facing currently used for geosynthetic wall construction. These facing elements are also known as modular block wall (MBW) units or modular block wall units. They are popular because of their aesthetic appeal, widespread availability, and relative low cost (Berg, 1991). A broken block, or natural stone-like, finish is the most popular SRW face finish.

SRWs are relatively small, squat concrete units, specially designed and manufactured for retaining wall applications. The units are typically manufactured by a dry casting process and weigh 15 to 50 kg each, with 35 to 50 kg units routinely used. The nominal depth (dimension perpendicular to wall face) of SRWs usually ranges between 0.2 and 0.6 m.   Unit heights typically vary between 100 and 200 mm for the various manufacturers. Exposed face length typically ranges between 200 and 600 mm.

SRWs may be manufactured solid or with cores. Full height cores are typically filled with aggregate during erection. Units are normally dry-stacked (*i.e.*, without mortar) in a running bond configuration. Vertically adjacent units may be connected with shear pins or lips.

The vertical connection mechanism between SRWs also contributes to the connection strength between the geosynthetic reinforcement and the SRWs. Connection strength should be addressed in design, and may control the maximum allowable tensile load in a given layer of reinforcement. Therefore, the reinforcement design strength and vertical spacing of layers is specific to the particular combination of SRW and geosynthetic reinforcement utilized. Geogrids, both stiff and flexible, are the common reinforcing elements of SRW-faced walls. Geotextiles have been used in walls with SRW facings, but to a limited extent. A detailed description of SRW units, and design with these units, is presented by Simac et al. (1993) and is summarized by Bathurst and Simac (1994).

Wrap-around facings are commonly used: i) for temporary structures; ii) for walls that will be subject to significant post-construction settlement; iii) where aesthetic requirements are low; and iv) where post-construction facings are applied for protection and aesthetics. The geosynthetic facing may be left exposed for temporary walls if the geosynthetic is stabilized against ultraviolet light degradation. A consistent vertical spacing of reinforcement, and therefore wrap height, of 0.3 to 0.5 m is typically used. A sprayed concrete facing is usually applied to permanent walls to provide protection against ultraviolet exposure, potential vandalism, and possible fire. Precast concrete or wood panels may also be attached after construction.

Geotextiles are commonly used in wrap-around-faced MSE walls. With the proper ultraviolet light stabilizer, these structures perform satisfactorily for a few years. They should be covered by a permanent facing for longer-term applications. Geogrids are also used for wrap facings, though a geotextile, an erosion control blanket, or sod is required to retain fill soil. Alternatively, rock or gravel can be used in the wrap area and a filter placed between the stone and fill soil. Secondary, biaxial geogrid can be used to provide the face wrap for the primary, uniaxial soil reinforcing elements for geogrids. With the proper ultraviolet light stabilizer specified, geogrids can be left uncovered for a number of years; reportedly for design lives of 50 years or more for heavy, stiff geogrids (Wrigley, 1987).

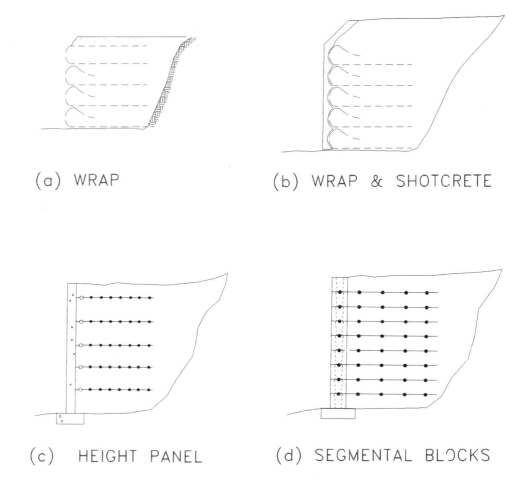

(a) WRAP

(b) WRAP & SHOTCRETE

(c) HEIGHT PANEL

(d) SEGMENTAL BLOCKS

Figure 9-13   *Possible geosynthetic reinforced wall facings: (a) geosynthetic facing - temporary wall; (b) geosynthetic facing protected by shotcrete; (c) full-height precast concrete panels; and (d) modular concrete units.*

Segmental concrete panels, similar to panels used to face metallic reinforcing systems, have been used to face geosynthetic reinforced walls. However, this facing is currently used in only a few states.

Stiff, polyethylene geogrids are exclusively used for precast concrete panel faces, where tabs of the geogrid are cast into the concrete and field-attached to the soil reinforcing geogrid layers. Flexible, polyester geogrids are not used because casting them into wet concrete would expose the geogrids to a high-alkaline environment.

Full-height concrete panels have also been used to face geosynthetic reinforced walls, although this facing is not very popular. It is used in only where aesthetics of full-height panels are specifically desired. Similar to the articulated panels, stiff, polyethylene geogrids are exclusively used for precast concrete panel faces where tabs are cast into the panel.

Timber facings are also used for geosynthetic reinforced walls. Timber-faced walls are normally used for low-to moderate-height structures, landscaping, or maintenance construction. Geotextiles and geogrids are used with timber facings.

Gabion facings can also be used for wall structures with geogrid or geotextile reinforcing elements. A geotextile filter is typically used between the back face of the gabion baskets and the fill soil to prevent soil from piping through the gabion stones.

The success of a geosynthetic wall is highly dependent on the type of facing system used and the care with which it is designed and constructed. Aesthetic requirements often determine the type of facing systems. Anticipated deflection of the wall face, both laterally and downward, may place further limitations on the type of facing system selected. Tight construction specifications and quality inspection are necessary to insure that the wall face is constructed properly; otherwise an unattractive wall face, or a wall face failure, could result.

Two failures of geosynthetic walls reported in the literature were in fact due to the failures of the wall facing systems (Richardson and Behr, 1988; Burwash and Frost, 1991).

Facings constructed as the wall is constructed must either allow the geosynthetic to deform freely during construction without any buildup of stress on the face, or the facing connection must be designed to take the stress. Although most wall design methods assume that the stress at the face is equal to the maximum horizontal stress in the reinforced backfill, research suggests that some stress reduction does occur near the face. Christopher, et al. (1990) recommend that the stress at the face at any depth below the top of the wall be varied linearly from approximately 75% of the maximum geosynthetic stress at the top of the wall to 100% of this stress at the base of the wall. This guideline appears reasonable and is therefore recommended.

Higher face connection stresses are possible due to bending at the connection caused by settlement of the wall. This can be a problem especially if full-height precast panels (Figure 9-13.c) are used, or if compaction at the wall face is poor. Heavy compaction at the wall face can also create high connection stresses. It is best to use lightweight hand-compactors within 1 m of the wall face. In any event, the level of compaction influences facing stresses.

The long-term strength of the facing connections must also be considered in facing design, because it is at the face where the wall environment (moisture, pH, temperature, and abrasion) is often most severe. If metallic connections are utilized, corrosion protection should be considered. If geosynthetic connections are utilized, installation damage, creep and chemical durability must be considered. A case in point is the wall at Poitiers, France, where polyester reinforcement strips, which were connected directly to the concrete facing, lost 45% of their strength after 17 years (Leflaive, 1988). The concrete created a highly alkaline environment and hydrolysis of the polyester resulted. Had corrective measures not been taken, the wall facing system would have surely failed.

Methods used to calculate required pullout length may also be used to calculate the required face embedment length, L_o, as shown in Figure 9-11 and Design Step 10. A minimum embedment length of about 1 m should be specified due to constructability requirements. Constructability requirements would also dictate that the vertical spacing of reinforcement be limited to a maximum of 0.4 m, if a single layer forming system is used to construct walls with a geosynthetic face wrap. Greater spacings could be used if a multi-layer external forming bracing system is utilized.

Connections for facing systems installed on geosynthetic face wrapped walls after all wall settlement and lateral movement are complete need only be designed to resist overturning, seismic, and gravity forces on the facing itself.

9.6　MATERIAL PROPERTIES

9.6-1　Reinforced Wall Fill Soil

Gradation: All soil fill material used in the reinforced section should be free draining and reasonably free from organic or other deleterious materials. Specifications presented in Table 9-2 are recommended.

Table 9-2
Reinforced Soil Fill Requirements

Sieve Size	Percent Passing
19 mm[1]	100
4.75 mm	100 - 20
0.425 mm	0 - 60
0.075 mm	0 - 15

Plasticity Index (PI) $\leq$ 6

Soundness: magnesium sulfate soundness loss < 30% after 4 cycles

NOTE:
1. The maximum size can be increased up to 100 mm, provided tests are performed to evaluate geosynthetic strength reduction due to installation damage (see Appendix F).

Chemical Composition (Berg, 1993): The chemical composition of the fill and retained soils should be assessed for effect on durability of reinforcement (pH, chlorides, oxidation agents, etc.). Some soil environments posing potential concern when using geosynthetics are listed in Appendix F. Soils with pH $\geq$ 12 or pH $\leq$ 3 should not be used in reinforced walls and a pH range $\geq$ 3 to $\leq$ 9 is recommended. Specific supporting test data should be required if pH > 9.

Compaction (Christopher et al., 1990): Soil fill should be compacted to 95% of optimum dry density (γ_d) and $\pm$ 2 % of optimum moisture content, w_{opt}, according to the standard Proctor test. Compacted lift heights of 200 to 300 mm are recommended for granular soils. Conventional compaction control density measurements should be performed for backfills where a majority of the material passes a 2.0 mm sieve. For coarse gravelly backfills, use either relative density for compaction control or a method type compaction specification for backfill placement. The latter is appropriate if the backfill contains more than 30% of 19 mm or larger materials.

Shear Strength: Peak shear strength parameters determined using direct shear or triaxial tests should be used in the analysis. However, the use of assumed shear values based on local experience may be acceptable for some projects. Verification of site soil type(s) should be completed following excavation or identification of borrow pit, as applicable.

Unit Weights: Dry unit weight for compaction control, moist unit weight for analyses, and saturated unit weight for analyses (where applicable) should be determined for the fill soil. The unit weight γ can be determined from the standard Proctor test or alternatively, a vibratory type relative density test may be used. For a conservative determination of γ, soak the material overnight and allow it to drain a few minutes before weighing it in a mold of known volume. The unit weight value of should be consistent with the design angle of internal friction, ϕ.

9.6-2 Geosynthetic Reinforcement

Geosynthetic reinforcement systems consist of geogrid or geotextile materials arranged in horizontal planes in the backfill to resist outward movement of the reinforced soil mass. Geogrids transfer stress to the soil through passive soil resistance on grid transverse members and through friction between the soil and the geogrid's horizontal surfaces (Mitchell and Villet, 1987). Geotextiles transfer stress to the soil through friction. Geosynthetic design strength must be determined by testing and analysis methods that account for the long-term geosynthetic-soil stress transfer and durability of the full geosynthetic structure. Long-term soil stress transfer is characterized by the geosynthetic's ability to sustain long-term load in-service without excessive creep strains. Durability factors include site damage, chemical degradation, and biological degradation. These factors may cause deterioration of either the geosynthetic tensile elements or the geosynthetic-soil stress transfer mechanism.

Allowable Tensile Strength (Berg, 1993): Allowable tensile strength (T_a) of the geosynthetic shall be determined using a partial factor of safety approach (Bonaparte and Berg, 1987). The procedure presented in Appendix F is derived from Task Force 27 guidelines for geosynthetic reinforced soil retaining walls (AASHTO, 1990), the Geosynthetic Research Institute's Methods GG4a and GG4b - Standard Practice for Determination of the Long Term Design Strength of Geogrids (1990, 1991), and the Geosynthetic Research Institute's Method GT7 - Standard Practice for Determination of the Long Term Design Strength of Geotextiles (1992). See also Appendix F.

For geosynthetic reinforced walls, the long-term allowable geosynthetic strength, T_a, is:

$$T_a = \frac{T_{ult}}{RF}$$

with:

$$RF = RF_{CR} \times RF_{ID} \times RF_{CD} \times RF_{BD} \times RF_{JNT}$$

where:

T_a	=	allowable geosynthetic tensile strength,(kN/m), for use in stability analyses;
T_{ult}	=	ultimate geosynthetic tensile strength,(kN/m);
RF_{CR}	=	partial factor of creep deformation, ratio of T_{ult} to creep-limiting strength, (dimensionless);
RF_{ID}	=	partial factor of safety for installation damage, (dimensionless);
RF_{CD}	=	partial factor of safety for chemical degradation, (dimensionless);
RF_{BD}	=	partial factor of safety for biological degradation, used in

environments where biological degradation potential may exist, (dimensionless); and

RF_{JNT} = partial factor of safety for joints (seams and connections), (dimensionless).

For geosynthetic walls, the safe design strength, T_d, is:

$$T_d = \frac{T_a}{FS}$$

where:

T_d = long-term safe design strength (kN/m); and
FS = overall factor of safety against failure (dimensionless).

Evaluation of long-term allowable strength, and, therefore, of safe design strength, for various geosynthetic reinforcement materials has been a problem for some public work agencies, as discussed in Section 9.12 Implementation.

As noted in the reinforced slopes chapter, RF values for durable geosynthetics in non-aggressive, granular soil environments range from 2.5 to 7. Appendix F suggests that a default value of RF = 7 may be used with geosynthetic products that have not been fully tested or evaluated, for design of routine, non critical structures which meet the soil, geosynthetic and structural limitations listed in the appendix. However, as indicated by the range of RF values, there is a potential to significantly reduce the reinforcing requirements and the corresponding cost of the structure by obtaining a reduced RF from test data.

Soil-Reinforcement Interaction: Two types of soil-reinforcement interaction coefficients or interface shear strengths must be determined for design: pullout coefficient, and interface friction coefficient. Pullout coefficients are used in stability analyses to compute the mobilized tensile force of each reinforcement layer. Interface friction coefficients are used to check the factor of safety against outward sliding of the entire reinforced mass. Procedures for quantifying these two interaction properties are presented in Appendix F.

The ultimate pullout resistance, P_r, of the reinforcement per unit width (Christopher et al., 1990) is given by:

$$P_r = 2 \cdot F^* \cdot \alpha \cdot \sigma'_v \cdot L_e$$

where:

$2\,L_e$	=	the total surface area per unit width of the reinforcement in the resistance zone behind the failure surface
L_e	=	the embedment or adherence length in the resisting zone behind the failure surface
F^*	=	the pullout resistance (or friction-bearing-interaction) factor
α	=	a scale effect correction factor
σ'_v	=	the effective vertical stress at the soil-reinforcement interfaces

For preliminary design and in absence of specific geosynthetic test data, F^* may be conservatively taken as:

$$F^* = 2/3 \tan \phi \quad \text{for geotextiles, and}$$

$$F^* = 0.8 \tan \phi \quad \text{for geogrids.}$$

9.7   DESIGN EXAMPLE

DEFINITION OF DESIGN EXAMPLE

- Project Description: An existing highway on compacted earthfill up to 4 m high is to be widened to provide for extra traffic lanes; because of natural and legal barriers, no additional right-of-way can be obtained.

- Type of Structure: Geosynthetic reinforced retaining wall
  Type: wrapped face (Figure 9-2a)

- Type of Application: Geosynthetic reinforcement

- Alternatives: Reinforced concrete cantilever retaining walls
  Bin wall and similar gravity structures
  Proprietary reinforcing systems with metallic reinforcement

GIVEN DATA

- Geometry -   wall height: up to 4 m
  wall length: up to 6 m
  face batter: 1:20
  surcharge : 0.75 m granular fill

- Soils
  Backfill: clean gravelly sand (SW); average w_n = 6%; nonplastic;
  standard Proctor $\gamma_{d\ max}$ = 20 kN/m^3; w_{opt} = 9.7 %;
  at 95% of standard Proctor, ϕ_r' = 35° and c_r' = 0;
  sand contains no deleterious chemical or biological agents; pH = 7

Foundation: Subsurface exploration indicates that the soils found along the alignment of highway are stiff glacial clayey till or dense sands and gravels outwash. They are non-frost susceptible and non swelling. Groundwater table is sufficiently deep that it will not influence wall performance. Seismic conditions at the site are well known. Drainage behind the wall will be not be a problem because of the clean nature of the proposed backfill soils. Runoff from above will be taken care of in the design of the roadway and pavement above the wall.

REQUIRED

Design geosynthetic-reinforced retaining wall.

DEFINE

A. Geosynthetic function(s):

B. Geosynthetic properties required:

C. Geosynthetic specification:

SOLUTION

A. Geosynthetic function(s):

Primary   -   reinforcement

Secondary   -   filtration and drainage

B. Geosynthetic properties required (see Table 1-2):

tensile characteristics (strength, modulus, creep)
interface shear strength (soil-geosynthetic friction angle)
survivability properties
apparent opening size (AOS) and permeability or permittivity

C. Geosynthetic specifications:

(To be determined after design.)

DESIGN

Design geosynthetic reinforced retaining wall.

STEP 1. SCOPE, DIMENSIONS, AND EXTERNAL LOADS

A. Wall height, H = 4 m

B. Wall length, L: to be determined by stability requirements

C. Face batter angle: 1:20

D. External loads: uniform surcharge load, q = 0.75m x 19 kN/m^3 = 14.25 kPa

E. Facing: Wrapped

F. Spacing requirement: maximum s_i 0.5 m; assume 0.4 m

G. Environmental conditions: either benign or taken care of in design

H. Service life: 120 yr

STEP 2. PROPERTIES OF FOUNDATION SOILS

Required soil properties for strength and compressibility of the foundation were determined by appropriate laboratory and in situ tests performed as part of the soil exploration and testing program. Groundwater table location and drainage conditions were given.

STEP 3. PROPERTIES OF BACKFILL AND RETAINED FILL SOILS

A. w_n = 6 %, non-plastic gravelly sand (SW) backfill

B. Use 95 % Standard Proctor: γ_d = 19 kN/m^3

C. ϕ' = 35°

D. pH neutral; no adverse chemical or biological activity

STEP 4. DESIGN FACTORS OF SAFETY

Use minimums and recommendations, as given in Section 9.3-2, Step 4, for both external and internal stability. Assume partial factors of safety for the geosynthetic reinforcement as discussed in Appendix F. For installation damage and creep, use FS = 1.5 each; for combined chemical and biological, use FS = 1.1.

STEP 5. DETERMINE PRELIMINARY WALL DIMENSIONS

A. Assume L = 0.7H = 2.8 m

B. Wall embedment, H_1 : 0.45m (minimum)

STEP 6. DEVELOP INTERNAL AND EXTERNAL LATERAL EARTH PRESSURE DIAGRAMS FOR REINFORCED SECTION

Use USFS "tie-back wedge" method; consider the given soil properties and uniform surcharge. From Equation 9-1, $K_a = \tan^2 (45° - \phi/2) = 0.27$

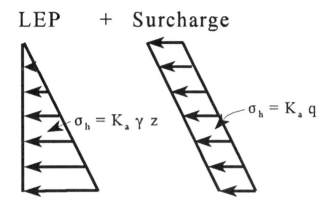

STEP 7. CHECK EXTERNAL WALL STABILITY

See Section 9.3-3 (commentary on the design procedure). Assume for this example that factors of safety for sliding, bearing capacity, overturning, and deep-seated stability are met or exceeded. Seismic stability is not a problem at this site.

STEP 8. SETTLEMENT

Assume that a detailed settlement analysis of the foundation indicates that estimated settlements are tolerable and that the operational criteria for the roadway are met.

STEP 9. CALCULATE MAXIMUM HORIZONTAL STRESS AT EACH LEVEL OF THE REINFORCEMENT

A. $\sigma_v = \gamma z$

B. $q = \gamma h_s$

C. $\sigma_h = K_a (\gamma z + q)$

From assumption of the average lift thickness of 0.4 m and the embedment depth of 0.45 m, a cross-section of the reinforced wall section can be drawn. Also shown are the surcharge fill height of 0.75 m and the assumed Rankine failure plane inclined at $45° + \phi_r'/2$. See also Figure 9-5.

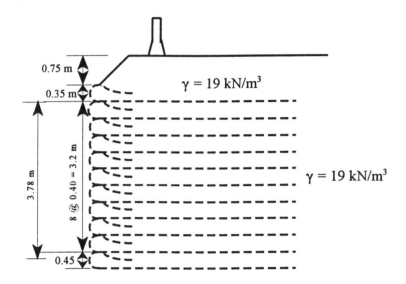

STEP 10.   CHECK INTERNAL STABILITY AND DETERMINE REINFORCEMENT
REQUIREMENTS

A. Determine the long-term design strength T_d for the assumed layer spacing $s_i \approx 0.4$ m.
See above figure. To determine the maximum tensile force, use Equation 9-6 and Equation C
from Step 9 above. The bottom most layer will have the greatest horizontal earth pressure.
Or for z = 3.78 m (midpoint of that layer),

$$T_{max} = 0.45 \text{ m} \{0.27 [19.0 (3.78) + 19.0 (0.75)]\} = 10.5 \text{ kN/m}$$

$$T_a = T_d = T_{max} (1.5) = 15.7 \text{ kN/m}$$

Note: minimum FS for internal stability is 1.5 (from Step 4). Apply reduction factors as
discussed in Section 9.6-2 and Appendix F. Or

$$T_{ult} = T_a (RF) = T_a [RF_{CR} \text{ x } RF_{ID} \text{ x } RF_{CD} \text{ x } RF_{BD}] = T_a [2.5]$$

$$= 15.7 [2.5] = 38.9 \text{ kN/m}$$

From Appendix E, Table E1, the most likely geosynthetics that should be suitable
reinforcement materials are the heavier woven monofilament, fibrillated tape, and
multifilament geotextiles and HDPE and PET geogrids.

B. Determine the embedment or pullout length L_e required behind the assumed Rankine
failure plane, as indicated in Figures 9-5, 9-11, and in Step 9 above. Use Equations 9-7 and
9-8. Factor of safety against pullout should be at least 1.5, and a minimum pullout length of
0.9 to 1 m is recommended.

$$L_e = \frac{0.35(0.27)[19(0.35) + 19(0.75)])}{2(19)(0.35)\frac{2}{3}\tan\phi} = 0.32m$$

Use minimum L_e = 0.9 or 1 m. Total length of geosynthetic back of the wall face L = H/tan 62.5° + 0.9 m = 3.65 m /tan 62.5° + 0.9 m = 1.9 + 0.9 = 2.8 m. For construction simplicity, use a constant length of 2.8 m throughout the full height of the wall.

For the wrapped face overlap length L_o, use Equation 9-8. L_o = 0.32 m. Use minimum embedment length of 0.9 or 1 m, although it is recognized that this is very conservative.

The final step in determining reinforcement requirements is to insure that the minimum survivability requirements are met. Be sure that *High Survivability* geotextile or geogrids are selected--see Tables 7-2 and 7-3.

STEP 11.   BACKFILL DRAINAGE AND SURFACE RUNOFF CONTROL

See Section 9.3-3 for comments on this aspect of the design. For geosynthetics properties requirements, use Table 5-4.

STEP 12.   LATERAL DISPLACEMENT ESTIMATES

See Section 9.4.

STEP 13.   PREPARE PLANS AND SPECS

See Section 9.9.

9.8   COST CONSIDERATIONS

At the FHWA-Colorado Department of Highways Glenwood Canyon geotextile test walls (Bell et al., 1983), the cost of the geotextile was only about 25% of the wall's total cost. Therefore, some conservatism on geosynthetic strength or on vertical spacing is not necessarily excessively expensive. A major part of the Glenwood Canyon costs involved the hauling and placement of backfill, as well as shotcrete facing. In some situations, especially where contractors are unfamiliar with geosynthetic reinforcement, artificially high unit costs have been placed on bid items such as the shotcrete facing, which effectively has made the reinforced soil wall uneconomical. The geosynthetic reinforcement is approximately 15% of the total in-place cost of highway reinforced walls with SRW facing and select granular soil fill. A cost comparison for reinforced versus other types of retaining walls is presented in Figure 9-14. For low walls, geosynthetics are usually less expensive than conventional walls and metallic geosynthetic wall systems. At the time of its construction, the Rainier Avenue wall was the highest geotextile wall ever constructed (Allen et al., 1992). It was unusually economical, partially because, as a temporary structure, no special facing was used. Permanent facing on a wall of that height would have increased its cost by $50/m^2$ or more.

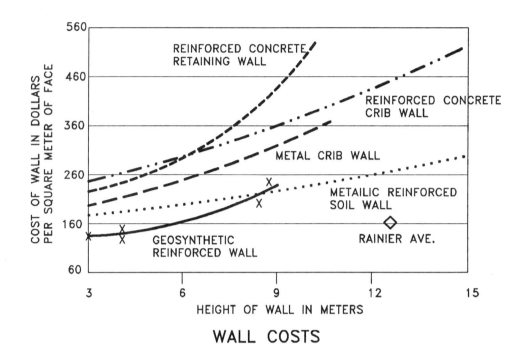

Figure 9-14 Cost comparison of reinforced systems.

Other factors impacting cost comparison include site preparation; facing cost, especially if precast panels or other special treatments are required; special drainage required behind the backfill; instrumentation; etc.

9.9   COST ESTIMATE EXAMPLES

9.9-1   Geogrid, SRW-Faced Wall

A preliminary cost estimate for a geosynthetic reinforced wall is needed to assess its viability on a particular project. Therefore, a rough design is required to estimate fill and soil reinforcement quantities. The project's scope is not fully defined, and several assumptions will be required.

STEP 1. Wall description

> The wall will be approximately 200 m long, and varies in exposed wall height from 4 to 6 meters. A gradual slope of 5H:1V will be above the wall. The wall will have a nominal (*e.g.*, < 3°) batter. Seismic loading can be ignored.

> An SRW facing will be specified. The geosynthetic will be a geogrid. Reinforced wall fill will be imported. Wall fill soils are not aggressive and pose no specific durability concerns.

STEP 2. Foundation soil

Wall will be founded on a competent foundation, well above the estimated water table. The in situ soils are silty sands, and an effective friction angle of 30° can be assumed for conceptual design. A series of soil borings along the proposed wall alignment will be completed prior to final design.

STEP 3. Reinforced fill and retained backfill properties

A well-graded gravely sand, with 20 mm maximum size, will be specified as wall fill, as it is locally available at a cost of approximately $3 per 1,000 kg delivered to site. An effective angle of friction of 34° and a unit weight of 20 kN/m^3 can be assumed. The fill is nonaggressive, and a minimum durability partial safety factor can be used.

The retained backfill will be on-site silty sand embankment material. An effective angle of friction of 30° and a unit weight of 19 kN/m^3 can be assumed.

STEP 4. Establish design factors of safety.

For external stability, use minimums of:
sliding	1.5
bearing capacity	2
overturning	2
overall stability	1.5

For internal stability, use minimums of:
FS = 1.5 against reinforcement failure
FS = 1.5 against pullout failure

STEP 5. Determine preliminary wall dimensions.

Average exposed wall height is approximately 5 m. An embedment depth of 0.45 (the minimum recommended) should be added to the exposed height. Total design height is 5.45 m.

Assume L/H ratio of approximately 0.7. Use an L = 0.7 (5.45) ≈ 4 m.

STEP 6. Develop earth pressure diagrams.

Use a tie-back wedge approach, but for conceptual design ignore the relatively flat 5:1 slope and the Meyerhof- type distribution of external load.

For internal stability $K_a = \tan^2(45 - \phi/2) = 0.28$

For external stability $K_a = \tan^2(45 - \phi/2) = 0.33$

STEP 7. Check external stability.

By observation and experience, it is assumed that the L/H ratio of 0.7 will provide adequate external safety factors for the project conditions.

STEP 8. Estimate settlement.

Again, by observation and experience, settlement is not a problem for these project conditions.

STEP 9. Calculate horizontal stress at each layer of reinforcement.

Not required for conceptual design; see next step.

STEP 10. Check internal stability and determine reinforcement requirements.

Lateral load to be resisted by the geogrid is equal to:

$$\tfrac{1}{2}\,K_\alpha\,\gamma\,H^2 \;=\; \tfrac{1}{2}\,(0.28)\,(20\ \text{kN/m}^3)\,(5.45\ \text{m})^2 \;=\; 83\ \text{kN/m}$$

Assuming 100% geogrid coverage in plan view, the geogrids must safely carry 83 kN/m per unit width of wall. Assume a geogrid with a long-term allowable strength of 20 kN/m will be used. The safe design strength of the geogrid is therefore equal to:

$$T_d \;=\; T_a\,/\,FS \;=\; 20\,/\,1.5 \;=\; 13.3\ \text{kN/m}$$

The approximate number of geogrid layers required is equal to:

$$83 \div 13.3 \;=\; 6.2$$

Round this number up and add an additional layer for conceptual design to account for practical layout considerations, 5:1 slope, and inclusion of Meyerhof-type distribution of external load with final design. Therefore, assume 8 layers of geogrid will be used.

COST ESTIMATE:

Material Costs:

Leveling Pad - 200 m ($10 / m) = $2,000

Reinforced wall fill -

200 m (4 m) (5.45 m) (20 kN/m^3) = 87,200 kN $\Rightarrow$ 87,200,000 N $\div$ 9.8 = 8,900,000 kg

8,900,000 kg ($3 / 1,000 kg) $\approx$ $ 27,000

Geogrid soil reinforcement -

8 layers (4 m) (200 m) = 6,400 m^2

From the range presented in Appendix E, assume a material cost, delivered to site, of $2.50 / m^2

6,400 m^2 ($2.50 / m^2) = $16,000

SRW face units -

From local market, SRWs range in cost from $50 to $70 / m^2

Assume a cost of $60 / m^2

200 m (5.45 m) ($60 / m^2) $\approx$ $65,000

Gravel drain fill within or behind the SRW units -

Assume 0.3 m thickness required. Assume a cost of $7.50 per compacted m^3.

200 m (5.45 m) (0.3 m) ($7.50 / m^3) = $3,000

Engineering and Testing Costs:

A line-and-grade specification will be used. Based upon previous projects, assume cost of design engineering, soil testing, and site assistance will be approximately $10 per m^2.

200 m (5.45 m) ($10 / m^2) $\approx$ $11,000

Installation Costs:

Based upon previous bids, assume cost to install will be approximately $50 per m^2

200 m (5.45 m) ($50 / m^2) $\approx$ $55,000

TOTAL ESTIMATED COST:

Materials + Engineering/Site Assistance + Installation =

($2,000 + $27,000 + $16,000 + $65,000 + $3,000) + $11,000 + $55,000 = $179,000

{Check: This is equal to an installed cost of $164 / m^2, which is reasonable based upon past experience.} Based upon this cost estimate, the geosynthetic reinforced wall option is the most economical for this project. Therefore, it is recommended that final design proceed using a geosynthetic reinforced wall. Note that estimate does not include site preparation, placement of random backfill, or final completion items (*e.g.*, seeding, railings).

9.9-2   Geotextile Wrap Wall

A preliminary cost estimate for a temporary wall is needed to assess its viability on a particular project. Therefore, a rough design is required to estimate fill and soil reinforcement quantities. The project scope is not fully defined, and several assumptions will be required.

STEP 1. Wall description.

The temporary wall will be approximately 50 m long and approximately 8 m high. A flat slope and no traffic will be above the wall. Seismic loading can be ignored.

A wrap-around facing will be used and an ultraviolet-stabilized geotextile specified. Thus, a gunite or other type of protective facing for this temporary structure will not be required. Wall fill soils are not aggressive and pose no specific durability concerns.

STEP 2. Foundation soil.

Wall will be founded on a competent foundation that overlies a soft compressible layer of soil. Details of foundation bearing capacity and global stability do not have to be addressed for this conceptual cost estimate, but will be addressed in final design. The in situ soils are silts, and an effective friction angle of 28° can be assumed for conceptual design.

STEP 3. Reinforced fill and retained backfill properties.

A well-graded sand will be specified as wall fill, as it is locally available at a cost of approximately \$2 per 1,000 kg delivered to site. An effective angle of friction of 32° and a unit weight of 19.5 kN/m^3 can be assumed. The fill is nonaggressive, and a minimum durability partial safety factor can be used.

The retained backfill will be on site silt embankment material. An effective angle of friction of 28° and a unit weight of 18.5 kN/m^3 can be assumed.

STEP 4. Establish design factors of safety.

For external stability, use minimums of:
sliding	1.5
bearing capacity	2
overturning	2
overall stability	1.5

For internal stability, use minimums of:

FS = 1.5 against reinforcement failure

FS = 1.5 against pullout failure

STEP 5. Determine preliminary wall dimensions.

Average exposed wall height is approximately 8 m. An embedment distance of 0.45 (the minimum recommended) should be added to the exposed height. Design height is 8.45 m.

Assume L/H ratio of approximately 0.7. Use an $L = 0.7 (8.45) \approx 6$ m.

STEP 6. Develop earth pressure diagrams.

Use a tie-back wedge approach.

For internal stability $K_a = \tan^2(45 - \phi/2) = 0.31$
For external stability $K_a = \tan^2(45 - \phi/2) = 0.36$

STEP 7. Check external stability.

By observation and experience, it is assumed that the L/H ratio of 0.7 will provide adequate external safety factors for the project conditions. {This assumption will checked in final design.}

STEP 8. Estimate settlement.

Again, by observation and experience, settlement is likely not a problem for these project conditions. Settlement will be quantified during final design.

STEP 9. Calculate horizontal stress at each layer of reinforcement.

Not required for conceptual design; see next step.

STEP 10. Check internal stability and determine reinforcement requirements.

Lateral load to be resisted by the geotextile is equal to:
$$\tfrac{1}{2} K_a \gamma H^2 = \tfrac{1}{2} (0.31) (19.5 \text{ kN/m}^3) (8.45 \text{ m})^2 = 216 \text{ kN/m}$$

Assuming 100% geotextile coverage in plan view, the geotextiles must safely carry 216 kN/m per unit width of wall. Assume a geotextile with a long-term allowable strength of 20 kN/m will be used. The safe design strength of the geotextile is therefore equal to:
$$T_d = T_a / FS = 20 / 1.5 = 13.3 \text{ kN/m}$$

The approximate number of geotextile layers required is equal to:
$$216 \div 13.3 = 16.2$$

Round this number up and add an additional layer for conceptual design to account for practical layout considerations with final design. Assume a vertical spacing of 0.5 m will be used. Therefore, 17 layers of geotextile will be used.

COST ESTIMATE:

Material Costs:

Reinforced wall fill -
50 m (8.45 m) (6 m) (20 kN/m^3) = 50,700 kN ÷ 50,700,000 N ÷ 9.8 = 5,173,500 kg
5,173,500 kg ($2 / 1,000 kg) ≈ $ 10,000

Geotextile soil reinforcement (include face area and wrap-tail length) -
17 layers (6 + 0.5 + 1.5 m) (50 m) = 6,800 m^2
From the range presented in Appendix E, assume a material cost, delivered to site, of $2 / m^2
6,800 m^2 ($2 / m^2) ≈ $14,000

Engineering and Testing Costs:

A line-and-grade specification will be used. Based upon previous projects, assume cost of design engineering, soil testing, and site assistance will be approximately $30 per m^2, because of the height and relatively low total area of wall that will be constructed.

50 m (8.45 m) ($30 / m^2) ≈ $13,000

Installation Costs:

Based upon previous bids, assume cost to install will be approximately $60 per m^2

50 m (8.45 m) ($60 / m^2) ≈ $25,000

TOTAL ESTIMATED COST:

Materials + Engineering/Site Assistance + Installation =
$24,000 + $13,000 + $25,000 = $62,000

{Check: This is equal to an installed cost of $147 / m^2, which is reasonable based the small size of this project and upon past experience.} Based upon this cost estimate, the geosynthetic reinforced wall option is the most economical for this project. Therefore, it is recommended that final design proceed using a geosynthetic reinforced wall.

9.10 SPECIFICATIONS

9.10-1 Geogrid, SRW-Faced Wall

The following example was obtained from New York D.O.T. It is a special provision, from a specific project, for materials and construction of a geogrid-SRW reinforced soil wall.

ITEM 17554.02 MECHANICALLY STABILIZED SEGMENTAL BLOCK RETAINING WALL SYSTEM (EXTENSIBLE REINFORCEMENT)

DESCRIPTION

Construct a Mechanically Stabilized Segmental Block Retaining Wall System (Extensible Reinforcement), (MSSBRWS) where indicated on the plans.

A MSSBRWS consists of an un-reinforced concrete or compacted granular leveling pad, facing and cap units, backfill, underdrains, geotextiles, and an extensible reinforcement used to improve the mechanical properties of the backfill.

Other definitions that apply within this specification are:

A. Leveling Pad An un-reinforced concrete or compacted granular fill footing or pad which serves as a flat surface for placing the initial course of facing units.

B. Facing Unit A segmental precast concrete block unit that incorporates an alignment and connection device and also forms part of the MSSBRWS face area. A corner unit is a facing unit having two faces.

C. Alignment and Connection Device

Any device that is either built into or specially manufactured for the facing units, such asshear keys, leading/trailing lips, or pins. The device is used to provide alignment and maintain positive location for a facing unit and also provide a means for connecting the extensible reinforcement.

D. Extensible Reinforcement

High density polyethylene, polypropylene or high tenacity polyester geogrid mats of specified lengths which connect to the facing unit and are formed by a regular network of integrally connected polymer tensile elements with apertures of sufficiently large size to allow for mechanical interlock with the backfill.

E. Unit Fill Well-graded aggregate fill placed within and/or contiguous to the back of the facing unit.

F. Cap Unit A segmental precast concrete unit placed on and attached to the top of the finished MSSBRWS.

G. Backfill   Material placed and compacted in conjunction with extensible reinforcement and facing units.

H. Underdrain   A system for removing water from behind the MSSBRWS.

I. Geotextile   A permeable textile material used to separate dissimilar granular materials.

MATERIALS

Not all materials listed are necessarily required for each MSSBRWS. Ensure that the proper materials are supplied for the chosen system design.

A. Leveling Pad

    1. For MSSBRWS greater than or equal to 4.6 m in total height, supply a leveling pad conforming to the following:

        a. Un-reinforced Concrete

        Supply concrete conforming to Section 501 (Class A Concrete).

    2. For MSSBRWS less than 15 feet in total height, supply a leveling pad conforming to one of the following:

        a. Un-reinforced Concrete

        Supply concrete conforming to Section 501 (Class A Concrete).

        b. Granular Fill

        Supply select granular fill conforming to Subsection 203-2.02C (Select Granular Fill and Select Structure Fill).

        c. Crushed Stone

        Supply crushed stone conforming to Section 623 (Screened Gravel, Crushed Gravel, Crushed Stone, Crushed Slag), Item 623.12, Crushed Stone (In-Place measure), Size Designation 1.

B. Facing and Cap Units

Supply units fabricated and conforming to Subsection 704-07 (Precast Concrete Retaining Wall Blocks). Notify the Director, Materials Bureau, of the name and address of the units' fabricator no later than 14 days after contract award.

C. Alignment and Connection Devices

Supply devices conforming to the designer-supplier's Installation Manual.

D. Extensible Reinforcement

Supply reinforcement manufactured and conforming to Sub-section 725-03 (Geogrid Reinforcement).

E. Unit Fill

Supply fill conforming to material and gradation requirements for Type CA-2 Coarse Aggregate under Sub-section 501-2.02, B.2 (Coarse Aggregate).

F. Cast-in-place Concrete

Supply concrete conforming to Section 501 (Class A Concrete).

G. Backfill

Supply backfill material as shown on the plans and conforming to Subsection 203-1.08 (Suitable Material). Backfill material must come form a single source, unless prior written approval for use of multiple sources is obtained from the Director, Geotechnical Engineering Bureau.

Stockpile backfill material conforming to the current Soil Control Procedure (SCP) titled "Procedure for the Control of Granular Materials."

1. Material Test Procedures

The State will perform procedures conforming to the appropriate Departmental publications in effect on the date of advertisement of bids. These publications are available upon request to the Regional Director, or the Director, Geotechnical Engineering Bureau.

2. Material Properties

    a. Gradation

        Stockpiled backfill material must meet the following gradation requirements:

Table 17554-2

Sieve Size Designation	Percent Passing by Weight
64 mm	100
6.5 mm	30 - 100
0.425 mm	0 - 60
0.075 mm	0 - 15

    b. Plasticity Index.
        The Plasticity Index must not exceed 5.
    c. Durability.

    The Magnesium Sulfate Soundness loss must not exceed 30 percent.

H. Geotextile

Supply geotextile material conforming to Section 207 (Geotextile), Item 207.03, Geotextile Underdrain.

I. Drainage

Supply underdrain and geotextile material as shown on the plans:

    1. Underdrain Pipe

        Supply optional underdrain pipe conforming to Section 605 (Underdrains).

    2. Geotextile Underdrain

        Supply geotextile underdrain conforming to Section 207 (Geotextile), Item 207.03, Geotextile Underdrain.

J. Identification Markers

Supply identification markers conforming to the designer-supplier's Installation Manual.

K. Basis of Acceptance

Accept cast-in-place concrete conforming to Section 501 (Portland Cement Concrete), Class A.

Accept other materials by manufacturer's certification. The State reserves the right to sample, test, and reject certified material conforming to the Departmental written instructions.

Only approved MSSBRWS designer-suppliers appearing on the attached Approved List of Products will be acceptable for use under this item.

Obtain all necessary materials (except backfill, underdrains, geotextiles, and cast-in-place concrete) form the approved designer-supplier. Upon award of the contract, notify the Deputy Chief Engineer, Technical Services (DCETS) of the name and address of the chosen designer-supplier. Once designated the designer-supplier shall not be changed.

Obtain from the designer-supplier and submit to the DCETS for approval, the MSSBRWS design and installation procedure. All designs must be stamped by a Professional Engineer licensed to practice in New York State. The DCETS requires 20 working days to approve the submission after receipt of all pertinent information. Begin work only after receiving DCETS approval.

Submit shop drawings and proposed methods for construction to the Engineer for written approval at least 30 working days before starting work. Shop drawings must conform to the size and type requirements given in Subsection 718-01 (Prestressed Concrete Units (Structural)) under Drawing Types, Subparagraph 2.A (Working Drawings, Size and Type).

Supply on-site technical assistance from a representative of the designated designer-supplier during the beginning of installation until such time as the Engineer determines that outside consultation is no longer required.

Provide the Engineer with two copies of the designated designer-supplier Installation Manual two weeks before beginning construction.

A. Excavation, Disposal and MSSBRWS Area Preparation

Excavate, dispose and prepare the area on which the MSSBRWS will rest conforming to Section 203 (Excavation and Embankment), except as modified here:

    1.   Grade and level, for a width equaling or exceeding the reinforcement length, the area on which the MSSBRWS will rest. Thoroughly compact this area to the Engineer's satisfaction. Treat all soils found unsuitable, or incapable of being satisfactorily compacted because of moisture content, in a manner directed by the Engineer, in conjunction with recommendations of the Regional Soils Engineer.

    2.   Remove rock to the limits shown on the plans.

    3.   Excavate the area for the leveling pad conforming to Section 206, (Trench, Culvert and Structure Excavation).

B. Facing and Cap Unit Storage and Inspection

Handle and store facing and cap units with extreme care to prevent damage. The State will inspect facing and cap units on their arrival at the work site and prior to their installation to determine any damage that may have occurred during shipment. Facing and cap units will be considered damaged if they contain any cracks or spalls and/or heavy combed areas with any dimension greater than 25 mm. The State will reject any damaged facing and cap units. Replace rejected units with facing and cap units acceptable to the Engineer.

C. Facing Unit Erection

1. Provide an un-reinforced concrete or compacted granular fill leveling pad as shown on the plans.

   a. Place concrete in conformance with Subsection 555-3, (Construction Details). The Engineer may waive any part of Subsection 555-3 that he determines is impractical.

   b. Place compacted granular fill in conformance with Subsection 203-3.12 (Compaction).

2. Install by placing, positioning, and aligning facing units in conformance with the designer-supplier's Installation Manual, unless otherwise modified by the contract documents or the Engineer, and check that requirements of Table 17554-4 are not exceeded. After placement, maintain each facing unit in position by a method acceptable to the Engineer.

3. Correct all misalignments of installed facing units that exceed the tolerances allowed in Table 17554-4 in a manner satisfying the Engineer.

Table 17554-4

Vertical control	6 mm over a distance of 3 m
Horizontal location control	13 mm over a distance of 3 m
Rotation from established plan wall batter	13 mm over 3 m in height

4. Control all operations and procedures to prevent misalignment of the facing units. Precautionary measures include (but are not limited to) keeping vehicular equipment at least 1 m behind the back of the facing units. Compaction equipment used within 1 m of the back of the facing units must conform to Subsection 203-3.12B.6 (Compaction Equipment for Confined Areas).

D. Unit Fill

    1.   Place unit fill to the limits indicated on the plans. Before installing the next course of facing units, compact the unit fill in a manner satisfying the Engineer and brush clean the tops of the facing units to ensure an even placement area.

    2.   Protect unit fill from contamination during construction.

E. Extensible Reinforcement

    1.   Before placing extensible reinforcement, backfill placed and compacted within a 1 m horizontal distance of the back of facing units must be no more than 25 mm above the required extensible reinforcement elevation. Backfill placed and compacted beyond the 1 m horizontal distance may be roughly graded to the extensible reinforcement elevation.

    2.   Place extensible reinforcement normal to facing units unless indicated otherwise on the plans. The Engineer will reject broken or distorted extensible reinforcement. Replace all broken or distorted extensible reinforcement.

    3.   Install extensible reinforcement within facing units conforming to the designer-supplier's Installation Manual. Pull taut and secure the extensible reinforcement before placing the backfill in a manner satisfying the Engineer.

F. Backfill

    1.   Place backfill materials (other than rock) at a moisture content less than or equal to the Optimum Moisture Content. Remove backfill materials placed at a moisture content exceeding the Optimum Moisture Content and either rework or replace, as determined by the Engineer. Determine Optimum Moisture Content in conformance with Soil Test Methods for compaction that incorporate moisture content determination. Use Soil Test Methods in effect on the date of advertisement of bids. Cost to rework or replace backfill materials shall be borne by the Contractor.

    2.   Place granular backfill material in uniform layers so that the compacted thickness of each layer does not exceed 0.25 m or one block height, whichever is less. Compact each layer in conformance with Subsection 203-3.12 (Compaction). The Engineer will determine by visual inspection that proper compaction has been attained.

9.10-2 Segmental Retaining Wall Unit

The following material specification for modular concrete units is from the National Concrete Masonry Association, Design Manual for Segmental Retaining Walls (Simac et al., 1993).

Section _____
SEGMENTAL RETAINING WALL UNITS

PART 1: GENERAL

1.01   Description

    A. Work includes furnishing and installing segmental retaining wall (SRW) units to the lines and grades designated on the construction drawings or as directed by the Architect/Engineer. Also included is furnishing and installing appurtenant materials required for construction of the retaining wall as shown on the construction drawings.

1.02   Related Work

    A. Section _____ - Site Preparation
    B. Section _____ - Earthwork
    C. Section _____ - Drainage Aggregate
    D. Section _____ - Geosynthetic Reinforcement {delete if not applicable}

1.03   Reference Standards

    A. ASTM C 90 - Load Bearing Concrete Masonry Units
    B. ASTM C 140 - Sampling and Testing Concrete Masonry Units
    C. ASTM D 698 - Moisture Density Relationship for Soils, Standard Method
    D. NCMA TEK 50A - Specifications for Segmental Retaining Wall Units
    E. NCMA SRWU-1 - Determination of Connection Strength between Geosynthetics and Segmental Concrete Units
    F. NCMA SRWU-2 - Determination of Shear Strength between Segmental Concrete Units
    G. NCMA Design Manual for Segmental Retaining Walls
    H. Where specifications and reference documents conflict, the Architect/Engineer shall make the final determination of applicable document.

1.04   Certification

    A. Contractor shall submit a notarized manufacturer's certificate prior to start of work stating that the SRW units meet the requirements of this specification.

1.05   Delivery, Storage and Handling

A. Contractor shall check the materials upon delivery to assure that specified type, grade, color and texture of SRW unit has been received.

B. Contractor shall prevent excessive mud, wet concrete, epoxies, and like material which may affix themselves from coming in contact with the materials.

C. Contractor shall protect the materials from damage. Damaged material shall not be incorporated into the reinforced soil wall.

1.06   Measurement and Payment

A. Measurement of SRW units is on a vertical square foot basis.

B. Payment shall cover supply and installation of SRW units along with appurtenant and incidental materials required for construction of the retaining wall as shown on the construction drawings. It shall include all compensation for labor, materials, supplies, equipment and permits associated with building these walls.

C. Quantity of retaining wall as shown on plans may be increased or decreased at the direction of the Architect/Engineer based on construction procedures and actual site conditions.

D. The accepted quantities of SRW units will be paid for per vertical square foot in place (total wall height). Payment will be made under:

Pay Item	Pay Unit
Segmental Retaining Wall Units	SQ FT

PART 2: MATERIALS

2.01   Segmental Retaining Wall Units

A. SRW units shall be machine formed concrete blocks specifically designed for retaining wall applications.

B. SRW units shall meet the following architectural requirements:
    1. Color of units shall be _____. {insert}
    2. Finish of units shall be _____. {insert split-faced, smooth, striated, etc.}
    3. Unit faces shall be of _____ geometry. {insert rounded, straight, offset, etc.}
    4. Maximum and minimum face area per unit shall be _____ and _____, respectively. {insert values or delete if not a requirement}
    5. Units shall be erected with a _____ configuration. {insert running or stacked bond}
    6. All units shall be sound and free of cracks or other defects that would interfere with the proper placing of the unit or significantly impair the strength or permanence of the construction. Cracking and excessive chipping may be grounds for rejection.

C. SRW units shall meet the following structural requirements:
    1. Concrete used to manufacture SRW units shall have a minimum 28 day compressive strength of {3000 psi} in accordance with ASTM C 90. The

concrete shall have adequate freeze/thaw protection with a maximum moisture absorption rate, by weight, of: i) 8% in southern climates; or ii) 6% in northern climates.

2. Units shall be positively interlocked to provide a minimum shear capacity of a_u = {400 lb/ft} and λ_u = {30°} as tested in accordance with NCMA SRWU-1.

3. Units shall provide a minimum connection strength between it and the geosynthetic reinforcement of a_{cs} = {200 lb/ft} and λ_{cs} = {40°} as tested in accordance with NCMA SRWU-2, if required.

4. SRW units molded dimensions shall not differ more than ±1/8 inch from that specified, except height which shall be ±1/16 inch.

D. SRW units shall meet the following constructability and geometric requirements:

1. Units shall be capable of attaining concave and convex curves.

2. Units shall be positively engaged to the unit below so as to provide a minimum of 3/32 inch horizontal setback per vertical foot of wall height. True vertical stacked units will not be permitted.

3. Units shall be positively engaged to the unit below so as to provide a maximum of _____ inch horizontal setback per vertical foot of wall height. {This controls amount of useable property at top of wall. Specify value or delete if not applicable.}

2.02   Levelling Pad and Unit Fill Material

A. Material for footing shall consist of compacted sand or gravel and shall be a minimum of 6 inches in depth.

B. Fill for units shall be the free draining gravel or drainage fill, see Section _____ Drainage Aggregate.

C. Do not run mechanical vibrating plate compactors on top of the units. Compact unit fill be running hand-operated compaction equipment just behind the unit. Compact to minimum 95% standard Proctor density (ASTM D 698) or 90% of modified Proctor density (ASTM D 1557).

2.03   Drainage Aggregate

A. Drainage layer materials shall be the free draining gravel or drainage fill, see Section _____ - Drainage Aggregate.

B. Vertical drainage layer behind the wall face shall be placed no less than 1 ft^3 per 1 ft^2 of wall face.

2.04   Infill Soil

A. The infill soil material shall be free of debris and consist of either of the following inorganic soil types according to their USCS designations (GP, GW, SW, SP, SM, ML, CL). The maximum particle size shall be 4 inches. There shall be less than 20% by weight of particles greater than 1½ inches, maximum 60% by weight passing the #200 sieve and PI<20.

B. The infill soil shall be compacted in maximum 8 inches compacted lifts to the following minimum densities (percentage of the maximum standard Proctor) (ASTM D 698): i) fine grained (ML-CL, SC, SM) soils to a minimum or 95%; and ii) coarse grained (GP, GW, SW, SP) soils to a minimum of 98%.

2.05   Common Backfill

    A. Soil placed behind the infill can be any inorganic soil with a liquid limit less than 50 and plasticity index less than 30, or as directed by the Engineer.

    B. Backfill shall be compacted to a minimum 90% of maximum standard Proctor density (ASTM D 698).

PART 3: EXECUTION

3.01   Excavation

    A. Contractor shall excavate to the lines and grades shown on the project grading plans. Contractor shall take precautions to minimize over-excavation. Over-excavation shall be filled with compacted infill material, or as directed by the Engineer/Architect, at the Contractor's expense.

    B. Architect/Engineer will inspect the excavation and approve prior to placement of bearing pad material.

    C. Excavation of deleterious soils and replacement with compacted infill material, as directed by the Architect/Engineer, will be paid for at the contract unit prices, see Section ____ - Excavation..

    D. Over-excavated areas in front of wall face shall be filled with compacted infill material at the Contractor's expense, or as directed by the Architect/Engineer.

    E. Contractor shall verify location of existing structures and utilities prior to excavation. Contractor shall ensure all surrounding structures are protected from the effects of wall excavation.

3.02   Levelling Pad Construction

    A. Levelling pad shall be placed as shown on the construction drawings with a minimum thickness of 6 inches.

    B. Foundation soil shall be proofrolled and compacted to 95% of standard Proctor density and inspected by the Architect/Engineer prior to placement of levelling pad materials.

    C. Soil levelling pad material shall be compacted to provide a level hard surface on which to place the first course of units. Compaction will be with mechanical plate compactors to 95% of maximum Proctor density (ASTM D 698).

    D. Levelling pad shall be prepared to insure intimate contact of retaining wall unit with pad.

3.03   Segmental Unit Installation

    A. First course of SRW units shall be placed on the bearing pad. The units shall be checked for level and alignment. The first course is the most important to insure accurate and acceptable results.

B. Insure that units are in full contact with base.

C. Units are placed side by side for full length of straight wall alignment. Alignment may be done by means of a string line or offset from base line to a molded finished face of the SRW unit. Adjust unit spacing for curved sections according to manufacturer's recommendations.

D. Install shear connectors (if applicable).

E. Place unit fill (if applicable).

F. Place and compact fill behind and within units.

G. Clean all excess debris from top of units and install next course. Ensure each course is completely filled prior to proceeding to next course.

H. Lay each successive course ensuring that shear connectors are engaged.

I. Repeat procedures to the extent of the wall height.

J. Uppermost row of SRW units or caps shall be glued to underlying units with an adhesive, as recommenced by the manufacturer.

END OF SECTION

9.10-3 Geotextile Wrap Around Wall

The following example of a special provision for materials and construction of a geotextile-reinforced soil retaining wall was obtained from the Washington State D.O.T.

GEOTEXTILE RETAINING WALL
November 15, 1994

Description

The Contractor shall construct geotextile retaining walls in accordance with the details shown in the Plans, these specifications, or as directed by the Engineer.

Quality Assurance

The base of the excavation shall be completed to within plus or minus 75 mm of the staked elevations unless directed by the Engineer. The external wall dimensions shall be placed to within plus or minus 50 mm of that staked on the ground. The reinforcement layer vertical spacing and overlap distance shall be completed to within plus or minus 25 mm of that shown in the Plans.

The completed wall(s) shall meet the following tolerances:

1. Vertical tolerances (design batter) and horizontal tolerances (alignment) for the geotextile face shall not exceed 75 mm when measured along a 3 m straight edge for permanent walls and 130 mm for temporary walls. The face batter and alignment measurements shall be made at the midpoint of each wall layer.

2. The overall vertical tolerances shall not exceed 50 mm per 3 meters of wall height for permanent walls and 80 mm per 3 meters of wall height for temporary walls.

3. The maximum outward bulge of the geotextile face between backfill reinforcement layers shall not exceed 100 mm for permanent walls and 150 mm for temporary walls.

Submittals

The Contractor shall submit to the Engineer, a minimum of 14 calendar days prior to beginning construction of each wall, detailed plans for each wall as well as other information. As a minimum, the submittals shall include the following:

1. Detailed wall plans showing the actual lengths proposed for the geotextile reinforcing layers and the locations of each geotextile product proposed for use in each of the geotextile reinforcing layers.

2. The Contractor's proposed wall construction method, including proposed forming systems, types of equipment to be used and proposed erection sequence.

3. Geotextile certificate of compliance and samples for the purpose of source approval as required elsewhere in these Special Provisions.

4. Details of geotextile wall corner construction as required elsewhere in these Special Provisions.

5. If the wall is permanent, the shotcrete mix design with compressive strength test results, and the method proposed for shotcrete wall face finishing and curing.

Approval of the Contractor's proposed wall construction details and methods shall not relieve the Contractor of his responsibility to construct the walls in accordance with the requirements of these specifications.

Materials

Geotextiles and Thread for Sewing

The material shall be woven or non-woven geotextile consisting only of long chain polymeric filaments or yarns formed into a stable network such that the filaments or yarns retain their position relative to each other during handling, placement, and design service life. At least 95 percent by weight of the long chain polymers shall be polyolefins, polyesters, or polyamides. The material shall be free of defects and tears. The geotextile shall conform to the properties as indicated in Tables 1 and 2. The geotextile shall be free from any treatment or coating which might adversely alter its physical properties after installation.

Thread used for sewing shall be high strength polypropylene, polyester, or Kevlar thread. Nylon threads will not be allowed. The thread used must also be resistant to ultraviolet radiation if the sewn seam is exposed at the wall face.

Geotextile Properties

Table 1: Minimum properties required for geotextiles used in geotextile retaining walls.

Geotextile Property	Test Method[2]	Geotextile Property Requirements[1] Woven/Nonwoven
Water Permittivity	ASTM D4491	.02 sec.[-1] min.
AOS	ASTM D4751	.84 mm max. (#20 sieve)
Grab Tensile Strength, min. in machine and x-machine direction	ASTM D4632	900 N/530 N min.
Grab Failure Strain, in machine and x-machine direction	ASTM D4632	$< 50\% / \geq 50\%$
Seam Breaking Strength[3]	ASTM D4632 and ASTM D4884 (adapted for grab test)	700 N/430 N min.
Puncture Resistance	ASTM D4833	280 N/220 N min.
Tear Strength, min. in machine and x-machine direction	ASTM D4533	280 N/220 N min.
Ultraviolet (UV) Radiation Stability	ASTM D4355	70% Strength Retained min., after 500 Hrs. in weatherometer

[1]All geotextile properties are minimum average roll values (i.e., the average test results for any sampled roll in a lot shall meet or exceed the values shown in the table).

[2]The test procedures used are essentially in conformance with the most recently approved ASTM geotextile test procedures, except for geotextile sampling and specimen conditioning, which are in accordance with WSDOT Test Methods 914 and 915, respectively. Copies of these test methods are available at the WSDOT Headquarters Materials Laboratory in Tumwater.

[3]Applies only to seams perpendicular to the wall face.

Table 2: Wide strip tensile strength required for the geotextile used in geotextile retaining walls.

Wall Location	Vertical Spacing of Reinforcement Layers	Reinforcement Layer Distance from Top of Wall	Minimum Tensile Strength Based on ASTM D4595[1,2]	
			Geotextile Polymer Type	
			Polyester	Polypropylene/ Polyethylene
*****	*****	*****	(Project Specific)	(Project Specific)

[1]Wide strip geotextile strengths are minimum average roll values (i.e., the average test results for any sampled roll in a lot shall meet or exceed the values shown in the table). These wide strip strength requirements apply only in the geotextile direction perpendicular to the wall face.

[2]The test procedures used are essentially in conformance with the most recently approved ASTM geotextile test procedures, except for geotextile sampling and specimen conditioning, which are in accordance with WSDOT Test Methods 914 and 915, respectively. Copies of these test methods are available at the WSDOT Headquarters Materials Laboratory in Tumwater.

Shotcrete Wall Facing

Shotcrete shall be the application of one or more layers of pneumatically placed concrete to soil, geosynthetic, concrete, or steel surfaces.

Plain shotcrete is defined as a Portland cement concrete mix containing admixtures to provide quick set, high early strength, and satisfactory adhesion, which is conveyed through a hose and pneumatically projected at high velocity onto a surface.

The wet-mix process consists of thoroughly mixing all ingredients except accelerating admixtures but including the mixing water, introducing the mixture into the delivery equipment and delivering it by positive displacement or compressed air to the nozzle. The mixed shotcrete is then air jetted from the nozzle at high velocity onto the surface in the same manner as for the dry mix process. The accelerator for the wet-mix is added to the shotcrete mixture in such a way that the quantity can be properly regulated and the material uniformly dispersed throughout the shotcrete when it is placed.

Portland cement shall be Type II as specified in Section 9-01.2(1). Air entrainment shall be 6.0 percent (plus or minus 1 1/2 percent) to comply with the requirements of Section 9-23.6.

Aggregate for shotcrete shall meet the following gradation requirements:

Sieve Size	Percent Passing by Weight
12.7 mm	100
9.5 mm	90 to 100
No. 4	70 to 85
No. 8	50 to 70
No. 16	35 to 55
No. 30	20 to 35
No. 50	8 to 20
No. 100	2 to 10
No. 200	0 to 2.5

Water for mixing and curing shall be clean and free from substances which may be injurious to concrete or steel. Water shall also be free of elements which would cause staining.

Reinforcement shall be as shown in the Plans and shall comply with the requirements of Section 9-07.

Proportioning Concrete

Shotcrete shall be proportioned to produce a 28 MPa compressive strength at 28 days. The shotcrete mix design and method of placement proposed for use at the jobsite shall be submitted to the Engineer by the Contractor at least 14 calendar days prior to beginning shotcrete placement. The Contractor shall also include evidence within this submittal that the proposed shotcrete mix design and method of placement will produce the required compressive strength at 28 days. The Contractor must receive notification from the Engineer that the proposed mix design and method of placement is acceptable before shotcrete placement can begin.

No admixture shall be used without the permission of the Engineer. If admixtures are used to entrain air, to reduce water-cement ratio, to retard or accelerate setting time, or to accelerate the development of strength, the admixtures shall be used at the rate specified by the manufacturer and approved by the Engineer.

Shotcrete Testing

The Contractor shall make shotcrete test panels for evaluation of shotcrete quality, strength, and aesthetics. Both preproduction test panels for approval and nozzlemen prequalification, as well as production test panels, shall be prepared by the Contractor. Any cores obtained for the purpose of shotcrete strength testing shall have the following minimum dimensions:

a. The core diameter shall be at least 3 times the maximum aggregate size, but not less than 50 mm.

b. The core height shall be 1.5 times the core diameter, but not less than 75 mm.

The cores will be obtained and tested in accordance with AASHTO T 24. Cores removed from the panel shall be immediately wrapped in wet burlap and sealed in a plastic bag. Cores shall be clearly marked to identify from where the cores were taken and whether they are for pre-production or for production testing. If for production testing, the section of the wall represented by the cores shall be clearly marked on the cores. Cores shall be delivered to the Engineer within 2 hours of coring. The remainder of the panel(s) shall become the property of the Contractor.

1. Pre-production testing - The Contractor shall prepare at least one 1.0 meter by 1.0 meter panel for each mix design for evaluation and testing of shotcrete quality and strength. The Contractor shall make one additional 1.2 meter by 1.2 meter qualification panel for evaluation and approval of the Contractor's proposed method for shotcrete, installation, finishing, and curing. Both the 1.0 meter and the 1.2 meter panels shall be constructed using the same methods and initial curing used to construct the shotcrete wall, except that the 1.0 meter panel shall not include wire reinforcement. The 1.0 meter panel shall be constructed to the minimum thickness necessary to obtain the core samples of the required dimensions. The 1.2 meter panel shall be constructed to the same thickness as proposed for the wall face. Production shotcrete work shall not begin until satisfactory test results are obtained and the panels are approved by the Engineer.

2. Production testing - The Contractor shall make at least one 1.0 meter by 1.0 meter panel for each section of wall shot or as many as directed by the Engineer. A section is defined as one day's placement. The production panels shall be constructed using the same methods and initial curing used to construct the shotcrete wall, but without wire reinforcement. The panels shall be constructed to the minimum thickness necessary to obtain core samples of the required dimensions. If the production shotcrete is found to be unsuitable based on the results of the test panels, the section(s) of the wall represented by the test panel(s) shall be repaired or replaced to the satisfaction of the Engineer at no expense to the Contracting Agency.

Shotcrete Coloration For Facing Alternate C

If facing Alternate C is required, the Contractor shall provide shotcrete coloration for finishing the sculptured shotcrete to match the color of the natural surroundings. Approval of the final appearance of the coloration will be based on the pre-production test panel. Approval of the long-term properties of the coloration material shall be based on a manufacturer's certification which verifies the following to be true about the product:

1. Resistance to alkalies in accordance with ASTM D 543.
2. Demonstrates no change in coloration after 1,000 hours of testing in accordance with ASTM D 822.
3. Does not oxidize when tested in accordance with ASTM D 822.
4. Demonstrates resistance to gasoline and mineral spirits when tested in accordance with ASTM D 543.

Additionally, the certification shall provide the product name, proposed mix design and application method, and evidence of at least one project where the product, using the proposed mix and application method, was applied and which has provided at least five years or more of acceptable durability and color permanency.

Wall Backfill Material

All backfill material used in the reinforced soil zone of the geotextile wall shall be free draining, free from organic or otherwise deleterious material and shall conform to the gradations for Gravel Borrow as specified in Section 9-03.14. The material shall be substantially free of shale or other soft, poor durability particles. The material shall have magnesium sulfate soundness loss of less than 30 percent after four cycles. The Contractor shall provide the Engineer with test results from an independent laboratory for the magnesium sulfate soundness loss test. The test shall be conducted in accordance with ASTM C 88.

The backfill material shall meet the following chemical requirements for permanent walls:

Property or Chemical	Test Method	Allowable Quantity
Resistivity	AASHTO T288-91I	Greater than 3,000 ohm-cm
pH	AASHTO T289-91I	Polypropylene/Polyethylene: 5 to 10
		Polyester: 5 to 9
Chlorides	AASHTO T291-91I	Less than 100 mg/kg
Sulfates	AASHTO T290-91I	Less than 200 mg/kg

For temporary walls, the only backfill chemical property requirements which apply are as follows: soil pH shall be between 3 and 11, and soil resistivity shall be greater than 1,000 ohm-cm.

Geotextile Approval and Acceptance
Source Approval
The Contractor shall submit to the Engineer the following information regarding each geotextile proposed for use:

Manufacturer's name and current address,
Full Product name,
Geotextile structure, including fiber/yarn type, and
Geotextile polymer type(s).

If the geotextile source has not been previously evaluated, a sample of each proposed geotextile shall be submitted to the Headquarters Materials Laboratory in Tumwater for evaluation. After the sample and required information for each geotextile type have arrived at the Headquarters Materials Laboratory in Tumwater, a maximum of 14 calendar days will be required for this testing. Source approval will be based on conformance to the applicable values from Tables 1 and 2. Source approval shall not be the basis of acceptance of specific lots of material unless the lot sampled can be clearly identified, and the number of samples tested and approved meet the requirements of WSDOT Test Method 914.

Geotextile Samples for Source Approval

Each sample shall have minimum dimensions of 1.5 meters by the full roll width of the geotextile. A minimum of 6 square meters of geotextile shall be submitted to the Engineer for testing. The geotextile machine direction shall be marked clearly on each sample submitted for testing. The machine direction is defined as the direction perpendicular to the axis of the geotextile roll.

The geotextile samples shall be cut from the geotextile roll with scissors, sharp knife, or other suitable method which produces a smooth geotextile edge and does not cause geotextile ripping or tearing. The samples shall not be taken from the outer wrap of the geotextile nor the inner wrap of the core.

Acceptance Samples

Samples will be randomly taken by the Engineer at the jobsite to confirm that the geotextile meets the property values specified.

Approval will be based on testing of samples from each lot. A "lot" shall be defined for the purposes of this specification as all geotextile rolls within the consignment (i.e., all rolls sent to the project site) which were produced by the same manufacturer during a continuous period of production at the same manufacturing plant and have the same product name. After the samples and manufacturer's certificate of compliance have arrived at the Headquarters Materials Laboratory in Tumwater, a maximum of 14 calendar days will be required for this testing. If the results of the testing show that a geotextile lot, as defined, does not meet the properties required in Tables 1 and 2, the roll or rolls which were sampled will be rejected. Two additional rolls for each roll tested which failed from the lot previously tested will then be selected at random by the Engineer for sampling and retesting. If the retesting shows that any of the additional rolls tested do not meet the required properties, the entire lot will be rejected. If the test results from all the rolls retested meet the required properties, the entire lot minus the roll(s) which failed will be accepted. All geotextile which has defects, deterioration, or damage, as determined by the Engineer, will also be rejected. All rejected geotextile shall be replaced at no expense to the Contracting Agency.

Certificate of Compliance

The Contractor shall provide a manufacturer's certificate of compliance to the Engineer which includes the following information about each geotextile roll to be used:

Manufacturer's name and current address,
Full product name,
Geotextile structure, including fiber/yarn type,
Geotextile polymer type(s),
Geotextile roll number, and
Certified test results.

Approval Of Seams

If the geotextile seams are to be sewn in the field, the Contractor shall provide a section of sewn seam before the geotextile is installed which can be sampled by the Engineer.

The seam sewn for sampling shall be sewn using the same equipment and procedures as will be used to sew the production seams. The seam sewn for sampling must be at least 2 meters in length. If the seams are sewn in the factory, the Engineer will obtain samples of the factory seam at random from any of the rolls to be used. The seam assembly description shall be submitted by the Contractor to the Engineer and will be included with the seam sample obtained for testing. This description shall include the seam type, stitch type, sewing thread type(s), and stitch density.

Construction Requirements

Geotextile Roll Identification, Storage, and Handling

Geotextile roll identification, storage, and handling of the geotextile shall be in conformance to ASTM D 4873. During periods of shipment and storage, the geotextile shall be kept dry at all times and shall be stored off the ground. Under no circumstances, either during shipment or storage, shall the materials be exposed to sunlight, or other form of light which contains ultraviolet rays, for more than five calendar days.

Wall Construction

Excavation for the wall shall be in accordance with the requirements of Section 2-09 and in conformity to the limits and construction stages shown in the Plans.

The base for the wall shall be graded to a smooth, uniform condition free from ruts, potholes, and protruding objects such as rocks or sticks. The geotextile shall be spread immediately ahead of the covering operation. The Contractor shall take precaution to direct surface runoff from adjacent areas away from the wall construction site.

Wall construction shall begin at the lowest portion of the excavation and each layer shall be placed horizontally as shown in the Plans. Each layer shall be completed entirely before the next layer is started. Geotextile splices transverse to the wall face will be allowed provided the minimum overlap is 0.3 meters or the splice is sewn together. Geotextile splices parallel to the wall face will not be allowed. The geotextile shall be stretched out in the direction perpendicular to the wall face to ensure that no slack or wrinkles exist in the geotextile prior to backfilling.

Under no circumstances shall the geotextile be dragged through mud or over sharp objects which could damage the geotextile. The fill material shall be placed on the geotextile in such a manner that a minimum of 150 mm of material will be between the vehicle or equipment tires or tracks and the geotextile at all times. Particles within the backfill material greater than 75 mm in size shall be removed. Turning of vehicles on the first lift above the geotextile will not be permitted. End-dumping fill directly on the geotextile will not be permitted without approval of the Engineer.

Should the geotextile be torn, punctured, or the overlaps or sewn joints disturbed as evidenced by visible geotextile damage, subgrade pumping, intrusion, or distortion, the backfill around the damaged or displaced area shall be removed. The damaged geotextile section shall be replaced by the Contractor with a new section of geotextile at no expense to the Contracting Agency.

If geotextile seams are to be sewn in the field or at the factory, the seams shall consist of one row of stitching unless the geotextile where the seam is to be sewn does not have a selvage edge. If a selvage edge is not present, the seams shall consist of two parallel rows of stitching, or shall consist of a J-seam, Type Ssn-1, using a single row of stitching. The two rows of stitching shall be 25 mm apart with a tolerance of plus or minus 13 mm and shall not cross except for restitching. The stitching shall be a lock-type stitch. The minimum seam allowance, i.e., the minimum distance from the geotextile edge to the stitch line nearest to that edge, shall be 40 mm if a flat or prayer seam, Type Ssa-2, is used. The minimum seam allowance for all other seam types shall be 25 mm. The seam, stitch type, and the equipment used to perform the stitching shall be as recommended by the manufacturer of the geotextile and as approved by the Engineer.

The seams shall be sewn in such a manner that the seam can be inspected readily by the Engineer or his representative. The seam strength will be tested and shall meet the requirements stated herein.

A temporary form system shall be used to prevent sagging of the geotextile facing elements during construction. A typical example of a temporary form system and sequence of wall construction required when using this form are shown in the Plans.

Soil piles or the manufacturer's recommended method, in combination with the forming system, shall be used as needed to hold the geotextile in place until the specified cover material is placed.

The wall backfill shall be placed and compacted in accordance with the wall construction sequence shown in the Plans. The minimum compacted backfill lift thickness of the first lift above each geotextile layer shall be 150 mm. The maximum compacted lift thickness anywhere within the wall shall be 250 mm.

Each layer shall be compacted to 95 percent of maximum density. The water content of the wall backfill shall not deviate above the optimum water content by more than 3 percent. Sheepsfoot rollers, other rollers with protrusions, and full-size vibratory rollers will not be allowed. Small vibratory rollers will be allowed with the approval of the Engineer. Compaction within 1 meter of the wall face shall be achieved using light mechanical tampers approved by the Engineer and shall be done in a manner to cause no damage or distortion to the wall facing elements or reinforcing layers.

If corners must be constructed in the geotextile wall due to abrupt changes in alignment of the wall face as shown in the Plans, the method used to construct the geotextile wall corner(s) shall be submitted to the Engineer for approval at least 14 calendar days prior to beginning construction of the wall. The corner must provide a positive connection between the sections of the wall on each side of the corner such that the wall backfill material cannot spill out through the corner at any time during the design life of the wall. Furthermore, the corner must be constructed in such a manner that the wall can be constructed with the full geotextile embedment lengths shown in the Plans in the vicinity of the corner in both directions.

The method of ending a geotextile wall layer at the top of the wall where changes in wall top elevation occur shall also be submitted for approval with the wall corner details submittal. The end of each layer at the top of the wall must be constructed in a manner which prevents wall backfill material from spilling out the face of the wall throughout the life of the wall. If the profile of the top of the wall changes at a rate of 1:1 or steeper, this change in top of wall profile shall be considered to be a corner. Also, wall angle points with an interior angle of less than 150 degrees shall be considered to be a corner.

Shotcrete Wall Facing Construction
If the wall is to be a permanent structure, the entire wall face shall be coated with a reinforced shotcrete facing as detailed in the Plans and as described herein.

Alignment Control
Non-corroding alignment wires and thickness control pins shall be provided to establish thickness and plane surface. Alignment wires shall be installed at corners and offsets not established by formwork. The Contractor shall ensure that the alignment wires are tight, true to line, and placed to allow further tightening. Alignment wires shall be removed after wall construction is complete.

Qualifications of Contractor's Personnel
All nozzlemen on this project shall have had at least one year of experience in the application of shotcrete. Each nozzleman will be qualified, by the Engineer, to place shotcrete on this project after successfully completing one test panel for each shooting position which will be encountered, and the results of the 7-day strength tests are known. No shotcrete shall be placed by personnel without these specific qualifications.

The Contractor shall notify the Engineer not less than 2 days prior to the shooting of a qualification panel. The mix design for the shotcrete shall be the same as that slated for the wall being shot.

Qualification will be based on a visual inspection of the shotcrete density, void structure, and finished appearance along with a minimum 7-day compressive strength of 17 MPa determined from the average test results from two cores taken from each panel.

If shotcrete finish Alternate B or C is specified by the Plans, all shotcrete crew members shall have completed at least three projects in the last five years where such finishing, or sculpturing and texturing, of shotcrete was performed. Evidence of this experience, in addition to approval of the shotcrete finish qualification panel, will form the basis for qualification of nozzlemen to perform the work as specified by shotcrete finish Alternate B or C. Shotcreting shall not begin until this qualification process is completed.

Placing Wire Reinforcement
Reinforcement of the shotcrete shall be placed as shown in the Plans. The wire reinforcement shall be securely fastened to the No. 3 epoxy coated rebar so that it will be 25 to 40 mm from the wall face at all locations. Wire reinforcement shall be lapped 1.5 squares in all directions.

Shotcrete Construction Requirements

A clean, dry supply of compressed air sufficient for maintaining adequate nozzle velocity for all parts for the work and for simultaneous operation of a blow pipe for cleaning away rebound shall be maintained at all times. Thickness, method of support, air pressure, and rate of placement of shotcrete shall be controlled to prevent sagging or sloughing of freshly-applied shotcrete.

The shotcrete shall be applied from the lower part of the area upwards so that rebound does not accumulate on the portion of the surface that still has to be covered. Surfaces to be shot shall be damp but not have free standing water. No shotcrete shall be placed on dry, dusty, or frosty surfaces. The nozzles shall be held at a distance and at an angle approximately perpendicular to the working face so that rebound will be minimal and compaction will be maximized. Shotcrete shall emerge from the nozzle in a steady uninterrupted flow. When, for any reason, the flow becomes intermittent, the nozzle shall be diverted from the work until a steady flow resumes.

Surface defects shall be repaired as soon as possible after initial placement of the shotcrete. All shotcrete which lacks uniformity; which exhibits segregation, honeycombing, or lamination; or which contains any dry patches, slugs, voids, or sand pockets, shall be removed and replaced with fresh shotcrete by the Contractor to the satisfaction of the Engineer at no additional expense to the Contracting Agency.

Construction joints in the shotcrete shall be uniformly tapered over a minimum distance of twice the thickness of the shotcrete layer and the surface of the joints shall be cleaned and thoroughly wetted before any adjacent shotcreting is performed. Shotcrete shall be placed in a manner which provides a shotcrete finish with a uniform texture and color across the finished construction joint.

The shotcrete shall be cured by applying a clear curing compound as specified in Section 9-23. The curing compound shall be applied immediately after final gunning. The air in contact with shotcrete surfaces shall be maintained at temperatures above 4° C for a minimum of 7 days. Curing compounds shall not be used on any surfaces against which additional shotcrete or other cementitious finishing materials are to be bonded unless positive measures such as sandblasting, are taken to completely remove curing compounds prior to the applications of such additional materials.

If field inspection or testing as directed by the Engineer indicates that any shotcrete produced by the Contractor fails to meet the requirements of this Specification, the Contractor shall immediately modify procedures, equipment, or system, as necessary and as approved by the Engineer to produce specification material. All substandard shotcrete already placed shall be repaired by the Contractor to the satisfaction of the Engineer at no additional expense to the Contracting Agency. Such repairs may include removal and replacement of all affected materials.

Shotcrete Finishing

The shotcrete face shall be finished using the alternate aesthetic treatment indicated in the Plans. The alternates are as follows:

Alternate A: After the surface has taken its initial set (crumbling slightly when cut), the surface shall be broom finished to secure a uniform surface texture.

Alternate B: Place shotcrete a fraction beyond the alignment wires and forms. Allow shotcrete to stiffen to the point where the surface will not pull or crack when screeded with a rod or trowel. Excess material shall then be trimmed, sliced, or scraped to true lines and grade. Alignment wires shall then be removed. The surface shall receive a steel trowel finish, leaving a smooth uniform texture and color. Once the shotcrete has cured, pigmented sealer shall be applied to the shotcrete face in accordance with the Special Provisions. The shotcrete surface shall be completed to within a tolerance of plus or minus 25 mm of true line and grade.

Alternate C: Shotcrete shall be hand-sculptured, colored, and textured to simulate the relief, jointing, and texture of the natural backdrop surrounding the wall. The ends and base of the wall shall transition in appearance as appropriate to more nearly match the color and texture of the adjoining roadway fill slopes. This may be achieved by broadcasting fine and coarse aggregates, rocks, and other native materials into the final surface of the shotcrete while it is still wet, allowing sufficient embedment into the shotcrete to become a permanent part of the surface.

Installation of Guard Rail Posts

Guard rail posts, if specified in the Plans, shall be installed after completion of the wall and prior to construction of the shotcrete wall facing. The posts shall be installed in a manner that will prevent bulging of the wall face and in a manner that will prevent ripping, tearing, or pulling of the geotextile reinforcement. Holes through the geotextile shall be the minimum size necessary for the post. The Contractor shall demonstrate to the Engineer prior to beginning installation of the guard rail posts that his installation method will not rip, tear, or pull the geotextile.

Measurement

Geotextile retaining wall will be measured by the square meter of face of completed wall. Shotcrete wall facing will be measured by the square meter of completed shotcrete wall facing.

Payment

The unit contract prices per square meter for "Geotextile Retaining Wall" and "Shotcrete Wall Facing", per 1,000 kg for "Gravel Borrow Incl. Haul", and per cubic meter for "Structure Excavation Class A" shall be full pay to complete the work in accordance with these Specifications, including compaction of the backfill material and the temporary forming system.

9.11   CONSTRUCTION PROCEDURES

9.11-1 Concrete Faced Wall (after Elias and Christopher, 1995)

The construction of a geosynthetic reinforced system with precast facing elements requires the following steps.

A. Prepare Subgrade

  1. The foundation should be excavated to the grade as shown on the plans.

  2. The excavated areas should be carefully inspected. Any unsuitable foundation soils should be compacted or excavated and replaced with compacted select backfill material.

  3. Foundation soil at the base of the wall excavation should be proofrolled with a vibratory or rubber-tired roller.

B. Leveling Pad

  1. A cast-in-place or precast concrete leveling pad should be placed at the foundation elevation for all reinforced structures with concrete (panel and SRW) facing elements. The unreinforced concrete pad is often only 0.3 m wide and 0.15 m thick. The purpose of the pad is to serve as a guide for facing panel erection and not to act as a structural foundation support.

C. Erection of Facing Units

  1. The first row of facing panels may be full- or half-height panels, depending upon the type of facing utilized. The first tier of panels must be shored up to maintain stability and alignment. For construction with SRWs, full-sized blocks are used throughout, with no shoring required.

  2. Erection of subsequent rows of facing panels proceed incremental with fill placement and compaction.

D. Backfill Placement and Compaction

  1. The backfill material should be placed over a compacted lift thickness as specified.
  2. The backfill material should be compacted to specified density, usually 95 to 100% of standard Protor maximum density.
  3. A key to good performance is consistent compaction. Wall fill lift thickness must be controlled, based upon specification requirements and vertical distribution of reinforcement elements (and incremental face unit height).

E. Reinforcement Placement

   1. The geosynthetic reinforcements are placed and connected to the facing units when the fill has been brought up to the level of the connection. Reinforcement is generally placed perpendicular to the back of the facing units.

F. Placement of Fill on Reinforcement

   1. Geosynthetic reinforcement shall be pulled taut and anchored prior to placing fill.

   2. Fill placement and spreading should prevent or minimize formation of wrinkles in the geosynthetic. Wrinkles or slack near the facing connection should be prevented as they can result in differential movement of the wall face.

   3. A minimum thickness of 150 mm of fill must be maintained between the tracks of the construction equipment and the reinforcement at all times.

9.11-2 Geotextile Wrap Around Wall

Construction procedures for geosynthetic MSE walls are straightforward. Experience of the U.S. Forest Service, the New York Department of Transportation, and the Colorado Department of Highways has been very valuable in developing technically feasible and economical construction procedures. These procedures, detailed in Christopher and Holtz (1985), as well as the other references, are outlined below.

A. Wall Foundation

   1. The foundation should be excavated to the grade as shown on the plans. It should be graded level for a width equal to the length of reinforcement plus 0.3 m.

   2. The excavated areas should be carefully inspected. Any unsuitable foundation soils should be compacted or excavated and replaced with compacted select backfill material.

   3. Foundation soil at the base of the wall excavation should be proofrolled with a vibratory or rubber-tired roller.

B. Placement of geosynthetic reinforcement

1. The geosynthetic should be placed with the principal strength (machine) direction perpendicular to the face of the wall. It should be secured in place to prevent movement during fill placement.

2. It may be more convenient to unroll the geosynthetic with the machine direction parallel to the wall alignment. If this is done, then the cross-machine design tensile strength must be greater or equal to the design tensile strength.

3. A minimum of 150 mm overlap is recommended along edges parallel to the reinforcement for wrapped-faced walls.

4. If large foundation settlements are anticipated, which might result in separation between overlap layers, then field-sewing of adjacent geotextile sheets is recommended. Geogrids should be mechanically fastened in that direction.

C. Backfill placement in reinforced section

1. The backfill material should be placed over the reinforcement with a compacted lift thickness of 200 mm or as determined by the Engineer.

2. The backfill material should be compacted to at least 95% of the standard Proctor maximum density at or below the optimum moisture. Alternatively, a relative density compaction specification could be used. For coarse, gravelly backfills, a method-type compaction specification is appropriate.

3. When placing and compacting the backfill material, avoid any folding or movement of the geosynthetic.

4. A minimum thickness of 150 mm of fill must be maintained between the wheels of the construction equipment and the reinforcement at all times.

D. Face construction and connections

1. Place the geosynthetic layers using face forms as shown in Figure 9-15, unless a precast propped panel facing is to be used.

2. When using temporary support of forms at the face to allow compaction of

the backfill against the wall face, form holders should be placed at the base of each layer at 1 m horizontal intervals. Details of temporary form work for geosynthetics are shown in Figure 9-15.

3. When using geogrids, it may be necessary to use a geotextile or wire mesh to retain the backfill material at the wall face (Figure 9-16).

4. A hand-operated vibratory compactor is recommended when compacting backfill within 0.6 m of the wall face.

5. The return-type method shown in Figure 9-13a can be used for facing support. The geosynthetic is folded at the face over the backfill material, with a minimum return length of 1 m to ensure adequate pullout resistance.

6. Apply facing treatment (shotcrete, precast facing panels, etc.). Figures 9-2 and 9-13 shows several facing alternatives for geosynthetic walls.

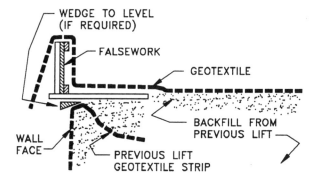

(1) PLACE FALSEWORK AND GEOTEXTILE
ON PREVIOUS LIFT

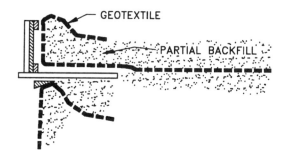

(2) PLACE/COMPACT PARTIAL BACKFILL
AND OVERLAP GEOTEXTILE

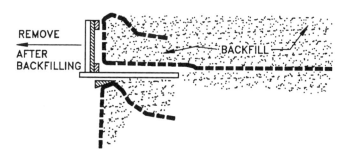

(3) PLACE/COMPACT REMAINDER OF
BACKFILL LIFT

Figure 9-15   *Lift construction sequence for geotextile reinforced soil walls
(Stewart et al., 1977).*

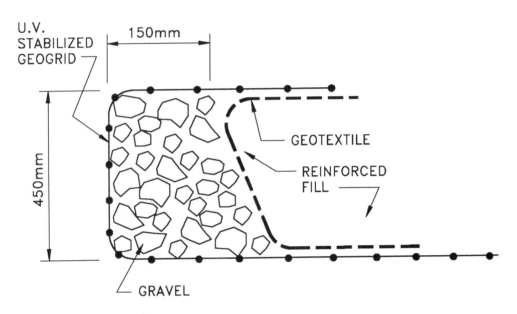

Figure 9-16   Typical face construction detail for vertical geogrid-reinforced retaining wall faces.

9.12   INSPECTION

As with all geosynthetic construction, and especially with critical structures such as geosynthetic reinforced walls and abutments, competent and professional construction inspection is absolutely essential for a successful project. The Engineer should develop procedures to ensure that the specified material is delivered to the project, that the geosynthetic is not damaged during construction, and that the specified sequence of construction operations are explicitly followed. Inspectors should use the checklist in Section 1.7. Other important details include construction of the wall face and application of the facing treatment to minimize the exposure of the geosynthetic to ultraviolet light.

Because geosynthetic reinforced retaining walls are sometimes considered experimental, they often are instrumented. In these cases, as a minimum, settlements and outward movements of the wall at its top should be determined by ordinary levels and triangulation surveys. Sometimes inclinometers and/or multiple-point extensometers are used for observing potential horizontal movements. On major projects, strain gages are placed on the geosynthetic to measure internal strains (Allen et al., 1992).

9.13   IMPLEMENTATION

There are two primary issues regarding implementation of geosynthetic wall technology in public works agencies. They are: determining who will complete and be responsible for final design; and specifying determination of the allowable and design tensile strengths of geosynthetic reinforcement.

Reinforced walls may be contracted on the basis of:
- in-house (agency) or consultant design with geosynthetic reinforcement, facing, drainage, and construction execution specified; or
- system or end-result design approach using approved systems with lines and grades noted on the drawings.

Both options are acceptable. The in-house or consultant option allows more facing and reinforcement options to be considered during design. This option requires an engineering staff or design consultant that is knowledgable about reinforced soil technology. This staff would also be valuable during construction, when questions and/or design modification requests arise.

The end-result approach, with sound specifications and prequalification of suppliers and materials, offers several benefits as well as some major disadvantages. Design of the structure is completed by trained and experienced engineers and the prequalified material components of geosynthetic and facing units have been successfully and routinely used together, which may not be the case for an in-house design with generic specification of these two components. Also, the system specification approach lessens the engineering requirements for an agency, and transfers some of a project's design cost to construction.

However, besides questions of liability, vendor-supplied designs typically assume a stable foundation and adequate overall slope stability, that no changes in materials or procedures will occur during construction, and that operational or environmental factors will not adversely influence wall performance or function. An excellent case history illustrating the drawbacks of a vendor-supplied design has been reported by Leonards, Frost, and Bray (1994). This case history should be required reading by all owners comtemplating purchase of a vendor-supplied designs.

Another issue difficult for many agencies and consultants is the evaluation and specification of the allowable and design tensile strengths of geosynthetic reinforcement. The procedure for evaluation, as summarized in this chapter and Chapter 8 on reinforced slopes, is from Berg (1993). This procedure is detailed in Appendix F.

This recommended procedure is based upon the assumption that materials will be prequalified and listed on an approved products list specification. The recommended requirements for supplier submissions and for agency review, and recommended delineation of responsibilities within a typical agency, are presented in Berg (1993). This procedure has been cumbersome for agencies that do not use approved products lists or that review and approve products based upon specific project submittals.

Because of these implementation problems, an alternative procedure for determining long-term allowable design strength of geosynthetic soil reinforcing elements is also presented in Appendix F. This alternative procedure is meant to complement, and not supersede, the detailed procedure of testing and evaluation of geosynthetic reinforcement materials.

9.13   REFERENCES

AASHTO (1992), Standard Specifications for Highway Bridges, Fifteenth Edition, with 1993 and 1994 Interims, American Association of State Transportation and Highway Officials, Washington, D.C.

AASHTO (1990), Design Guidelines for Use of Extensible Reinforcements (Geosynthetic) for Mechanically Stabilized Earth Walls in Permanent Applications, Task Force 27 Report - In Situ Soil Improvement Techniques, American Association of State Transportation and Highway Officials, Washington, D.C.

Allen, T.M., Christopher, B.R. and Holtz, R.D. (1992), Performance of a 12.6 m High Geotextile Wall in Seattle, Washington, Geosynthetic Reinforced Soil Retaining Walls, J.T.H. Wu Editor, A.A. Balkema, Rotterdam, pp. 81-100.

Allen, T.M. and Holtz, R.D. (1991), Design of Retaining Walls Reinforced with Geosynthetics, Geotechnical Special Publication No. 27, Proceedings of ASCE Geotechnical Engineering Congress, American Society of Civil Engineers, New York, Vol. II, pp. 970-987.

Allen, T.M. (1991), Determination of Long-Term Tensile Strength of Geosynthetics: A State-of-the-Art Review, Proceedings of Geosynthetics '91 Conference, Atlanta, GA, Vol. 1, pp. 351-380.

Bathurst, R. J. and Simac, M.R. (1994), Geosynthetic Reinforced Segmental Retaining Wall Structures in North America, Preprint of Special Lecture and Keynote Lectures of Fifth International Conference on Geotextiles, Geomembranes and Related Products, Singapore, September, 1994, pp. 31-54.

Bell, J.R., Barrett, R.K. and Ruckman, A.C. (1983), Geotextile Earth-\Reinforced Retaining Wall Tests: Glenwood Canyon, Colorado, Transportation Research Record 916, pp. 59-69.

Bell, J.R., Stilley, A.N. and Vandre, B. (1975), Fabric Retained Earth Walls, Proceedings of the 13th Annual Engineering Geology and Soils Engineering Symposium, Moscow, ID, April, pp. 271-287.

Berg, R.R. (1993), Guidelines for Design, Specification, & Contracting of Geosynthetic Mechanically Stabilized Earth Slopes on Firm Foundations, Report No. FHWA-SA-93-025, Federal Highway Administration, Washington, D.C., 87 p.

Berg, R.R. (1991), The Technique of Building Highway Retaining Walls, Geotechnical Fabrics Report, Industrial Fabrics Association International, St. Paul, MN, July, pp. 38-43.

Bonaparte, R. and Berg, R.R. (1987), Long-Term Allowable Tension for Geosynthetic Reinforcement, Proceedings of Geosynthetics '87 Conference, Volume 1, New Orleans, LA, pp. 181-192.

Burwash, W.J. and Frost, J.D. (1991), Case History of a 9 m High Geogrid Reinforced Retaining Wall Backfilled with Cohesive Soil, Proceedings of Geosynthetics '91 Conference, Atlanta, GA, Vol. 1, pp. 485-493.

Carroll, R.G. and Richardson, G.N. (1986), Geosynthetic Reinforced Retaining Walls, Proceedings of The Third International Conference on Geotextiles, Austria, Vienna, Vol. II, pp. 389-394.

Cedergren, H.R. (1989), Seepage, Drainage, and Flow Nets, Third Edition, John Wiley and Sons, New York, 465 p.

Christopher, B.R., Gill, S.A., Giroud, J.P., Juran, I. Scholsser, F., Mitchell, J.K. and Dunnicliff, J. (1990), Reinforced Soil Structures, Volume I. Design and Construction Guidelines, Federal Highway Administration, Washington, D.C., Report No. FHWA-RD--89-043, 287 p.

Christopher, B.R. and Holtz, R.D. (1989), Geotextile Design and Construction Guidelines, Federal Highway Administration, National Highway Institute, Report No. FHWA-HI-90-001, 297 p.

Christopher, B.R. and Holtz, R.D. (1985), Geotextile Engineering Manual, Report No. FHWA-TS-86/203, Federal Highway Administration, Washington, D.C., March, 1044 p.

Claybourn, A.F. and Wu, J.T.H. (1993), Geosynthetic-Reinforced Soil Wall Design, Geotextiles and Geomembranes, Vol. 12, No. 8, pp. 707-724.

Elias, V. and Christopher, B.R. (DRAFT - 1995), Demo 82 Manual, Federal Highway Administration, Washington, D.C.

GRI Test Method GG4a (1990), Determination of Long-Term Design Strength of Stiff Geogrids, Geosynthetic Research Institute, Drexel University, Philadelphia, PA, March.

GRI Test Method GG4b (1991), Determination of Long-Term Design Strength of Flexible Geogrids, Geosynthetic Research Institute, Drexel University, Philadelphia, PA, January.

GRI Test Method GT7 (1992), Determination of Long-Term Design Strength of Geotextiles, Geosynthetic Research Institute, Drexel University, Philadelphia, PA, April.

Holtz, R.D., Allen, T.M. and Christopher, B.R. (1991), Displacement of a 12.6 m High Geotextile-Reinforced Wall, Proceedings of the Tenth European Conference on Soil Mechanics and Foundation Engineering, Florence, Italy, Vol. 2, pp. 725-728.

Jaky, J.(1948), Earth Pressure in Silos, Proceedings of the Second International Conference on Soil Mechanics and Foundation Engineering, Rotterdam, Vol. I, pp.103-107.

Leflaive, E. (1988), Durability of Geotextiles: The French Experience, Geotextiles and Geomembranes, Vol. 7, Nos. 1 & 2, Elsevier Applied Sciences, England, pp.23-31.

Leonards, G.A., Frost, J.D., and Bray, J.D. (1994), Collapse of Geogrid-Reinforced Retaining Structure, Journal of Performance of Constructed Facilities, American Society of Civil Engineers, Vol. 8, No. 4, pp. 274-292.

Mitchell, J.K. and Villet, W.C.B. (1987), Reinforcement of Earth Slopes and Embankments, NCHRP Report No. 290, Transportation Research Board, Washington, D.C.

Perloff, W.H. and Baron, W. (1976), Soil Mechanics: Principles and Applications, Ronald Press, 504 p.

Richardson, G.N. and Behr, L.H., Jr. (1988), Geotextile-Reinforced Wall: Failure and Remedy, Geotechnical Fabrics Report, Industrial Fabrics Association International, Vol. 6, No. 4, July-August, pp. 14-18.

Simac, M.R., Bathurst, R.J., Berg, R.R. and Lothspiech, S.E. (1993), Design Manual for Segmental Retaining Walls, National Concrete Masonry Association, Herdon, VA, 336 p.

Steward, J., Williamson, R. and Mohney, J. (1977), Guidelines for Use of Fabrics in Construction and Maintenance of Low-Volume Roads, USDA, Forest Service, Portland, OR. Also reprinted as Report No. FHWA-TS-78-205.

Terzaghi, K. and Peck, R.B. (1967), Soil Mechanics in Engineering Practice, John Wiley & Sons, New York, 729 p.

Vrymoed, J. (1990), Dynamic Stability of Soil Reinforced Walls, Transportation Research Record 1242, pp. 29-38.

Wrigley, N.E. (1987), Durability and Long-Term Performance of 'Tensar' Polymer Grids for Soil Reinforcement, Materials Science and Technology, Vol. 3, , pp. 161-170.

Wu, T.H. (1975), Retaining Walls, Chapter 12 in Foundation Engineering Handbook, Winterkorn and Fang, Editors, Von Nostrand Reinhold, pp. 402-417.

U.S. Department of the Navy (1982), Soil Mechanics, Design Manual 7.1, Naval Facilities Engineering Command, Alexandria, VA.

U.S. Department of the Navy (1982), Foundations and Earth Structures, Design Manual 7.2, Naval Facilities Engineering Command, Alexandria, VA.

10.0 GEOMEMBRANES AND OTHER GEOSYNTHETIC BARRIERS

10.1 BACKGROUND

Barriers are used in earthwork construction to control movement of water, other liquids and sometimes vapors. Barriers are used to waterproof structures, to prevent moisture changes beneath roadways, to contain water and wastes, and to support other applications in transportation works. The function of these barriers is to either prevent damage to highway pavements and structures or to contain water or waste materials. Barriers must be engineered to perform their intended function for the particular application and project being designed.

Traditional barriers, or liners, are field-constructed of soil or aggregate-based materials. Thick compacted clay layers, cast-in-place concrete, and asphalt concrete are used to construct liners. Another conventional liner material is geomembranes, which have been used in transportation applications for more than forty years. The U.S. Bureau of Reclamation has been using geomembranes in water conveyance canals since the 1950s (Staff, 1984). Other types of geosynthetic barriers have also been used in transportation applications. These include thin-film geotextile composites, geosynthetic clay liners, and field-impregnated geotextiles.

While soil or aggregate-based liners are well-suited to some applications, geomembrane and other geosynthetic barriers are more appropriate for other projects. Suitability may be defined during design and with due consideration to material availability, long-term performance, and cost. For example, rigid concrete and asphalt liners or semi-stiff compacted clay liners are not well-suited to sites where barriers are subject to foundation settlements; conversely a geomembrane which has adequate flexibility would be suitable.

10.2 GEOSYNTHETIC BARRIER MATERIALS

Geosynthetic barrier materials can be classified as geomembranes, thin-film geotextile composites, geosynthetic clay liners, or field-impregnated geotextiles. Materials within each classification are reviewed herein, starting with a general material definition. The components, manufacturing processes, resultant product characteristics, and typical dimensions are presented. Several test standards are available to quantify property values.

10.2-1 Geomembranes

The term geomembrane is defined as a very low-permeability synthetic membrane liner or barrier used with any geotechnical engineering-related material to control fluid migration in a man-made project, system, or structure (ASTM D 4439). However, within this document the term geomembrane will be used to specifically describe materials which are manufactured of continuous polymeric sheets. Commonly available geomembranes are manufactured of the polymers listed in Table 10-1.

Table 10-1
Common Types of Geomembranes

POLYMER TYPE	ABBREVIATION	AVAILABLE	
		WITHOUT SCRIM REINFORCEMENT	WITH SCRIM REINFORCEMENT
Chlorinated Polyethylene	CPE	✓	✓
Chlorosulfonated Polyethylene	CSPE	✓	✓
Ethylene Interpolymer Alloy	EIA		✓
High-Density Polyethylene	HDPE	✓	
Polypropylene	PP	✓	✓
Polyvinyl Chloride	PVC	✓	✓
Very Low-Density Polyethylene	VLDPE	✓	

The HDPE and VLDPE geomembranes are supplied in roll form. Widths of approximately 4.6 to 10.5 m are available. Roll lengths of 200 to 300 m are typical, although custom roll lengths are available. These materials are generally available in sheet thicknesses of 1.0, 1.5, 2.0, and 2.5 mm. These PE geomembranes are usually of singular (*i.e.,* not composite) manufacture. However, thick coextruded PE composites are available with light-colored heat reflective surfaces, electrical conductive surfaces (for leakage testing), or a VLDPE layer sandwiched between an upper and lower layer of HDPE.

Geomembranes manufactured of CPE, CSPE, PVC, and PP are supplied in large panels, accordion-folded onto pallets. The panels are fabricated by factory-seaming of rolls, typically 1.4 to 2.5 m in width. Panels as large as 1,800 m^2 are available. These geomembranes may be of singular manufacture or composites with a fabric scrim incorporated to modify the mechanical properties (Table 10-2).

Geomembranes are relatively impermeable materials - *i.e.*, all materials are permeable, but the permeability of geomembranes (on the order of 10^{-14} m/s) is significantly lower than that of compacted clays. Hence, geomembranes are sometimes referred to as being impermeable, relative to soil. Theoretically, multiple layers of geomembranes and drains can be utilized to construct impermeable structures (Giroud, 1984).

Leakage, and not permeability, is the primary concern when designing geomembrane containment structures. Leakage can occur through poor field seams, poor factory seams, pinholes from manufacture, and puncture holes from handling, placement, or in-service loads. Leakage of geomembrane liner systems is minimized by attention to design, specification, testing, quality control (QC), quality assurance (QA) of manufacture, and QC and QA of construction.

Geomembrane materials are better-defined than other geosynthetic barriers, due to their widespread use in environmental applications and their historical use in other applications. Manufacturing QC/QA standards, index property test methods, performance test methods, design requirements, design detailing, and construction QC/QA are well-established (Daniel and Koerner, 1993).

10.2-2 Thin-Film Geotextile Composites

The moisture barrier commonly used in roadway reconstruction is a thin-film geotextile composite. These composites are used to prevent or minimize moisture changes in pavement subgrades, as discussed in Section 10.3. Two styles of composites, as illustrated in Figure 10-1, are available.

One commercially available composite consists of a very lightweight PP nonwoven geotextile sandwiched between two layers of PP film (Figure 10-1a). This product has a mass per unit area of 140 g/m^2 and is available in roll widths of 2.3 m and roll lengths of 91 m.

Another commercially available composite consists of a polyethylene (PE) film sandwiched between two layers of nonwoven PP geotextiles (Figure 10-1b). This product has a mass per unit area of 300 g/m^2 and is available in roll widths of 3.65 m and roll lengths of 91 m. Similar products are available with PVC, CSPE, and PP geomembranes as the core.

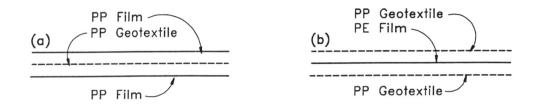

Figure 10-1   *Thin-film geotextile composites: (a) PP film / PP geotextile / PP film; and (b) PP geotextile - PE film - PP geotextile.*

10.2-3 Geosynthetic Clay Liners

Geosynthetic clay liners (GCLs) are another type of composite barrier materials. A dry bentonite clay soil is supported between two geotextiles or on a geomembrane carrier, as illustrated in Figure 10-2. Geotextiles used above and below the dry clay may or may not be connected with threads or fibers, to increase the in-plane shear strength of a hydrated GCL.

Approximately 5 kg/m^2 of *dry* sodium bentonite is used in the manufacture of GCLs. The bentonite is at a moisture content of 6 to 20% in its *dry* condition. This *dry* bentonite hydrates and swells upon wetting, creating a very-low permeability barrier. The fully hydrated bentonite typically will have a permeability in the range of 1 to 5 x 10^{-11} m/sec.

GCLs are supplied in roll form. Widths of approximately 4.1 to 5.3 m are available. Roll lengths of 30 to 60 m are typical, although custom roll lengths may be used for large projects.

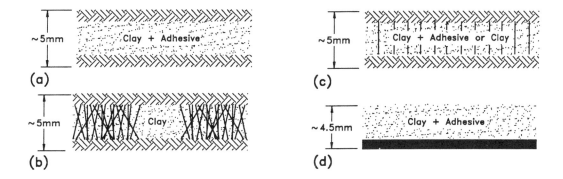

Figure 10-2   Geosynthetic clay liners: (a)   geotextile / bentonite clay / geotextile; (b)   stitched bonded geotextile GCL; (c) needle punched geotextile GCL; and (d) bentonite clay / PE geomembrane (after Koerner, 1994).

10.2-4 Field-Impregnated Geotextiles

Impregnated geotextiles are also used as moisture and liquid barriers. The coating treatment is applied in the field, after the geotextile is deployed and anchored. A nonwoven geotextile is used with a variety of coatings, including asphalt, rubber-bitumen, emulsified asphalt, or polymeric formulations. The coating may be proprietary. The geotextile's type and mass per unit area will be a function of the coating treatment, although use of lightweight nonwoven geotextiles, in the range of 200 to 400 g/m^2, is common. Heavier-weight, nonwoven geotextiles may be used to provide gas venting, if gas potential exists on a site.

The barrier is formed as sprayed-on liquid solidifies into a seam-free membrane. Although sprayed-on membranes are seam-free, bubbles and pinholes may form during installation and can cause performance problems. Proper preparation of the geotextile (*i.e.*, clean and dry) to be sprayed is important. These types of barriers have been used in canals, small reservoirs, and ponds for water control. Water storage applications have used air-blown asphalt coatings. (Matrecon, 1988)

Engineers also use field-impregnated geotextiles to provide moisture control in friable roadway soils. Pavement application of a barrier is called membrane encapsulated soil layers (MESLs).

10.3   APPLICATIONS

Geomembranes and other geosynthetic barriers are used in wide variety of applications for transportation construction and maintenance. Geomembrane transportation applications, as summarized by Koerner and Hwu (1989), are summarized below. The applications are noted as representing regular use or limited use of geomembranes and other synthetic barriers in highway works.

- Control of vertical infiltration of moisture into a subgrade of expansive soil. This minimizes the change in soil water content and subsequent volume changes. Placement of the geosynthetic barrier is illustrated in Figure 10-3. Thin-film geotextile composites or geomembranes are often used in this application.

- Control of horizontal infiltration of moisture into a subgrade of expansive soil. This minimizes change in soil water content and volume changes. Placement of the barrier is illustrated in Figure 10-4. Depth of moisture barrier is approximately 450 to 600 mm beneath estimated swell depth, or a typical total depth of 1.5 to 2.5 m. Thin-film geotextile composites or geomembranes are usually used in this application, although geomembranes and field-impregnated geotextiles are also used.

- Maintenance of water content of frost-sensitive soils with a horizontally placed barrier. This application is illustrated in Figure 10-5, as the MESL previously described. Thin-film geotextile composites or geomembranes are usually used in this application.

- Waterproofing of tunnels, as illustrated in Figure 10-6. Geomembranes (in conjunction with heavyweight nonwoven geotextiles) and GCLs are used in this application.

- Transport of water in canals lined with a geomembrane, as illustrated in Figure 10-7, or a GCL.

- Geomembranes are used for secondary containment of underground fuel storage tanks, as illustrated in Figure 10-8. GCLs and geomembranes are also used for secondary containment of above ground fuel storage tanks.

- Rest area waste water treatment lagoons may be lined with geomembranes.

- Sealing of berms for wetland mitigation.

- Storm water retention and detention ponds also may be lined with geomembranes, GCLs, or coated geotextiles.

- Geomembranes are used beneath structures as methane and radon gas barriers.

- Geomembranes are used for containment of waste, caustic soils (*e.g.*, pyritic soils) and construction debris.

- Deicing salt and aviation deicing fluid runoff may be contained in geomembrane-lined facilities.

- Geomembranes, GCLs, and coated geotextiles may be used to waterproof walls and bridge abutments. Geomembranes may be used to prevent infiltration of corrosive deicing salt runoff into metallic MSE walls, as illustrated in Figure 10-9.

- Railroads use geosynthetic barriers to waterproof subgrades, to prevent upward groundwater movement in cuts, and to contain diesel spills in refueling areas.

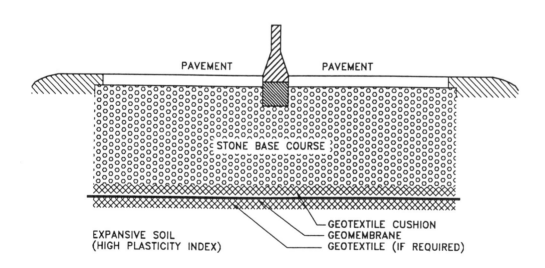

Figure 10-3   *Control of expansive soils (from Koerner and Hwu, 1989).*

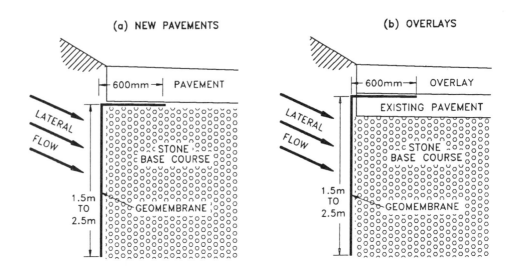

Figure 10-4   Control of horizontal infiltration of base.

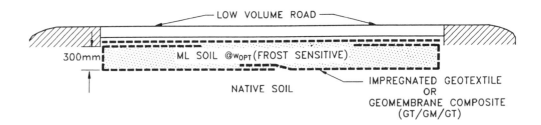

Figure 10-5   Maintenance of optimum water content (from Koerner and Hwu, 1989).

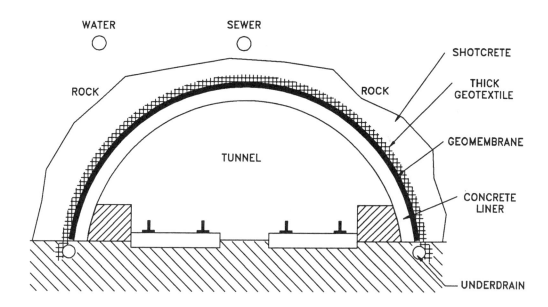

Figure 10-6   Waterproofing of tunnels (from Koerner and Hwu, 1989).

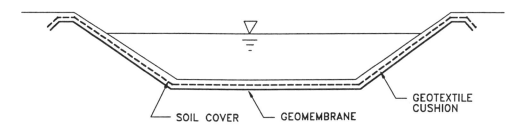

Figure 10-7   Water conveyance canals.

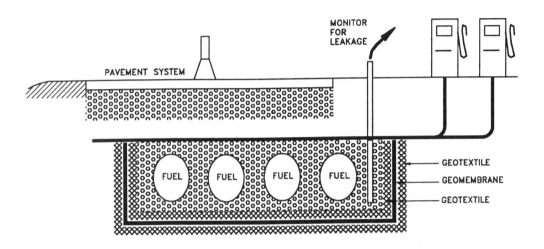

Figure 10-8   Secondary containment of underground fuel tanks (from
                Koerner and Hwu, 1989).

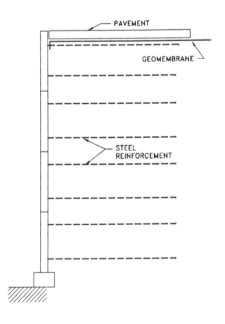

Figure 10-9   Waterproofing of walls (from Koerner and Hwu, 1989).

10.4   DESIGN CONSIDERATIONS

All geosynthetic barriers are continuous materials which are relatively impermeable as manufactured. However, to fulfill its barrier function the geosynthetic system must remain leak-proof and relatively impermeable when installed and throughout its design life. The following steps are part of the design process for geosynthetic barrier systems:

- define performance requirements;
- design for in-service conditions;
- durability design for project-specific conditions;
- design for installation, under anticipated project conditions;
- peer review (optional); and
- economic analysis.

10.4-1 Performance Requirements

The required function and performance of a geosynthetic barrier must be defined prior to design and material selection. The purpose of the barrier significantly affects the design and installation requirements. Some of the questions that should be asked regarding performance include the following.

- Is the barrier functioning as a primary or a secondary liner?
- Is a barrier less-permeable than adjacent soil required, or is an impermeable barrier system (liner(s) and drain(s)) required?
- What are the consequences of leakage?
- Can an acceptable leakage rate be defined?
- What is the anticipated life of the system, *i.e.*, is it temporary or permanent?

10.4-2 In-Service Conditions

Applications of a barrier vary, and the in-service exposure and stresses that the barrier must withstand likewise vary. Some of the questions that should be addressed regarding in-service conditions include the following.

- Will the barrier be placed within soil, or remain exposed to environmental elements throughout the design life?
- What environmental conditions (e.g., temperature variations, sunlight exposure, etc.) will the geosynthetic barrier be exposed to throughout its design life?
- Will the barrier be subject to deformation-controlled (*e.g.*, due to post-construction movements caused by settlement of underlying soil) stresses ?
- Will the geosynthetic barrier be subject to downdrag forces (*e.g.*, on a side slope of a surface impoundment) or to load-controlled stresses?
- Will the barrier system be exposed to varying stress levels due to fluctuating water loads?
- Will the barrier system result in a low-friction interface that must be analyzed?
- Will the barrier trap gases and/or liquids generated beneath the liner, and require venting?

- Are performance requirements such that abrasion or puncture protection is required?

10.4-3 Durability

Many geomembranes and other geosynthetic barriers in transportation applications contain nonpolluted water. As such, geosynthetic chemical durability is not normally a concern. However, chemical resistance is a concern when liquids such as fuel or other contaminants must be contained. The chemical resistance of candidate geosynthetic barriers, as well as of their components (if applicable), must be specifically evaluated when other than nonpolluted water is to be contained. The EPA 9090 test is available for such an assessment. Available geosynthetic barriers have a wide range of chemical resistance to various elements and compounds.

Resistance to ultraviolet light must be assessed for those applications where the barrier remains exposed over its design life. Oxidative or hydrolytic degradation potential also may be assessed. Biological degradation potential should also be checked. Degradation due to vegetation growth, burrowing animals, or microorganisms may be a concern. Biological degradation of materials beneath a liner can result in gas formation that must be vented around or through the liner.

10.4-4 Installation Conditions

Installation conditions are a design consideration for all geosynthetics in all applications. Installation of geomembranes and other geosynthetic barriers is a primary design consideration. Location and installation time of year can affect barrier material selection. Environmental factors such as temperature, temperature variation, humidity, rainfall, and wind must be considered. Some geosynthetics are more sensitive to temperature than others, and moisture and wind affect field-seaming ability. Barriers constructed of field-impregnated coated geotextiles must be placed during carefully defined weather conditions.

Placement, handling, and soil covering operations can also affect geosynthetic design and selection. The panel weight and size must be compatible with project requirements and constraints. The timing of soil placement over the liner may dictate ultraviolet light resistance requirements. And the geomembrane or other geosynthetic barrier must be capable of withstanding the rigors of installation.

The subgrade material, subgrade preparation, panel deployment method, overlying soil fill type, and placement and compaction of overlying fill soil all affect the geosynthetic barrier's survivability. Recommended properties of geomembrane barriers (Koerner, 1994) are presented in Table 10-2.

Table 10-2

Recommended Minimum Properties for General Geomembrane Installation Survivability

(from Koerner, 1994)

PROPERTY AND TEST METHOD	REQUIRED DEGREE OF SURVIVABILITY			
	Low[1]	Medium[2]	High[3]	Very high[4]
Thickness, mm - ASTM D 1593	0.63	0.75	0.88	1.00
Tensile (25 mm strip), kN/m - ASTM D 882	7	8.7	10.5	12.2
Tear (Die C), N - ASTM D 1004	33	45	67	90
Puncture, N - ASTM D 3787, modified	110	130	160	180
Impact, J - ASTM D 3998, modified	9	11	15	19

NOTES
1. Low refers to careful hand-placement on very uniform, well-graded subgrade with light loads of a static nature - typical of vapor barriers beneath building floor slabs.
2. Medium refers to hand- or machine-placement on machine-graded subgrade with medium loads - typical of canal liners.
3. High refers to hand- or machine-placement on machine-graded subgrade of poor texture with high loads - typical of landfill liners and covers.
4. Very high refers to hand- or machine-placement on machine-graded subgrade of very poor texture with high loads - typical of reservoir covers and liners for heap leach pads.

Geotextiles and other geosynthetics are often used with geomembranes to enhance the barrier's puncture resistance during installation and in-service. Geotextiles, and other geosynthetics, act as cushions and further prevent puncture of the geomembrane. The cushion can be placed below the geomembrane to resist rocks, roots, etc., in the subgrade, and/or above the geomembrane to resist puncture from subsequently placed fill or waste.

Selection of the most-effective geotextile will depend upon several factors (Richardson and Koerner, 1990), including:

- mass per unit area;
- geotextile type;
- fiber type;
- thickness under load;
- polymeric type; and
- geomembrane type and thickness.

The level of puncture protection provided by the geotextile is directly related to the mass per unit area.

10.4-5 Peer Review

A peer, or design quality assurance, review is recommended for landfill barrier systems (Rowe and Giroud, 1994). A peer review of a geosynthetic barrier structure for a transportation application may likewise be warranted, depending upon the critical nature of the structure, experience of design team, and project location and function. The goal of such a project review is to enhance the quality of the constructed project.

A peer review is recommended (Berg, 1993) for:
- projects where performance is crucial to public safety and/or the environment;
- projects that are controversial or highly visible;
- proposed designs that incorporate new materials or construction techniques;
- projects requiring state-of-the-art expertise;
- designs that lack redundancy in primary components;
- designs that have a poor performance record;
- projects with accelerated design and/or construction schedules; and
- projects with overlapping design and construction schedules.

10.4-6 Economic Considerations

Cost should be considered in design after function, performance, and installation design criteria are addressed. Material and in-place costs will obviously vary with the type of geosynthetic barrier and the quantity of barrier specified. In-place cost of geosynthetic barriers can vary from approximately $2.50/m^2$ to $16.00/m^2$. Cost of conventional compacted clay liners can vary between approximately $5.00/m^2$ to $30.00/m^2$ in-place.

10.5   INSTALLATION

A well-designed geosynthetic must be installed correctly to perform its function as a barrier. Handling and installation specifications for geomembrane and other geosynthetic barriers should, as a minimum, conform to the manufacturer's recommendations. Special project requirements should also be noted in the construction specifications and plans.

Geosynthetic barrier handling and storage requirements at the construction site should be specifically designated. Layout of the geosynthetic normally should be predetermined and documented on a roll or panel layout plan. The installer or geosynthetic supplier is normally required, by specification, to provide the layout plan.

Three areas of construction which are critical to a successful installation are:

- subgrade preparation;
- field seaming; and
- sealing around penetrations and adjacent structures.

The subgrade must provide support to the geosynthetic barrier and minimal point loadings. The subgrade must be well-compacted and devoid of large stones, sharp stones, grade stakes, etc., that could puncture the geosynthetic barrier. In general, no objects greater than 12 mm should be protruding above the prepared subgrade (Daniel and Koerner, 1993). Geotextiles are often used as cushions for geomembranes to increase puncture resistance, as previously discussed. Geotextiles and geocomposite drains are also used beneath geosynthetic barriers to vent underlying gas (e.g., from decomposing organic deposits) or relieve excess hydrostatic pressure.

The method of seaming is dependent upon the chosen geosynthetic material and the project design. Overlaps, of a designated length, are typically used for thin-film geotextile composites and geosynthetic clay liners. Geomembranes are seamed with thermal methods or solvents. Temperature, time, and pressure must be specified and maintained within tolerances for thermal seaming. With solvent seams, solvent application is important, because too much solvent can weaken the geomembrane and too little solvent can result in a weak or leaky seam. Pressure, or heat, is used in conjunction with solvents. Seaming procedures for a variety of geomembranes is detailed in several waste containment manuals (Daniel and Koerner, 1993; Landreth and Carson, 1991; Matrecon, 1988).

Construction details around penetrations and adjacent structures depend upon the chosen geosynthetic material and the project design. As such, they must be individually designed and detailed. Geosynthetic manufacturers and several waste containment manuals (Daniel and Koerner, 1994; Matrecon, 1988; Richardson and Koerner, 1988) can provide design guidance.

10.6 INSPECTION

Quality assurance (QA) and quality control (QC) are recognized as critical factors in the construction of geomembrane-lined waste containment facilities. QA and QC may or may not be as important for highway-related barrier works. The extent of QA and QC for highway barrier works should be project, and barrier product, specific.

10.6-1 Manufacture

Manufacturing quality control (MQC), normally performed by the geosynthetic manufacturer, is necessary to ensure minimum (or maximum) specified values in the manufactured product (Daniel and Koerner, 1993). Additionally, manufacturing quality assurance (MQA) programs are used to provide assurance that the geosynthetics were manufactured as specified. Quality of raw materials and of finished geosynthetic products are monitored in an MQA program. The MQA program may be conducted by the manufacturer, in a department other than manufacturing, or by an outside organization. Details on MQA and MQC for geomembrane and geosynthetic clay liners, along with other geosynthetic components, are presented in an EPA Technical Guidance Document (Daniel and Koerner, 1993).

10.6-2 Field

Construction quality assurance (CQA) and construction quality control (CQC) programs should be used for most geosynthetic barrier structure construction. CQC is normally performed by the geosynthetic installer to ensure compliance with the plans and specifications. CQA is performed by an outside organization to provide assurance to the owner and regulatory authority (as applicable) that the structure is being constructed in accordance with plans and specifications. Typically, for waste containment facilities, the CQA-performing organization is not the installer or designer, *i.e.*, it is a third-party organization. CQA may be performed by the transportation agency for highway works. Details on CQA and CQC for geomembrane and geosynthetic clay liners -- and for traditional compacted clay barriers -- are presented in an EPA Technical Guidance Document (Daniel and Koerner, 1993).

10.7   SPECIFICATION

Geosynthetic barrier specifications should contain the following components:
  • statement on purpose of barrier;
  • material specification for the barrier and all associated geosynthetic components, including component property requirements, product requirements, manufacturing quality control requirements, and manufacturing quality assurance;
  • shipping, handling, and storage requirements;
  • installation requirements;
  • requirements for sealing to and around penetrations and appurtenances;
  • seaming requirements, including pass/fail criteria;
  • anchoring requirements; and
  • statement on construction quality control and construction quality assurance.

Again, detailed information on geomembrane and GCL specifications is presented in waste management manuals (Daniel and Koerner, 1993; Matrecon, 1988).

As discussed under Installation, Section 10.5, subgrade preparation is crucial to a successful installation. Therefore, it is imperative that the accompanying subgrade preparation specification be written specifically for the geosynthetic barrier to be installed. The subgrade must be inspected and approved prior to placement of the geosynthetic barrier.

10.8　REFERENCES

Berg, R.R. (1993), Project Peer Review in Geosynthetic Engineering, Geotechnical Fabrics Report, Industrial Fabrics Association International, St. Paul, MN, April, pp. 34-35.

Daniel, D.E. and Koerner, R.M. (1993), Quality Assurance and Quality Control for Waste Containment Facilities, U.S. EPA Technical Guidance Document, EPA/600/R-93/182, Cincinnati, OH, 305 p.

Giroud, J.P. (1984), Impermeability: The Myth and a Rational Approach, Proceedings of the International Conference on Geomembranes, Vol. 1, Denver, June, pp. 157-162.

Koerner, R.M. (1994), Designing With Geosynthetics, 3rd Edition, Prentice-Hall Inc., Englewood Cliffs, NJ, 783 p.

Koerner, R.M. and Hwu, B-L (1989), Geomembrane Use in Transportation Systems, Transportation Research Record 1248, Transportation Research Board, Washington, D.C., pp. 34-44.

Landreth, R.E. and Carson, D.A. (1991), Inspection Techniques for the Fabrication of Geomembrane Field Seams, U.S. EPA Technical Guidance Document, EPA/530/SW-91/051, Cincinnati, OH, 174 p.

Matrecon, Inc. (1988), Lining of Waste Containment and Other Impoundment Facilities, U.S. EPA Technical Guidance Document, EPA/600/2-88/052, Cincinnati, OH, 1026 p.

Richardson, G.N. and Koerner, R.M., Editors (1990), A Design Primer: Geotextiles and Related Materials, Industrial Fabrics Association International, St. Paul, MN, 166 p.

Richardson, G.N. and Koerner, R.M. (1988), Geosynthetic Design Guidance for Hazardous Waste Landfill Cells and Surface Impounds, EPA Contract No. 68-03-3338, U.S. Environmental Protection Agency, Cincinnati, OH.

Rowe, R.K. and Giroud, J.P. (1994), Quality Assurance of Barrier Systems for Landfills, IGS News, International Geosynthetics Society, Vol. 10, No. 1, March, pp. 6-8.

Staff, C.E. (1984), The Foundation and Growth of the Geomembrane Industry in the United States, Proceedings of International Conference on Geomembranes, Denver, June, pp. 5-8.

11.0 GEOTEXTILES IN WASTE CONTAINMENT SYSTEMS

11.1 BACKGROUND AND APPLICATIONS

Geotextiles form an integral part of containment system design for both municipal and hazardous waste landfills and impoundments. The important roles that a geotextile can play in a successful containment design are illustrated in Figure 11-1. Within the three primary components of a waste containment system (the liner, the waste itself and the cap), specific geotextile applications and functions include:

- Containment Liner System
  - site stabilization and access for construction of the liner
  - protection of geomembrane liner against possible punctures
  - gas venting to prevent gas from collecting beneath the liner
  - leachate collection system filters
  - intermittent drainage between geomembrane sheets
  - leachate collection system filters
- Within Body of Waste
  - materials to provide daily cover during placement of waste
  - venting of gas within between layers of waste
  - internal drainage
- Caps and Covers
  - stabilization of soft or otherwise unstable landfill mass
  - to facilitate cap construction and minimize subsidence
  - gas transfer media for low gas production landfills
  - filters for open flow geonet or gravel gas vent layers
  - pore water dissipation and improvement of friction between geomembranes and cohesive soil
  - protection of geomembrane cover against possible punctures
  - filters and separators for geonet or aggregate drainage layers above the hydraulic barrier
  - separators between biotic barriers and surface topsoil layers
  - filters beneath erosion control armor stone on the cap surface and in drainage channels and spillways
  - reinforcement of cover soils

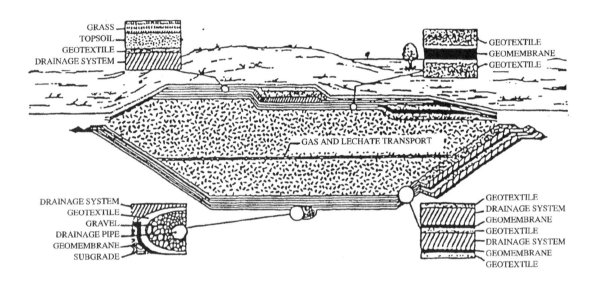

Figure 11-1   *Section through a typical landfill liner and cover illustrating the various uses of geotextiles.*

As can be seen from this list there are numerous potential geotextile applications, and several have already been discussed in earlier chapters. However, each application will have some what different selection and evaluation requirements, primarily due to the environment in which the geotextiles are placed as well as special considerations required for waste containment construction.

In each of the above applications, geotextiles have significant advantages over the classical granular soil layers they replace. This is because of
- Improved performance, if properly selected.
- Potential cost savings, especially if granular materials are relatively expensive.
- Substantial savings due to reduction in liner and cap thicknesses, which result in increased air space for disposal (potential for substantial cost benefits over and above any material cost saving.)
- Reduction in landfill settlement due to decreased weight of cap.

As with the graded granular soil layers that it replaces, the geotextile must be carefully selected based on the properties required to adequately perform its intended function in the specific application. Each application area is reviewed in this chapter and design and initial selection guidance is provided. In cases where applications are very similar to those discussed in earlier chapters, reference will be made to the appropriate design methodology in those chapters.

11.2   SITE STABILIZATION AND CONSTRUCTION ACCESS

Wet, soft or otherwise poor subgrade conditions may not provide a suitable foundation for construction of the liner system. In these cases, a geotextile may be used to expedite access road construction and to provide increased structural support for the hydraulic barrier and leachate collection system. In this application, the geotextile acts as a separator and serves with an initial granular support layer to stabilize soft or otherwise unstable subgrades. This application is well established for roadway construction, whose principles are equally applicable to providing a stable working platform for construction of the hydraulic barrier and to facilitate placement of the geomembrane and leachate collection system. If properly designed, the support layer may also function as a gas vent.

11.2-1 Design Approach
Design of geotextiles as separators is covered in Chapter 5. The principal geotextile requirements are: 1) adequate strength and durability to survive the installation, 2) appropriate filtration properties to retain the underlying soil, and 3) adequate drainage to

prevent pore pressure buildup in the subgrade (thus its strength improves with time). The following outline provides a review of the design considerations for proper selection of the geotextile.

- Stress Conditions
  - Construction equipment and ground pressure
  - Estimated traffic

- Input Parameters
  - Shear strength or CBR of subgrade soil .
  - Grain size and permeability of subgrade soil

- Design Method
  - Determine required aggregate thickness
  - Determine geotextile requirements to survive construction
  - Determine filtration requirements for geotextile (see Chapter 2)

- Controlling Geotextile Requirements
  - Grab Strength
  - Puncture Resistance
  - Tear Resistance
  - AOS
  - Permeability

11.3   GEOMEMBRANE PROTECTION

Impermeable geomembrane liners used in containment systems are relatively thin and thus can be easily damaged, both during installation of the liner system as well as after completion of construction. Adequate mechanical protection must be provided to resist both short-term equipment loads and long-term loads imparted by the waste. Experience has shown that geotextiles can play an important role in the successful installation and longer-term performance of geomembranes by acting as a cushion to prevent puncture damage to the geomembrane (see Figure 11-1). As such, geotextiles can be placed (1) below geomembranes to resist puncture and wear due to abrasion caused by sharp-edged rocks in the subgrade, and (2) above the geomembrane to resist puncture caused either by the drainage aggregate or direct contact with waste materials. Equally, during placement of intermittent and final covers, geotextiles can be placed below the geomembrane to reduce risk of damage by sharp objects in the landfill and above the geomembrane to prevent damage during placement of drainage aggregate or cover soil.

In geomembrane protection, geotextiles are used to augment and/or replace conventional protective soil layers. For operation of equipment directly over the geomembrane, protective-soil layers at the base and on the side slopes of the landfill can typically be reduced by as much as 150 mm through the use of a geotextile. Greater reductions can be achieved by careful management and control of construction sequence and cover material placement. Even with a geotextile, the minimum protective soil layer should be 300 mm thick. In transition areas, which include the toe of slopes, the protective layer thickness should be increased to allow for equipment turning and placement of protective soil cover in slope transition areas. It may be possible to eliminate, at least temporarily, the protective soil layer beneath the geomembrane and above the geomembrane in areas where equipment will not be operated. For example, placement of side slope cover soils may be delayed and placed in stages during placement of the waste to reduce downdrag forces on the geotextile and membrane.

The effectiveness of geotextiles for geomembrane protection is illustrated in Figures 11-2 and 11-3, which present the results of puncture tests on four types of geomembranes. The geomembranes were all 0.75 mm thick and used in combination with 200 g/m^2 and 400 g/m^2 polypropylene continuous filament needlepunched nonwoven geotextiles. Testing was accomplished according to a modified ASTM D 3738 test procedure in which a 8 mm diameter blunt steel piston was used. These results show the effectiveness of geotextiles in protecting the liner.

The selection of the most effective geotextile in this application will depend on several factors including:
- Mass per unit area,
- Type of geotextile,
- Type of fibre (continuous filament, staple filament)
- Thickness under load,
- Type of polymer, and
- Polymer composition and thickness of geomembrane.

Puncture protection is directly related to the mass per unit area of a given geotextile. Increasing this quality provides additional confidence that the geomembrane will not be damaged. Figure 11-4 illustrates the relative increase in puncture resistance with increased mass per unit area of a nonwoven geotextile.

The puncture resistance also varies with the type of geotextile. Figure 11-5 shows the results of puncture tests on several different geotextiles of similar mass per unit area. The figure indicates a distinct difference between the types of geotextiles evaluated.

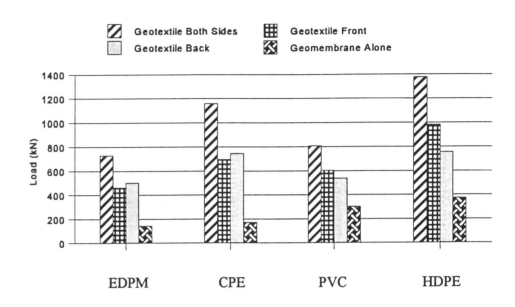

Figure 11-2   *Result of puncture test using 205 g/m² geotextiles on different geomembranes, all 76 mm thick.*

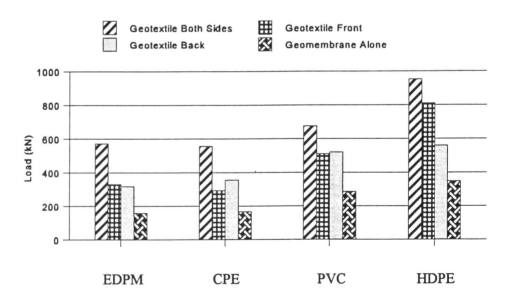

Figure 11-3   *Result of puncture test using 410 g/m² geotextiles on different geomembranes, all 76 mm thick.*

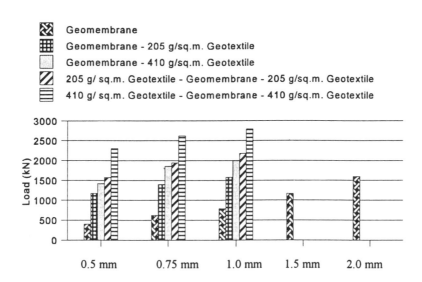

Figure 11-4   *Relative increase in puncture resistance with increased mass per unit area of a nonwoven geotextile.*

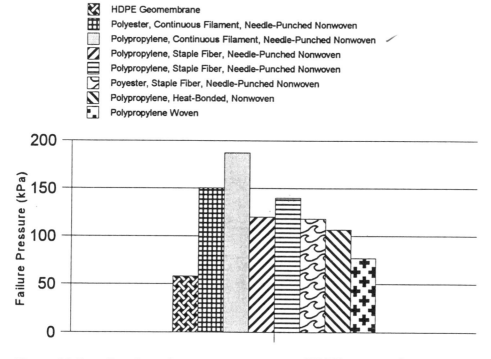

Figure 11-5   *Results of puncture tests on HDPE geomembranes using different geotextiles of a similar mass per unit area.*

Another important factor is time. The effectiveness of the geotextile will decrease with time due to creep under load. Unless long-term data is available to clearly show otherwise, a minimum factor of safety of 10 should be used to arrive at the long-term effectiveness based on short-term tests.

Although the above data can provide some guidance for initial selection of suitable geotextiles, there is a need for a consistent evaluation method to facilitate proper selection of the most suitable geotextile. ASTM Committee D35 is currently evaluating simplified procedures that simulate field conditions and provide a rapid assessment of geotextile effectiveness. Because point penetration has been found to be more important than rupture of the geotextile, the evaluation procedure uses a point loading condition to simulate sharp rocks as opposed to the blunt piston currently used to evaluate the puncture resistance of geotextiles. Until such a standard is developed, manufacturers should have test data available to clearly show the effectiveness of their geotextiles with different geomembranes under long-term point loads. In the absence of any other information, medium to heavy weight needlepunched nonwoven geotextiles with a mass per unit area of at least 250 g/m^2 are recommended.

11.3-1 Geomembrane Puncture Resistance Design

Design and selection of the geotextile for the specific geomembranes type and thickness consists of evaluating the local stress conditions. An equivalent point load is determined using the appropriate design curve (as illustrated below), then the factor of safety for long-term loading conditions is chosen. Finally, the geotextile is selected on the basis of its ability to provide the necessary puncture resistance.

STEP 1. Evaluate the stress conditions and determine the effective point load.
a. Determine size of gravel in subgrade or cover
b. Estimate ground pressure of construction equipment, σ_e
c. Determine the overburden pressure, $\sigma_p = \sigma_{pr} + \sigma_{pc}$
    σ_{pr} = Refuse depth x refuse unit weight, and
    σ_{pc} = Thickness of cap x cap unit weight
d. Select the greater stress condition s from (b) or (c) above, and use Figure 11-6 to determine the puncture force, P'

STEP 2. Determine design point load, P_L.
$P_L = FS_{long-term} \times FS_{design} \times P'$
where $FS_{long-term} > 10$, or chosen from long-term test data, and
$FS_{design} > 2$ for hazardous waste and >1 for caps

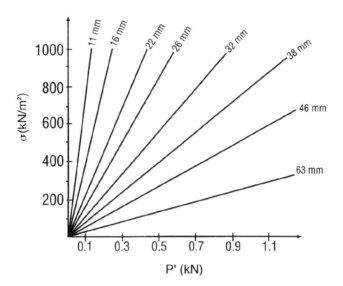

Figure 11-6   *Puncture force on a stone as a function of normal overburden pressure.*

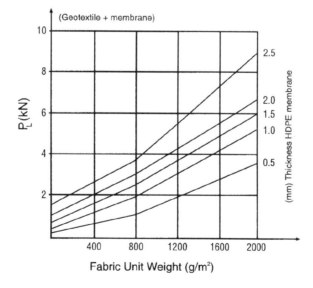

Figure 11-7   *Example design chart for continuous filament polypropylene needlepunched nonwoven geotextiles on HDPE geomembranes (Werner, et al., 1990).*

STEP 3. Based on point load resistance test data, select mass per unit area required for each type of geotextile considered. As an example , see Figure 11-7.

11.4 FILTERS FOR LEACHATE COLLECTION SYSTEMS

Geotextiles are commonly used as filters in leachate collection systems. They are an essential part of the collector drain and are used to encapsulate the gravel and pipe components. They are also used in the drainage blankets placed above the primary liner and in between the primary and secondary liners, whether as part of a geocomposite drainage system (*e.g.*, geonets) or adjacent to sand or gravel drain layers. When used in these applications, the geotextile acts as a filter allowing leachate to enter the collection system while preventing soil or waste from entering and clogging the system. The design of geotextile filters for conventional drains is covered in detail in Chapter 2, and these procedures are equally appropriate for leachate filtration. In addition to the geotextile selection requirements covered in Chapter 2, chemical and biological considerations are important for leachate collection filter design.

The main chemical and biological considerations are the potential for chemical precipitants or microbes to be trapped in the pore spaces of the filter and reduce its flow capacity. Also, there is a potential for microbes, once trapped, to grow and clog the filter. A U.S. EPA study by Koerner and Koerner (1991) to evaluate the clogging potential of various filter materials showed that for the filter to be effective, the dissolved solids and microorganisms (found to be on the order of 0.07 mm or less) must be permitted to pass through the filter. Although their study was performed under extreme biological conditions, the results indicated that very open structures (as found in needle-punched nonwoven geotextiles and open monofilament wovens) offered the highest biological clogging resistance; flow reductions over a 6-month period were less than one order of magnitude in most cases. Tests performed with tightly woven and heat bonded geotextiles showed substantially reduced flow rates (several orders of magnitude). However, in all cases, geotextiles performed substantially better than sand filters.

11.4-1 Filter Design

The design approach presented in Chapter 2 followed three basic principles for geotextile design and selection:

1. If the larger pores in the geotextile filter are smaller than the largest particles of soil (or waste), these particles will not pass the filter. As with graded granular filters, the larger particles of soil form a filter bridge or *cake* on the geotextile, which in turn filters the smaller particles of the soil or waste. Thus, the soil or waste is retained and particle movement and piping is prevented.

2. If the smaller openings in the geotextiles are sufficiently large so that the smaller particles of soil and biological matter are able to pass through the filter, then the geotextile will not clog.

3. A large number of openings should be present in the geotextile so that proper flow can be maintained even if some of the openings later become clogged.

The following design steps are appropriate:

Step 1.   Evaluate the critical nature of the application.

Step 2.   Determine the gradation of the material to be filtered; particularly determine D_{85}, D_{60}, D_{50}, D_{15}, and D_{10}

Step 3.   Determine the inflow requirements.
For leachate collection system design, the inflow will depend on the nature of the waste (*i.e.* unit weight, liquid content, and permeability), local rainfall conditions, size and side slope configuration, and duration of exposure (*i.e.* time period until final closure.) The Hydrologic Evaluation of Leachate Potential (HELP) model developed by the USEPA and USAE WES can be used to estimate the inflow requirements.

Step 4.   Determine the geotextile requirements.
   A. Retention criteria
       $AOS < B\ D_{85}$
   B. Evaluate permeability requirements
       $k_g > 10\ k_s$ (This is a critical application!)
   C. Evaluate clogging potential
       $AOS > 3\ D_{15}$ for $C_u > 4$
       Porosity > 80%
       When in doubt run gradient ratio tests.
   D. Biological activity
       Active: $k_g > 100\ k_s$
       Inactive: $k_g > 10\ k_s$

Step 5.   Determine the maximum gravel size in the drainage layer and its placement procedures.

To perform effectively, the geotextile must also survive the installation process (survivability criteria). For a detailed discussion of its development and background along with recommended geotextile filtration criteria, see Chapter 2, especially Table 2-2.

11.5   WATER AND GAS TRANSMISSION

Relatively thick needlepunched nonwoven geotextiles have a three-dimensional fiber structure and a high percentage of air voids that allow lateral free flow of liquids and gases in the plane of the product. As such, they can be used as a drainage layer in the following situations:

-    Between the two geomembranes of a double lining system, in order to drain any leakage that may occur through the primary (top) liner.
-    Between the geomembrane and soil, or between the geomembrane and waste material, to drain the slope, seepage from the waste, or groundwater. (Moisture between a geomembrane and underlying soil can significantly decrease the interface friction angle.)
-    Beneath a geomembrane liner in a liquid containment system to divert gases from beneath the system that can accumulate due to organics in the underlying soils.
-    Beneath intermittent and final landfill cover systems over the waste to act as a gas transmission media and to divert gases to the collection system.

11.5-1 Water Transport Design

Design and selection of the geotextile will depend on the magnitude of anticipated seepage, the normal stress anticipated over the geotextile, and the geotextile characteristics including thickness, mass per unit area, and fiber type (continuous or staple filament). Figure 11-8 illustrates the transmissivity response curve of 200, 260 and 400 g/m^2 polypropylene continuous filament needlepunched nonwoven geotextiles as a function of normal stress.

Using Darcy's Law, in situ transmissivity can now be checked to determine if it is sufficient for the desired water discharge quantity in question:

$$Q/B = \theta \times i$$

where

Q = water flow quantity,
B = width of geotextile,
θ = transmissivity, and
i = hydraulic gradient.

The gradient i is dependent on the water pressure present and on the landfill configuration. In the least complicated situation for drainage along a slope of inclination angle β, the gradient i = sin β.

If greater transmissivity is required than can be provided by heavyweight needlepunched nonwoven geotextiles, geonets or geocomposites must be installed.

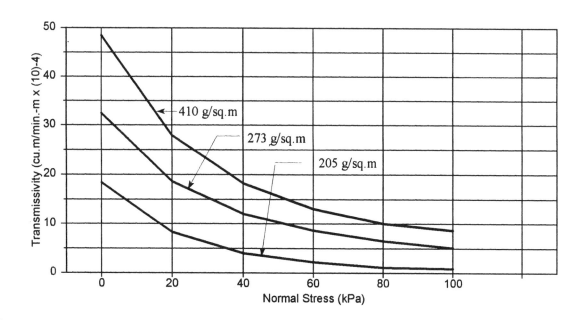

Figure 11-8   *Transmissivity of three needlepunched nonwoven geotextiles as a function of normal stress.*

Gas Transport Design. Gas generation may be sufficiently low for certain landfills (such as controlled sanitary disposal facilities) to consider using the lateral transmission potential of needlepunched nonwoven geotextiles as a gas transfer media. Design and selection of the geotextile will depend upon the magnitude of anticipated gas generated, the normal stress anticipated on the geotextile, and geotextile characteristics including thickness, mass per unit area, and fiber type (continuous or staple filament).

Although gas transmission rates through the plane of the geotextiles are not commonly available, gas flow can be estimated from water transmissivity values obtained from ASTM D 4716, a standard test for the water permeability of geotextiles; these values are commonly reported in manufacturers' literature. Typical water transmissivity values for various needlepunched nonwoven geotextiles under typical final cover loads range from 1 to 10×10^{5} m^{2}/ sec for medium to heavy weight geotextiles, respectively. Under comparable conditions, gas transmissivity is approximately two orders of magnitude greater than water transmissivity (Koerner et al., 1984). Therefore, it is sufficient to base the design on water transmissivity, as there will be sufficient gas transmissivity available for the venting of gases, even with some moisture in the geotextile. A conservative gas flow estimate will also allow for long term compression of the geotextile. Using standard flow equations, anticipated final cover side slopes, and distance to outlet vents, in situ transmissivity can then be checked to determine if it is sufficient for the desired gas discharge quantity. The geotextile must also meet filtration requirements reviewed in the previous section based on the potential for fine grained soils to enter the layer and reduce its flow.

If greater transmissivity is required than can be provided by heavyweight needlepunched nonwoven geotextiles, drain laterals can be constructed with geonets, slotted pipe, gravel or strip drains placed at regular intervals to increase the flow capacity and facilitate venting.

11.5-2 Gas Transport Design

Step 1.   Evaluate vertical stress conditions. Use overburden pressure from depth of refuse or thickness of cap.

Step 2.   Estimate gas generation rate per unit width of drain, Q'.

Step 3.   Calculate the transmission requirement, q, for the geotextile.
$$\theta > Q'/\sin b$$
where b is the inclination angle of the slope.

Step 4.   Prepare specifications based on controlling geotextile requirements:

Transmissivity, q, under given pressure (use the results of ASTM D 4716 with a factor of safety of 1, or direct long-term gas flow measurements with a factor of safety of 100).

- Mass/unit area.
- Constructibility requirements.
- Durability requirements.

11.6   DAILY COVERS

In the municipal solid waste landfills, refuse must be covered at the end of each working day to mitigate problems such as blowing refuse, rodents and odor. Commonly, a layer of soil 150 to 300 mm thick is used for this purpose. Daily cover materials should:

- minimize dust;
- minimize odors;
- minimize blowing litter;
- minimize fire hazards;
- minimize infiltration of precipitation;
- control flies, mosquitoes, rodents, and other vermin;
- be aesthetically acceptable;
- be conveniently handled and placed and be fully functional under extremes of climate expected at the site; and
- allow migration of leachate and gas within the landfill.

Specially modified geotextiles can be used as a low cost replacement for soil daily covers. In addition, these products increase landfill capacity through a decrease in daily cover thickness. The following discussion on these alternate daily covers is adapted from Rohr, et al. (1990).

Generally, geosynthetics used for daily cover are fabricated into panels before being transported to the working face. At the end of each day, a panel is attached to the landfill equipment using ropes or chains attached to sleeves, slits, or knots in the geosynthetic corners or edges. The panel is then dragged over the working face until it covers the face. Materials such as soil, sandbags, or tires are placed around the perimeter of the panel to keep it in place. The next day, the panel is removed by dragging it off the working face with landfill equipment, and depending on space limitations, it can be rolled or folded as required. The cycle of placement and removal of a geosynthetic panel for an average 1000 m² working face takes a total of approximately 30-45 min and requires, in general, one or two laborers and two pieces of landfill equipment. Manual applications of the geosynthetic daily cover materials is also an option for covering small areas or for geosynthetics requiring more care in handling.

So far, our experience with geotextiles for this application has been very good. Geotextiles are durable and perform well the key requirements of a daily cover. Based on experience with UV stabilized geotextiles, a panel can be expected to serve approximately five or six reuses in winter, while during the summer and in mild weather conditions, one panel may provide up to 15 reuses.

Geosynthetics for daily cover should be allowed by regulatory agencies provided they meet the required performance criteria. Alternative products can be tried at operating landfills and evaluated over extended periods of time as to their economics, practicality, and overall environmental efficiency.

11.7   OTHER SELECTION CONSIDERATIONS

Other design and selection considerations include chemical and biological resistance, as well as adequate ultraviolet. stability for the specific application conditions.

Chemical and Biological Compatibility. Polypropylene and polyethylene polymers are highly resistant to chemical and biological degradation. Manufacturers should provide specific chemical compatibility information (including test results) about the efficacy of their products in environments similar to those anticipated at the site in question. If the potential exists for a caustic environment (*e.g.*, high pH conditions), polyesters can potentially degrade. Acidic conditions in the presence of metals can affect the additives used in polypropylenes. Table 11-1 provides initial guidance for polymer selection. Should there be any questions as to the effect of liquid wastes on the geomembrane/geotextile system, the system components must be tested for long-term chemical compatibility using approved test methods. ASTM is currently developing test methods for field and laboratory immersion procedures based on EPA Method 9090.

If geotextiles are to be exposed to sunlight for any length of time, consideration must also be given to ultraviolet protection, as all polymers degrade when exposed to sunlight. ultraviolet stability is an especially critical consideration when geotextiles are used in the construction of containment systems. It is not uncommon for the geotextiles used to remain exposed to sunlight at these sites for several months. The rate of degradation will depend on:
-    The base polymer of the geotextile;
-    Ultraviolet stabilizers added to the base polymer;
-    Geotextile thickness and fiber type;
-    Time of exposure;
-    Geographic latitude;
-    Altitude; and
-    Season of exposure.

Table 11-1
Environmental Properties and Chemical Resistance of Common Polymers used for (a) Geotextiles (ICI, 1981), and (b) Geomembranes (Koerner, 1994).

(a)

Environmental Properties of the Polymers used in Geotextiles

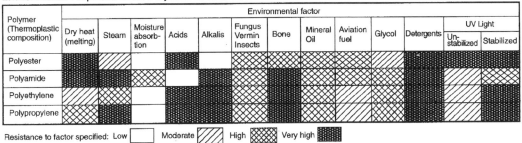

Resistance to factor specified: Low ☐ Moderate ⊘ High ⊠ Very high ▨

(b)

Chemical	Butyl rubber 100°F	158°F	Chlorinated polyethylene (CPE) 100°F	158°F	Chloro-sulfonated polyethylene (CSPE) 100°F	158°F	Elasticized polyolefin 100°F	158°F	Epichlorohydrin rubber 100°F	158°F	Ethylene propylene diene monomer (EPDM) 100°F	158°F	Polychloroprene (neoprene) 100°F	158°F	polyethylene 100°F	158°F	Polyvinyl chloride (PVC) 100°F	158°F
General:																		
Allphatic hydrocarbons			X	X			X		X	X			X	X	X	X		
Aromatic hydrocarbons							X		X	X			X	X	X	X		
Chlorinated solvents	X	X					X		X	X	X		X		X	X		
Oxygenated solvents	X	X					X		X		X	X	X	X	X	X		
Crude petroleum solvents			X	X			X		X	X			X	X	X	X		
Alcohols	X	X	X	X			X		X	X	X	X	X	X	X	X	X	X
Acids:																		
Organic	X	X	X	X	X		X		X		X	X	X	X	X	X	X	X
Inorganic	X	X	X	X	X		X		X		X	X	X	X	X	X	X	X
Bases:																		
Organic	X	X	X	X	X		X		X	X	X	X	X	X	X	X	X	X
Inorganic	X	X	X	X	X		X		X	X	X	X	X	X	X	X	X	X
Heavy metals	X	X	X	X	X		X	X	X	X	X	X	X	X	X	X	X	X
Salts	X	X	X	X	X		X	X	X	X	X	X	X	X	X	X	X	X

Geomembrane type

* X = generally good resistance.

Figure 11-9 shows the relative degradation of various types of polymers and stabilizers used in the manufacture of geotextiles. Requirements for the geotextile should be rigorously specified, with support data requested from manufactures. The current standard for U.V. evaluation of geotextiles is based on ASTM D 4355, and requirements usually range from 70% to 90% strength retention after 500 hours of U.V. exposure under the conditions specified in the test.

Mechanical Properties. Other selection considerations include mechanical properties such as friction angle between the geomembrane-geotextile, components, strength of the geotextile to resist tensile loading, and properties of the geotextile needed to survive construction and installation. Each of these considerations are application specific, and the reader is referred to standard texts such as Koerner (1990) and to the manufactures of geomembranes and geotextiles for design assistance.

11.8   SPECIFICATIONS

All engineers know that good specifications are essential for the success of any project. Good specifications are even more critical for projects in which geosynthetics are used in waste containment systems. As geotextile performance can be critical to the successful performance of the containment system, each geotextile requirement should be clearly specified along with adequate quality control/quality assurance requirements. This is necessary to assure that geotextiles delivered to the site indeed have the properties required to perform their intended function, and that they are properly installed. When chemical compatibility is of concern, control of the polymer is equally important for geotextiles as it is for geomembranes. Avoid the use of "or equal" specifications, as this can result in the selection of a geotextile by the contractor which may have completely different properties than intended by the designer.

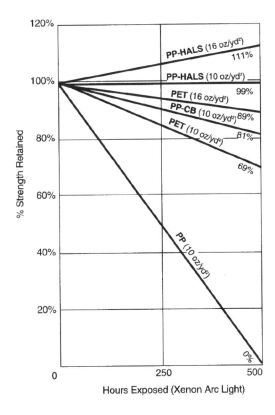

Polypropylene (HALS)	**PP-HALS**
Polyester (No UV Stabilizers)	**PET**
Polypropylene (Carbon Black)	**PP-CB**
Polypropylene (No UV Stabilizers)	**PP**

Figure 11-9   Relative strength retained of nonwoven geotextiles of different polymers and stabilizers after exposure to UV radiation according to ASTM D 4355.

The following sample specifications were developed by Wayne and Christopher (1993).

1.0   *Geotextile specifications for waste containment applications.*

    1.1 Scope of Work

        A. The contractor shall furnish all labor, materials, tools, supervision, transportation, and installation equipment necessary for the installation of the geotextile (filter, separator, cushion) layers, as specified herein, as shown on the drawings and in accordance with the construction quality assurance (COA) plan.

        B. The contractor shall be prepared to install the geotextile (filter, separator, cushion) in conjunction with earthwork and other components of the containment system.

    1.2 References

        The current ASTM Standards on Geosynthetics, sponsored by ASTM Committee D-35, apply to these specifications.

    1.3 Materials: Requirements for Geotextiles

    A. Physical Properties

        The geotextile shall be manufactured using a (list here acceptable polymer types and any special requirements such as additives, UV stabilizers, etc.). Manufacturer's Certification shall be required from the manufacturer stating the geotextile has been inspected for broken needles. This inspection shall be via metal detectors permanently installed on-line at the production facility.

        The geotextile shall meet or exceed the properties listed in Table. 11-2. Unless noted otherwise, all numerical values indicate minimum average roll values (i.e., test results from any sampled roll in the lot when tested in accordance with ASTM D 4759-88 shall meet or exceed the minimum average roll value listed). All strength values are for the weaker principal direction.

    B. Submittals

        Prior to bid, the contractor shall submit the manufacturer's specification information which indicates the property values of the products that are to be submitted for use on the project. A sample (0.3 m x 0.3 m) of each geotextile will be submitted along with the specification data.

Table 11-2   Geotextile Properties

PROPERTY	TEST METHOD	UNITS
Mass Per Unit Area	ASTM D 5199	g/m^2
Wide Strip Tensile	ASTM D 4595	kN/m
Trapezoidal Tear	ASTM D 4533	N
Puncture Resistance	ASTM D 4833	N
Permittivity	ASTM D 4491	sec^{-1}
Grab Tensile	ASTM D 4632	N
Grab Elongation	ASTM D 4632	%
AOS	ASTM D 4751	mm
UV Resistance 500 hrs	ASTM D 4355	% Strength Retained
Seam Strength	ASTM D 4884	% efficiency

The successful bidder shall provide written certification, signed by a responsible party (employed by the manufacturer) stating that the geotextile(s) shipped to the project meets or exceeds the minimum average roll values in Table 11-2. Upon request, the geotextile manufacturer will supply quality control/quality assurance test reports to the owners' representative prior to arrival of the geotextiles on site.

Manufacturer's quality control/quality assurance sampling and testing shall be conducted in accordance with ASTM D 4354 and applicable test standards as found in Table 11-2. Recommended manufacturing quality control/quality assurance tests and test frequencies are found in Table 11-3.

Additional supporting documentation: Upon request the contractor shall supply additional documentation which will include supporting test data and EPA 9090 test data to assist the engineer in determining the suitability of the geotextile for the intended application.

Table 11-3   Recommended Test Frequencies

PROPERTY	TEST FREQUENCY
Mass Per Unit Area	every 10,000 m²
Grab Tensile	every 10,000 m²
Apparent Opening Size	1 per production lot
Permeability	1 per production lot
Puncture Resistance	every 40,000 m²
Trapezoidal Tear	every 40,000 m²

1.4 Shipping, Storage and Handling

A. General

The geotextiles shall be shipped, stored and handled in accordance with ASTM D 4873 and as specified herein. The contractor will be responsible for shipping, storage and handling.

B. Roll Identification

Each roll will be labeled either by printing directly on the geotextile or tagged with a roll identification number, identification number, name of manufacturer, product grade, and physical dimensions. The label or tag information shall be affixed or attached to the roll at all times during deployment of the roll. The roll identification number and manufacturer name will also be marked on the protective covering.

C. Shipping

The geotextiles shall be shipped and stored in opaque protective covering. The contractor shall notify the engineer at least twenty-four hours prior to scheduled delivery. No materials will be unloaded without the presence of the owner's representative. Geotextiles delivered to the site shall be inspected for damage, unloaded and stored with minimal handling. The contractor will assist the owner's representative in conducting inventory, handling and sampling of the geotextiles at no additional cost to the owner.

D. Handling

No hooks, tongs or other sharp tools or instruments shall be used for handling geotextile rolls. Acceptable are the use of slings or a pole which extends a minimum of 0.3 m beyond each end to unload or handle individual rolls. The geotextiles shall not be dragged along the ground.

E. Site Storage

Geotextile rolls whose protective covering has been damaged shall be protected from ultraviolet (UV) light exposure, precipitation or other inundation, soil, mud, dirt, debris, puncture, cutting, or other damaging or deleterious conditions. The geotextiles shall not be stored directly on the ground.

1.5 Conformance Testing

A. Sampling Procedures

Upon delivery to the site, samples of the geotextiles will be removed and sent to a laboratory, chosen by the owner, for testing to evaluate conformance to the project specifications. Samples shall be selected by the owner's representative in accordance with the specifications. Samplers shall be taken from a portion of the roll that has not been damaged. Unless otherwise specified, samples shall be 1 m long by the roll width. Conformance testing as detailed in 1.5B shall be at the rate of two per lot or one per 10,000 m^2, whichever results in the greater number of samples.

B. Interpretation of Conformance Test Results

A test specimen shall be obtained from each conformance sample. The minimum number of specimens tested per conformance sample for each geotextile property will be determined in accordance with the respective ASTM standard. The average value will be calculated from the specimen test values of each sample and compared to the minimum average roll value of each tested geotextile property. A conformance sample that yields an average number which is less than the specified minimum average roll property will be recorded as a failure. If two conformance samples fail, all be recorded as a failure. If two conformance samples fail, all rolls within the sampled 10,000 m^2 or lot will be rejected for use on the project unless the manufacturer and owner agree that additional testing is necessary. If only one conformance test fails, the roll that yielded the failure will be rejected and a subsequent conformance sample from the same 10,000 m^2 or lot will be rejected for use on the project if the subsequent conformance test fails. If subsequent conformance test pass, the roll that yielded a failure is the only roll rejected from the project.

1.6 Geotextile Inspection

Prior to installation, the contractor shall visually inspect, all geotextile rolls for imperfections and possible damage. All defective rolls shall be marked and repaired in accordance with approved methods.

1.7 Installation

A. Placement

The geotextiles shall be installed as shown on the drawings, approved installation drawings, and as specified herein or in accordance with the approved installation procedures. No geotextile roll shall be installed without approval of the engineer.

The geotextile shall be cut only with an approved geotextile cutter.

The contractor will exercise extreme care during geotextile installation to prevent damage to the prepared supporting subgrade surface or to other installed geosynthetics. The contractor shall exercise care to prevent the entrapment of rocks, clods of earth or other matter which could damage the geotextile or other geosynthetic, clog the geotextile or hamper seaming. Any geotextile surface showing damage from penetration or distress caused by foreign objects shall be repaired or replaced.

No foot traffic will be allowed on the geotextile except with approved smooth-sole shoes. No vehicular traffic will be allowed on the geotextile. The contractor shall not use the geotextiles as a work area or storage area for tools and supplies.

The engineer will have the authority to order the immediate stoppage of work as a result of improper installation procedures or any reason that may cause defective installation. All changes to approved installation drawings and procedures shall be approved by the engineer prior to implementation.

Cleanup within the work area will be an ongoing responsibility of the contractor. Particular care will be taken to ensure that trash, tools, or other materials are not trapped beneath the geotextile.

B. Repairs

Geotextile repairs will be made with patches of the same geotextile material, using approved seaming systems, equipment and techniques. The patch size shall be 0.6 m larger in all directions than the area to be repaired. All corners must be rounded. All stitches shall be located no closer than 25 mm from the edge of the patch.

C. Joining Methods

Sewn geotextile panels will use UV stabilized, polypropylene thread. The thread color should contrast with the geotextile color to assist in inspection of the seam.

Geotextile seams shall be prayer or flat seams for nonwoven geotextiles. All seams will be formed by mating the edges of the geotextile panels and sewing them together with continuous stitches located a minimum of 75 mm from the edges. A two-thread type 401K double-lock stitch shall be used for all sewn work. Sewing methods will conform to the latest procedures recommended by the geotextile manufacturer. Thermal welding can also be performed on geotextiles with a mass per unit area greater than $270g/m^2$. For both forms of seaming, the contractor shall demonstrate that the seam efficiency meets the requirements of Table 1.

D. Cover

All geotextiles should be covered within fourteen days. If a geotextile will be left exposed for more than fourteen days, the manufacturer shall provide field UV test data and written recommendations for the maximum time of UV exposure prior to material shipment. Geotextiles which will be left exposed for more than fourteen days will be routinely sampled and tensile strength tests performed every four weeks. Samplers should be taken from the exposed geotextile with the sample areas repaired. If the tensile strength of the geotextile falls below 80% of the original values within the recommended exposure period, the manufacturer will replace the defective material. A limited warranty should be provided for the required exposure period.

E. Measurement and Payment

The geotextile will be measured by the square yard with payment based on square meters in place. This excludes overlaps and seam allowances (contract unit price per square meters in place). Prices quoted for geotextiles will include all installation charges for geotextile placement as shown on the plans. The contractor is responsible for any damage that occurs to the geotextile during installation and will replace the damaged geotextile at no additional cost.

11.9 REFERENCES

Hullings, D. and Koerner, R.M. (1991), Puncture Resistance of Geomembranes Using a Truncated Cone Test, Proceedings of Geosynthetics '91, Vol. 1, Atlanta, GA, USA, IFAI, pp. 273-285.

ICI (1981) A Guide to Test Procedures Used in the Evaluation of Civil Engineering Fabrics, ICI Fibres, U.K., 7 p.

Koerner, R.M. (1994) Designing with Geosynthetics, 3rd Edition, Prentice-Hall, Inc., Englewood Cliffs, NJ, 783 p.

Koerner, R.M. and Koerner G.R. (1991), Landfill Leachate Clogging of Geotextiles (and Soil) Filters, Cooperative Agreement No. CR-814965, Report to Risk Reduction Laboratory, Office of

Research and Development, U.S. Environmental Protection Agency, Cincinnati, Ohio, 148 p.

Koerner, R. M., Monteleone, M.J., Schmidt, J.R. and Roether, A.T. (1986), Puncture and Impact Resistance of Geosynthetics, Proceedings of the Third International Conference on Geotextiles, Vol. 32, Vienna, Austria, IFAI, pp. 677-681.

Koerner, R.M., Bove, J.A. and Martin, J.P. (1984) Water and Air Transmissivity of Geotextiles, International Journal of Geotextiles and Geomembranes, I 9(1), pp. 57-74.

Rohr, J.J., Lundell, C.M., Rueda, J., and Querio, A.J. (1990) The Use of Geosynthetics as Daily Cover at Solid Waste Landfills, Presented at the ASTSWMO 1990 Solid Waste Forum, Milwaukee.

Wayne, M. and Christopher, B.R. (1993), Manufacturing Quality Assurance and Construction Quality Control of Polypropylene Geotextiles, Proceedings of the 6th GRI Seminar on MOC/MOA and COC/COA of Geosynthetics, Koerner, R.M., ed., IFAI, St. Paul, MN, pp. 61-67.

Werner, G., Pühringer, G., and Frobel, R.K. (1990), Multiaxial Stress Rupture and Puncture Testing of Geotextiles, Proceedings of the Fourth International Conference on Geotextiles, Geomembranes, and Related Products, The Hague, Vol. 2, pp 765-770.

APPENDICES

Appendix A
GEOSYNTHETIC LITERATURE

A-1   1988 List (Holtz, R.D. and Paulson, J.N., Geotechnical News, Vol. 6, No. 1, March 1988, pp. 13-15)

Introduction

In only a very few years, geosynthetics have joined the list of possible solutions to a number of important civil engineering problems. Two important and increasingly critical examples come to mind immediately; the control of hazardous waste containment and soil reinforcement. In many cases, the use of a geosynthetic can significantly increase the structure's safety factor, improve performance, and reduce costs in comparison with conventional solutions.

The range of applications of geotextiles, geogrids, and related products is enormous. They are used in:

• filtration, drainage, and in erosion protection and control systems.
• stabilization of roadways and railroads on soft subgrades.
• reinforcement of retaining walls, earth and waste slopes, and embankments.

Geomembranes are primarily used as:
• pond and canal liners.
• barriers in hazardous and other waste containment systems.
• rehabilitation of old earth fill and concrete dams.
• environmental control in highways and other civil engineering construction.

Geotextiles, geogrids, and geomembranes can also be used together in a Geocomposite system to provide multiple functions. Common applications of geocomposites are in hazardous waste containment systems and as prefabricated drainage layers.

Even with the rapid recent growth in the use of these materials, it has been our experience that civil engineers often have difficulty obtaining what they consider to be reliable and impartial information about design, specifications, and construction with geosynthetics. This perceived lack of information is primarily a communications problem, because a relatively large body of information on geosynthetics has been published in books, journals, and conference proceedings.

This article presents some sources of information which are readily available to civil engineers contemplating design and construction with geosynthetics. Included are:
• books
• journals and other periodicals

- conference proceedings
- reports

In some cases, addresses where publications may be obtained are also given.

The field of geosynthetics is developing so rapidly that it can be difficult for the non-specialist to remain current. To assist in this regard, we also list organizations and technical committees concerned with geosynthetics. Watch for announcements of their meetings, symposia, and conferences, because that is where the latest information, research findings, and case histories are presented and discussed.

A. General References, Books and Manual

Bell, J.R., Hicks, R.G., et al, (1980 and 1982) "Evaluation of Test Methods and Use Criteria for Geotechnical Fabrics in Highway Applications," Oregon State University, Corvallis, Interim Report No. FHWA/RD-80/021, Final Report No. FHWA/RD-82.

Christopher, B.R. and Holtz, R.D. (1985) Geotextile Engineering Manual, STS Consultants Ltd., Northbrook, Illinois; Federal Highway Administration, Report FHWA-TS-86/203, 1044 p.

Giroud, J.P. (1984) Geotextiles and Geomembranes, Definitions, Properties and Design, I.F.A.I., Suite 450, 345 Cedar Street, St. Paul, MN 55101, 325 p.

Jones, C.J.F.P, (1985) Earth Reinforcement and Soil Structures, Butterworths, London, 183 p.

Kays, W.B. (1977) Construction of Linings for Reservoirs, Tanks and Pollution Control Facilities, John Wiley and Sons, 1977.

Koerner, R.M. and Welsh, J.P. (1980) Construction and Geotechnical Engineering Using Synthetic Fabrics, John Wiley and Sons, 267 p.

Koerner, R.M. (1986) Designing with Geosynthetics, Prentice-Hall, Inc., 424 p.

Rankilor, P.R. (1981) Membranes in Ground Engineering, John Wiley and Sons, 377 p.

Scott, J.D. and Richards, E.A. (1985) Geotextile and Geomembrane International Information Source, BiTech Publications, 902-1030 West Georgia Street, Vancouver, B.C., Canada, V6E 2Y3.

Steward, J., Williamson, R. and Mohney, J. (1977) "Guidelines for Use of Fabrics in Construction and Maintenance of Low-Volume Roads," USDA, Forest Service, Portland, Oregon. Also published by FHWA as Report No. FHWA-TS-78-205.

Veldhuijzen van Zanten, R. Ed. (1986) Geotextiles and Geomembranes in Civil Engineering, John Wiley, 658 p.

B. Periodicals and Journals

Specifically devoted to geosynthetics:

Geotechnical Fabrics Report, published by Industrial Fabrics Association International (IFAI), Suite 450, 345 Cedar Street, St. Paul, MN 55101, (612-22-2508). First issue Summer, 1983; now six issues/year; $30 (complimentary with membership in the North American Geosynthetics Society).

Geotextiles and Geomembranes, published by Elsevier Applied Science Publishers, Crown House, Linton Road, Barking, Essex IGII 8JU, England. First volume published in 1984; now four issues/year; $128 ($77 for members of the International Geotextile Society).

Useful technical articles on geosynthetics also occasionally appear in the standard geotechnical journals:
- Journal of Geotechnical Engineering (ASCE)
- Geotechnical Testing Journal - (ASTM)
- Canadian Geotechnical Journal (CGS)
- Transportation Research Record (TRB)

C. Proceedings of Conferences, Symposia, Etc.

1. International:
International Conference on the Use of Fabrics in Geotechnics (1977), Paris. Proceedings available from Ecole Nationale des Ponts et Chausees, 28 rue de Saints-Peres, Paris, France (2 Vols.).

Second International Conference on Geotextiles (1982), Las Vegas, NV. Proceedings available from IFAI (4 Vols.).

Third International Conference on Geotextiles (1986), Vienna. Proceedings available from IFAI (5 Vols.).

2. Canadian:

Proceedings of the 1st Canadian Symposium on Geotextiles (1980), Canadian Geotechnical Society, Calgary, Alberta.

Proceedings of the Second Canadian Symposium of Geotextiles and Geomembranes (1985), Edmonton, Alberta.

Seminar on the Use of Synthetic Fabrics in Civil Engineering (1981), Consulting Engineers of Ontario, Toronto, Ontario.

3. Other Geotextile Publications:

The Use of Geotextiles for Soil Improvement (1980) Preprint 80-177 ASCE National Convention, Portland, OR.

"Fabrics," Special Supplement to Civil Engineering (England), March, 1981.

"Engineering Fabrics in Transportation Construction," (1983) Transportation Research Record 916.

Symposium of Polymer Grid Reinforcement in Civil Engineering (1984) London, Available from Tensar Corp., Atlanta, GA.

"Geotextiles as Filters and Transitions in Fill Dams" (1986) Bulletin 55 International Commission on Large Dams; available from USCOLD, Denver, CO.

Proceedings of the Conference to Geosynthetics '87, (1987) New Orleans, Louisiana (2 Vols.), sponsored by IFAI, NAGS, and IGS. Proceedings available from IFAI (NOTE: Also contains papers on geomembranes.)

Fluet, J.E. (editor) (1987) Geotextile Testing and the Design Engineer ASTM Special Technical Publication 952.

4. Geomembranes:

National Conference on Management of Uncontrolled Hazardous Waste Substances (1981) Hazardous Materials Control Research Institute Silver Springs, MD.

Management of Uncontained Hazardous Waste Sites (1983), Hazardous Materials Control Research Institute, Silver Springs, MD.

International Conference on Geomembranes (1984) Denver. Proceedings available from IFAI.

Johnson, A.I., Frobel, R.K., Cavaiti, N.J., and Pettersson, C.B. (editors) (1985) Hydraulic Barriers in Soil and Rock, ASTM, STP 874, 332 p.

Van Zyl, D.J.A., Abt, S.R., Nelson, J.D. and Shepard, T.A. (editors) (1987) Geotechnical and Geohydrological Aspects of Waste Management, Lewis Publishers, Inc., Chelsea, Michigan, 312 p.

D. Manufacturer's Brochures, Product Specifications, and Design Manuals

A number of manufactures have brochures describing typical applications, and listing the general and index properties of their products. Some have also produced detailed design manuals, often written by reputable consulting engineers and professors. In some instances, they have technical assistance personnel on staff. This information and assistance is usually free upon request and should be considered along with the other information given above.

E. Other Information Sources

A number of societies and committees have been formed and are good sources of current information to keep you up to date. Groups interested in geosynthetics (chairpersons in parenthesis):

ASCE: Geotechnical Engineering Division Committee on Placement and Improvement of Soils (L.R. Anderson: 801-750-2775). For publication information, contact ASCE (212-705-7538).

ASTM: Committee on Geotextiles, Geomembranes, and Related Products (B.R. Christopher: 312-272-6520). For publication information, contact ASTM (215-297-5400)

TRB: Committee A2K07 on Geosynthetics (V.C., McGuffey: 518-457-4712). For publication information, contact TRB (202-334-3218).

ISSMFE: Committee on Geotextiles (J.P.Giroud: 305-736-5400).

International Geotextile Society (IGS) (J.P. Giroud: 305-736-5400).

North American Geosynthetic Society (NAGS) (J.E. Fluet: 305-736-5400).

A-2   1992 List (Cazzuffi, D. and Anzani, A., IGS News, Vol. 8, No. 1, March, 1992)

The list of publications given in this document has been compiled by the Members of the IGS Education Committee under the guidance of D. Cazzuffi and Anna Anzani. The list does not purport to be complete but is offered as a starting point for those readers interested in acquiring recognized high-quality publications on geotextiles, geomembranes and related products.

The list of reference documents include conference proceedings, textbooks and magazines and have been grouped into the following categories:
*   General Topics
*   Material Characteristics and Testing
*   Reinforcement Applications
*   Dams
*   Bank Protection
*   Waste Containment Applications.

In each category the reference documents have been listed in chronological order (except for General Topics, where Conference Proceedings have been grouped first, followed by textbooks and magazines).

This list has been reviewed by all Members of the IGS Education Committee.

General Topics

Proceedings of the International Conference on the Use of Fabrics in Geotechnics - First International Conference on Geotextiles (1977) - Paris (3 volumes, 532 pages). Order from: ENPC, Service Formation Continue, 28 Rue des Saints Peres, 75006 Paris, France.
(Price: US $ 50 plus postage)

Proceedings of the Second International Conference on Geotextiles (1982) - Las Vegas (4 volumes, 1024 pages). Order from: IFAI, 345 Cedar St., Suite 800, St. Paul, MN 55101, USA.
(Price: US $72 plus postage)

Proceedings of the Third International Conference on Geotextiles (1986) - Wien (5 volumes, 1550 pages).
Order from; IFAI, see address above (for North America).
(Price: US $ 128 plus postage)
Balkema, P.E. Box 1875, NL - 3000 BR Rotterdam, The Netherlands (for the rest of the world).
(Price: 300 dFl plus postage)

Proceedings of the Fourth International Conference on Geotextiles, Geomembranes, and Related Products (1990) - The Hague (2 volumes, 884 pages, plus a third volume in print).
Order from: Balkema, see address above.
(Price: 290 dFl plus postage)

Balkema, Old Post Road, Brookfield, VT 06036, USA (for North America). (Price: US $160 plus postage).

Proceedings of the International Conference on Geomembranes (1984) - Denver, (2 volumes, 511 pages). Order from: IFAI - see address above.
(Price: US $40 plus postage)

Proceedings of Geosynthetics '87 (1987) - New Orleans (2 volumes, 639 pages).
Order from: IFAI, see address above. (Price: US $50 plus postage)

Proceedings of Geosynthetics '89 (1989) - San Diego (2 volumes, 600 pages). Order from: IFAI, see address above. (Price: US $55 plus postage)

Proceedings of Geosynthetics '91 (1991) - Atlanta (2 volumes, 864 pages). Order from: IFAI, see address above. (Price: US $55 plus postage)

Rankilor, P.R. (1981), Membranes in Ground Engineering, John Wiley and Sons Ltd., Chichester, UK (377 pages).
Order from: J. Wiley, Baffins Lane, Chichester, West Sussex, PO191UD, UK.

Giroud, J-P. (1985), Geotextiles and Geomembranes, Definitions, Properties and Design, IFAI, St. Paul, MN, USA (404 pages). Order from: IFAI, see address above. (Price: US $49 plus postage).

van Zanten, R.V. - Editor (1986), Geotextiles and Geomembranes in Civil Engineering, John Wiley & Sons Ltd., Chichester, U.K. Order from: John Wiley, see address above.

John , N.W.M. (1987), Geotextiles. Blackie and son Ltd., Glasgow and London, UK.
Order from: Blackie and son Ltd., 7 Leicester Place, London W2CH 7BP, UK.

Koerner, R.M. (1990), Designing with Geosynthetics, Prentice Hall, Englewood Cliffs, NJ, USA (652 pages).
Order from: IFAI, see address above. (Price: US $70 plus postage).

Venkatappa, R. G. and Raju, G.V.S.S. - Editors (1990). Engineering with Geosynthetics, Tata McGraw-Hill, New Delhi, (316 pages). Order from: Tata McGraw-Hill Publishing Company Ltd., 4/12 Asaf All Road, New Delhi - 110 002 (Price: Rs. 215/-)

Hausmann, M.R. (1991), Engineering Principles of Ground Modifications, McGraw-Hill, New York, USA (632 pages).
Order from: McGraw-Hill Book Company, International Group, 1221 Avenue of the Americas, New York, NY 10020, USA.

Geotextiles and Geomembranes (Editor T.S. Ingold), an official journal of the IGS published by Elsevier in six issues per year. Order from: Elsevier Science Publisher Ltd., Crown House, Linton Road, Barking, Essex IG1 8JU, UK.
(Price £ 160/year) Elsevier Science Publishing Co., Inc., Journal Information Center, 655 Avenue of the Americas, New York, NY 10010, USA (for North American) (Price: US $296/year)

Material Characteristics and Testing

Fluet, J. - Editor (1987), Geotextile Testing and the Design Engineer, ASTM, Philadelphia, USA (192 pages). Order from: ASTM European Office, 27/29 Wilbury Way, Hitchin, Herts SG4 OSX, UK (for Europe). (Price: £ 25 plus postage) ASTM, 1915 Race Street, Philadelphia, PA 19103, USA.

Rilem (1988), Durability of Geotextiles, Chapman and Hall, London (230 pages). Order from: E & FN Spon. Marketing Dept., 2-6 Boundary Row, London SE18HN, UK. (Price: £ 29 plus postage).

Peggs, I.D. - Editor (1990), Geosynthetics, Microstructure and Performance, ASTM, Philadelphia, USA (170 pages) . Order from: ASTM European Office, see address above. (Price: $26 plus postage).

Rollin, A. and Rigo, J.M. - Editors (1991), Geomembranes, Identification and Performance Testing, Chapman and Hall, London (376 pages). Order from: E & FN Spon, see address above. (Price: £ 45 plus postage).

Reinforcement Applications

Jones, J.F.P. (1984), Earth Reinforcement and Soil Structures, Butterworths, London (184 pages). Order from: Butterworths, Borough Green, Sevenoaks, Kent TN158PH, UK (Price: £ 30 plus postage).

ASCE (1987), Soil Improvement - A Ten Year Up-To-Date, ASCE Geotechnical Special Publication No. 12. Order from: ASCE, 345 East 47th Street, New York, NY 10017, USA.

Jarrett, P.M. and McGown, A. - Editors (1987), The Application of Polymeric Reinforcements in Soil Retaining Structures. Kluwer Academic Publisher. Dordrecht, The Netherlands (638 pages). Order from: Kluwer Academic Publisher Group, P.O. Box 322, 3300 AH Dordrecht, NL (Price: 275 hFl plus postage). Kluwer Academic Publisher, 101 Philip Drive, Norwell, MA 02061, USA (Price: US $149 plus postage).

Mitchell, J.K, and Villet, W.C.B., (1987), Reinforcement of Earth Slopes and Embankments, National Cooperative Highway Research Program Report No. 290, Transportation Research Board.

Proceedings of the International Geotechnical Symposium: Theory and Practice of Earth Reinforcements, (1988) - Fukuoka (618 pages). Order from: Balkema, P.O. Box 1675, NL - 3000 BR Rotterdam, The Netherlands. (Price 120 hFl plus postage).
Balkema, Old Post Road, Brookfield, VT 06036, USA (for North America). (Price: US $59)

Rigo, J.M. and Degeimbre, R. - Editors (1989), Reflective Cracking in Pavements, Assessment and Control, Liege University. Order from: Universite de Liege, Inst. du Ganie Civil, Qual Banning, 6, B-4000 Ljege.

Shercliffe, D.A. - Editor (1990), Reinforced Embankments - Theory and Practice, Thomas

Telford, London (177 pages). Order from: Thomas Telford, Ltd., Thomas Telford House, 1 Heron Quay, London E144JD, UK.

McGown, A., Yoe, K.C. and Andrawes, K.Z. - Editors (1991), Performance of Reinforced Soil Structures, Thomas Telford, London (485 pages). Order from: Thomas Telford Ltd., see address above.

Dams

ICOLD (1986), Geotextiles as Filters and Transitions in Fill Dams, Paris (130 pages). Order from: Commission Internationale des Grands Barrages, 151, bd Haussmann, 75008 Paris, France. (Price: Ff. 100 plus postage).

ICOLD (1991), Watertight Geomembranes for Dams, State of the Art, Paris (140 pages). Order from: CIGB, see address above. (Price: Ff. 180 plus postage).

Bank Protection

Flexible Armoured Revetments Incorporating Geotextiles (1984), T. Telford, London (400 pages). Order from: Thomas Telford Ltd., see address above. (Price: £ 29.94 plus postage).

PIANC (1987), Guidelines for the Design and Construction of Flexible Revetments Incorporating Geotextiles for Inland Waterways, PIANC, Bruxelles (156 pages). Order from: Thomas Telford Ltd., see address above. (Price: £ 20 plus postage).

Memphill, R. W. and Bramley, M.E. (1989) Protection of River and Canal Banks, CIRIA - Butterworths (200 pages). Order from: Butterworths, see address above. (Price: £ 45 plus postage).

Waste Containment Applications

Proceedings of the Second International Landfill Symposium. Sardinia 89, (1989) - Porto Conte (Alghero) (2 volumes, 1,220 pages). Order from: CIPA, Via Palladio 25, I-20235, Milano. (Price: Lit 250000 plus postage)

Koerner, R.M. - Editor (1990), Geosynthetics Testings for Waste Containment Applications, ASTM, Philadelphia (386 pages). Order from: ASTM European Office, see address above (Price: £ 33 plus postage).

ISSFME - ETC (1991), Technics of Landfills and Contaminated Land. Technical Recommendations "GLC", ed. by the German Geotechnical Society for the ISSMFE; Ernst, Berlin (76 pages). Order from: Hans L. Jessberger, Ruhr University, Bochum, P.O. Box 102148, D-4630, Bochum 1, Germany.
Proceedings of the Third International Landfill Symposium, Sardinia 91 (1991) - Cagliari (2 volumes, 1,816 pages). Order from: CISA, Via Marengo 34, I-09123, Cagliari (Price: Lit 400000 plus postage).

Appendix B
GEOSYNTHETIC TERMS

apparent opening size (AOS, O_{95}) - a property which indicates the approximate largest particle that would effectively pass through a geotextile

blinding - condition whereby soil particles block the surface openings of a geotextile, thereby reducing hydraulic conductivity

California Bearing Ratio (CBR) - the ratio of (1) the force per unit area required to penetrate a soil mass with a 3-square-inch circular piston (approximately 2-inch diameter) at the rate of 0.05 inches/minute to (2) the force per unit area required for corresponding penetration of a standard material

clogging - condition where soil particles move into and are retained in the openings of a geotextile, thereby reducing hydraulic conductivity

cross-machine direction - the direction in the plane of the geosynthetic perpendicular to the direction of manufacture

filtration - the process of retaining soils while allowing the passage of water (fluid)

geocell - a three-dimensional comb-like structure, to be filled with soil or concrete

geocomposite - a geosynthetic material manufactured of two or more materials

geogrid - a geosynthetic formed by a regular network of tensile elements and apertures, typically used for reinforcement applications

geomembrane - an essentially impermeable geosynthetic, typically used to control fluid migration

geonet - a geosynthetic consisting of integrally connected parallel sets of ribs overlying similar sets of ribs, for planar drainage of liquids or gases

geosynthetic - a planar product manufactured from polymeric material used with soil, aggregate, or other geotechnical engineering materials

geotextile - a permeable geosynthetic comprised solely of textiles

index test - a test procedure which may contain a known bias but which may be used to establish an order for a set of specimens with respect to the property of interest

machine direction - the direction in the plane of the geosynthetic parallel to the direction of manufacture

permeability - the rate of flow of a liquid under a differential pressure through a material

permittivity - the volumetric flow rate of water per unit cross sectional area per unit head under laminar flow conditions, in the normal direction through a geotextile

Bibliography

Standard Terminology for Geosynthetics, ASTM D 4439, American Society for Testing and Materials, Philadelphia, PA, 1994, 3 p.

Frobel, R.K., *Geosynthetics Terminology - An Interdisciplinary Treatise*, Industrial Fabrics Association International, St. Paul, 1987, 126 p.

Appendix C
NOTATION AND ACRONYMS

C.1 NOTATION

a	=	radius of tire contact area
A	=	area
AOS	=	apparent opening size
b	=	a dimension; horizontal length of embankment slope
B	=	a coefficient; width of geosynthetic or embankment
c	=	undrained shear strength ("cohesion") in terms of total stresses
c'	=	effective stress strength parameter
c_a	=	soil-geosynthetic adhesion
c_v	=	coefficient of consolidation
C_c	=	compressive index
C_r	=	recompression index
C_u	=	uniformity coefficient, D_{60}/D_{10}
CBR	=	California Bearing Ratio
d	=	depth
D	=	grain size (subscript indicates percent smaller than); depth of embankment; thickness of soft layers
e	=	eccentricity
F*	=	the pullout resistance (or friction-bearing interaction) factor
FS	=	factor of safety
RF_{cr}	=	partial factor for creep deformation, ratio of T_{ult} to creep limiting strength,
RF_{ID}	=	partial factor for installation damage
RF_{CD}	=	partial factor for chemical degradation
RF_{BD}	=	partial factor for biological degradation
RF_{JNT}	=	partial factor for joints, seams, and connections
g	=	acceleration due to gravity
G_L	=	lower strength geogrid
G_H	=	higher strength geogrid
GVW	=	gross vehicle weight
H	=	head difference (gradient ratio test); embankment, slope or wall height
i	=	hydraulic gradient
k	=	coefficient of permeability
K	=	stress ratio; force coefficient
K_A	=	active earth coefficient of the retained backfill
L	=	length; length of reinforcement; length of failure arc
L_E	=	embedment length to resist pullout
M	=	moment

n	=	porosity
N	=	number of layers
N_c	=	bearing capacity factor for cohesive soils
O	=	opening size; subscript indicates percent smaller than
P_a	=	active earth pressure
P_b	=	resultant active earth pressure due to the retained backfill
PI	=	plasticity index
P_q	=	resultant active earth pressure due to the uniform surcharge
P_Q	=	resultant of live load
q	=	flow rate; surcharge load
q_a	=	allowable bearing capacity
q_{ult}	=	ultimate bearing capacity
Q_L	=	live load
R	=	radius of critical failure circle
R_v	=	resisting force (Meyerhof's approach)
S	=	vertical spacing between horizontal geogrid layer
SF	=	safety factor
t	=	thickness of geogrid
T	=	tensile strength of the geosynthetic
T_a	=	allowable tensile strength of the geosynthetic
T_d	=	design tensile strength of the geosynthetic (usually at a given strain)
T_{ult}	=	ultimate tensile strength of a geosynthetic
T_l	=	creep limit tensile strength of a geosynthetic
v	=	vertical
V_q	=	vertical force due to surcharge
w	=	water content
W	=	vertical force due to the weight of the fill
x	=	a dimension or coordinate
y	=	a dimension or coordinate
α	=	peak horizontal acceleration for seismic loading
β	=	slope of soil surface; angle of reinforcement force
γ	=	unit weight
Δ	=	change in some parameter or quantity
ε	=	strain
Θ	=	inclination of the wall face
μ	=	friction coefficient along the sliding plane, which depends on the location plane, *i.e.*, $\tan \phi_r$ or $\tan \phi_f$
$\mu*$	=	the pullout resistance of shearing friction between soil and geogrid
ψ	=	permittivity
θ	=	transmissivity; an angle; angle of failure plane
σ_h	=	horizontal stress

σ_o =   overburden stress

σ_p' =   preconsolidation stress

σ_v =   vertical stress

ϕ =   angle of internal friction

ϕ' =   effective angle of internal friction

τ =   shear resistance

C.2 ACRONYMS

AASHTO	American Association of State Highway and Transportation Officials
CPE	chlorinated polyethylene
CSPE	chlorosulfonated polyethylene
CQA	construction quality assurance
CQC	construction quality control
EIA	ethylene interpolymer alloy
EPA	U.S. Environmental Protection Agency
FHWA	U.S. Department of Transportation, Federal Highway Administration
HDPE	high-density polyethylene
GCL	geosynthetic clay liner
MSE	mechanically stabilized earth
MQA	manufacturing quality assurance
MQC	manufacturing quality control
MESL	membrane encapsulated soil layer
PP	polypropylene
PVC	polyvinyl chloride
QA	quality assurance
QC	quality control
SRW	segmental retaining wall (unit)
VLDPE	very low-density polyethylene
USFS	U.S. Department of Agriculture, Forest Service

Appendix D
REPRESENTATIVE[1] LIST OF GEOSYNTHETIC
MANUFACTURERS AND SUPPLIERS

Akzo Nobel Industrial Systems
Ridgefield Business Center
Suite 318, Ridgefield Court
Asheville, NC 28802
(704) 665 - 5050

Claymax Corporation
P.O. Box 88
Fairmount, GA 30139
(706) 337 - 5316

Hoechst Celanese Corp.
P.O. Box 5650
I-85 & Road 57
Spartanburg, SC 29304
(800) 845 - 7597

National Seal Company
1245 Corporate Blvd
Aurora, IL 60504
(708) 898 - 1161

Seaman Corporation
1000 Venture Blvd
Wooster, OH 44691
(615) 691 - 9476

Spartan Technologies
P.O. Box 1658
Spartanburg, SC 29304
(800) 638 - 1843

Tenax Corporation
4800 East Monument Street
Baltimore, MD 21205
(410) 522 - 7000

Amoco Fabrics and Fibers
900 Circle 75 Parkway, Suite 300
Atlanta, GA 30339
(404) 984 - 4444

Environmental Protection
P.O. Box 333
Mancelona, MI 49659-0333
(616) 587 - 9108

**JPS Elasomerics
Corporation**
395 Pleasant St.
Northhampton, MA 01060
(413) 586 - 8750

Nicolon Corporation
3500 Parkway Lane, Suite 500
Norcross, GA 30092
(800) 234 - 0484

Poly-Flex, Inc.
2000 West Marshall Drive
Grand Prairie, TX 75051
(214) 647 - 4374

Serrot Corporation
8383 Cassia Way
Henderson, NV 89014
(702) 566 - 4739

Staff Industries, Inc.
240 Chene Street
Detroit, MI 48207
(800) 526 - 1368

Tensar Earth Technologies
5775-B Glenridge Drive
Suite 450, Lakeside Center
Atlanta, GA 30328
(800) 836 - 7271

Bayex Inc.
14770 East Ave.
P.O. Box 390
Albion, NY 14411-0390
(800) 263 - 5715

Gundle Lining Systems
19103 Gundle Road
Houston, TX 77073
(800) 435 - 2008

LINQ Industrial Fabrics
2550 West 5th North Street
Summerville, SC 29483
(800) 543 - 9966

Palco Lining Inc.
2624 Hamilton Blvd
S. Plainfield, NJ 07080
(908) 898 - 6262

Reemay Inc.
70 Old Hickory Blvd.
Old Hickory, TN 37138
(800) 321 - 6271

SLT North America, Inc.
200 S. Trade Center Pkwy
Conroe, TX 77385
(713) 350 - 1813

Synthetic Industries
Construction Products
Division
4019 Industry Drive
Chattanooga, TN 37416
(800) 621 - 0444

Watersaver Company, Inc.
P.O. Box 16465
Denver, CO 80216-0465
(303) 289 - 1818

---

1. List is from the Industrial Fabrics Association International, Geotextile and Geomembrane Divisions membership lists, 1995.

GENERAL PROPERTIES AND COSTS OF GEOTEXTILES AND GEOGRIDS

TABLE E1 - GENERAL RANGE OF STRENGTH AND PERMEABILITY PROPERTIES[1,2] FOR REPRESENTATIVE TYPES OF GEOTEXTILES AND GEOGRIDS

Geotextile Type	Weight[3] (g/m²)	Ultimate Tensile[4] Strength (kN/m)	Strain at Ultimate[4] Tensile Strength (%)	Secant[4] Modulus at 10% Strain (kN/m)	Grab[5] Strength (N)	Puncture[6] Strength (N)	Burst[7] Strength (kPa)	Tear[8] Strength (N)	Equivalent[9] Darcy Permeability (m/sec)
Woven									
Monofilament-Polypropylene	120 - 240	16 - 70	20 - 40	70 - 260	700 - 2300	320 - 700	2700 - 4800	200 - 440	$10^{-4} - 10^{-2}$
Silt-Film and Fibrillated Tape	50 - 170	12 - 45	20 - 40	50 - 260	320 - 1600	80 - 600	1400 - 4800	200 - 1600	$10^{-4} - 10^{-3}$
Multifilament Polypropylene	240 - 760	35 - 210	15 - 40	175 - 700	700 - 6200	700 - 1100	4100 - 10400*	440 - 1800	$10^{-4} - 10^{-3}$
Multifilament-Polyester	140 - 710	25 - 350*	10 - 30	175 - 1050*	700 - 9000*	200 - 1400	3400 - 10400*	360 - 2300	$10^{-4} - 10^{-3}$
Nonwoven									
Continuous Filament-Melt Bonded	50 - 240	4 - 35	30 - 100	18 - 90	180 - 1800	80 - 440	550 - 3500	120 - 900	$10^{-4} - 10^{-2}$
Needlepunched (lightweight)	70 - 240	4 - 18	40 - 150	2 - 25	180 - 1100	200 - 550	1000 - 2700	120 - 700	$10^{-3} - 10^{-2}$
Needlepunched (heavyweight)	240 - 850	8 - 35	40 - 150	9 - 55	700 - 2300	440 - 1100	2000 - 6900	320 - 900	$10^{-4} - 10^{-2}$
Geogrid									
Polypropylene	140 - 240	8 - 35	10 - 20	90 - 230	n/a	n/a	n/a	n/a	> 10
High-Density Polyethylene	240 - 710	8 - 90	10 - 20	55 - 70	n/a	n/a	n/a	n/a	> 10
Polyester	240 - 710	35 - 140	5 - 15	350 - 2600	n/a	n/a	n/a	n/a	> 10

1. Data was obtained from numerous sources, in some cases estimated, and represents an average range. There may be products outside this range. No relation should be inferred between maximum and minimum limits for different tests.
2. Both directions
3. Method 1.1.84, Appendix B, FHWA Geotextile Engineering Manual, 1985
4. Wide Width Method, ASTM D-4595
5. ASTM D-4632
6. ASTM D-4833
7. ASTM D-3786
8. ASTM D-4533
9. ASTM D-4491
* Limited by test machine

TABLE E2
GENERAL DESCRIPTION OF GEOTEXTILES

Property	Description		
	LOW	MODERATE	HIGH
Tensile Strength (kN/m)	< 15	15 - 50	> 50
e @ ult.	< 20%	20% - 50%	> 50%
Burst (kPa)	< 1400	1400 - 3400	> 3400
Permeability, k (m/s)	< 0.0001	0.0001 - 0.001	> 0.001
Cost ($/m^2)	< $1	$1 - $2	> $2

TABLE E3
APPROXIMATE COST[1,2] RANGE OF GEOTEXTILES AND GEOGRIDS

Geosynthetic	Material Cost[1,2] ($/m^2)
Filtration Geotextiles - high survivability	1.25 - 1.75
Erosion Control Mats	3.50 - 6.00
Temporary Erosion Control Blankets	1.25 - 2.50
Roadway Geotextile Separators - high survivability	1.25 - 1.75
Asphalt Overlay Geotextiles	0.60 - 1.25
Geotextile Embankment Reinforcement[3]	2.50 - 12.00
Geogrid/Geotextile Wall and Slope Reinforcement[4,5] - per 15 kN/m long-term allowable strength	1.50 - 3.50

NOTES:
1. Typical costs for materials delivered on-site, for use in Engineer's estimate. Costs are exclusive of installation and Contractor's markup..
2. Installation cost of geosynthetics typically cost $0.30 to $0.90, except for very soft ground and underwater placement
3. Assumes design strength is based upon a 5% to 10% strain criteria with an ASTM D 4595 test.
4. Assumes allowable design strength is based upon a complete evaluation of partial safety factors, as detailed in Appendix F.
5. Material costs of $14.00 to $20.00 should be anticipated if the alternative procedure for determination of long-term design strength (Appendix F, section F.5) is used.

Appendix F
GEOSYNTHETIC REINFORCEMENT TENSILE AND
SOIL-INTERACTION STRENGTHS DETERMINATION

Table of Contents

F.1   BACKGROUND

Geosynthetic reinforcement systems consist of geogrids or geotextiles arranged in horizontal planes in the fill soil to resist outward movement of the composite reinforced soil mass. Geogrids transfer stress to the soil through passive soil resistance on transverse members of the grid and through friction between the soil and the geogrid's horizontal surfaces (Mitchell and Villet, 1987). Geotextiles transfer stress to the soil through friction. Geosynthetic design strength must be determined by testing and analysis methods that account for the long-term geosynthetic-soil stress transfer as well as the durability of the geosynthetic structure. The geosynthetic must be able to sustain long-term load in-service without excessive creep strains. Durability factors include installation damage, chemical degradation and biological degradation. These factors may cause deterioration of either the geosynthetic's tensile elements or the geosynthetic-soil stress transfer mechanism.

This appendix is divided into several sections. First, the partial factor of safety methodology for determining long-term allowable tensile strength is presented. Next, application of long-term allowable strength to design strength is reviewed for wall and slope applications. Then, a brief discussion on use of the partial factor of safety method and evaluation of supplier submittals by agencies is presented. Implementation obstacles are noted. An alternative procedure, less cumbersome to implement, for determination of an allowable strength is then presented. This simplified procedure is complementary to the more-detailed methodology, and is presented as a means to encourage use of geosynthetics in reinforced soil structures.

F.2   LONG-TERM ALLOWABLE TENSILE STRENGTH DERIVED WITH PARTIAL FACTORS (after Berg, 1993)

Long-term allowable tensile strength (T_a) of the geosynthetic should be determined using a partial factor of safety approach (Bonaparte and Berg, 1987). The procedure presented within this section is taken from Berg (1993). This procedure was derived from the Task Force 27 (AASHTO, 1990) guidelines for geosynthetic reinforced soil retaining walls, the Geosynthetic Research Institute's Methods GG4a and GG4b the Geosynthetic Research Institute's Method GT7 (1992).

For reinforced slopes and walls, the long-term allowable geosynthetic design strength, T_a, is:

$$T_a = \frac{T_{ult}}{RF_{CR} \; x \; RF_{ID} \; x \; RF_{CD} \; x \; RF_{BD} \; x \; RF_{JNT}}$$

or; alternatively this equation may be shown as:

$$T_a = \frac{T_{ult}}{RF}$$

with

$$RF = RF_{CR} \; x \; RF_{ID} \; x \; RF_{CD} \; x \; RF_{BD} \; x \; RF_{JNT}$$

where:

T_a = allowable geosynthetic tensile strength,(kN/m);

T_{ult} = ultimate geosynthetic tensile strength,(kN/m);

RF_{CR} = partial factor of creep deformation, ratio of T_{ult} to creep limiting strength, (dimensionless);

RF_{ID} = partial factor for installation damage, (dimensionless);

RF_{CD} = partial factor for chemical degradation, (dimensionless);

RF_{BD} = partial factor for biological degradation, used in environments where biological degradation potential may exist, (dimensionless); and

RF_{JNT} = partial factor for joints (seams and connections), (dimensionless).

F.2-1 Ultimate Strength

Ultimate strength values shall be based upon minimum average roll values (MARV) determined in accordance with ASTM D 4759 (1988). Ultimate strength for agency quality assurance (QA) purposes may be determined according to ASTM D 4595 Test Method for Tensile Properties of Geotextiles by the Wide-Width Strip Method (1986) or GRI:GG1 Geogrid Single Rib Tensile Strength (1988). The test procedure used to determine ultimate strength, however, must be the same as that used to define RF_{CR}.

F.2-2 Creep

Long-term tension-strain-time polymeric reinforcement behavior shall be determined from results of controlled laboratory creep tests conducted for a minimum duration of 10,000 hours for a range of load levels on samples of the finished product according to GRI:GG3 Tension Creep Testing of Geogrids (1991, 1991), or GRI:GT5 Tension Creep Testing of Geotextiles (1992). Samples shall be tested in the direction in which the load will be applied in use, in either a confined or unconfined mode (AASHTO, 1990). Confined creep testing shall use project-specific or representative backfill material and typical placement and confinement conditions.

The requirement for a 10,000-hour minimum creep test period for geogrids and geotextiles may be waived for a new product if it can be demonstrated that it is sufficiently similar to a proven 10,000 hour creep tested product of a similar nature. Product similarity must consider base resin, resin additives, product manufacturing process, product geometry, and creep response. Creep testing of all products, regardless of similarity, shall be conducted for a minimum of 1,000 hours. The 1,000 hour creep curves shall pattern very closely to the 1,000-hour portion of the 10,000- hour creep curves of the similar product.

Creep test data at a given temperature may be directly extrapolated over time up to one order of magnitude, in accordance with standard polymeric practices. Accelerated testing is required to extrapolate 10,000-hour creep test data to a minimum 75-year design life. Procedures for test acceleration are discussed in GRI:GG4 (1990, 1991) and GRI:GT7 (1992) Standards of Practice. Accelerated testing is used to extrapolate to a 75-year design life and to ensure that the failure mechanism, ex. ductile to brittle transition, does not change. Task Force 27 guidelines (1990) reference ASTM D 2837 (1990) as providing guidance on analytical procedures for time-temperature extrapolation of data. This ASTM procedure is for plastic pipes. Therefore, temperatures and failure modes quoted in this standard are not directly applicable to geosynthetic reinforcement materials.

For reinforced soil structures, total strain of the reinforcement shall be less than 10% over the design life of 75 years. Formulation of RF_{CR} -- ratio of ultimate strength to creep-limiting strength -- is defined in GRI:GG3 (1991, 1992) and GRI:GT5 (1992). A default factor for creep is not permitted for detailed and final design. However, an appropriate creep-reduction factor may be used for preliminary design only (AASHTO, 1990).

Additional judgement should be exercised when selecting an allowable total strain value for geosynthetics whose creep response is established with confined (in-soil) creep testing. A study on confined creep testing procedures and interpretation of test results was initiated in 1991 by the FHWA. Field performance of structures designed on the basis of confined creep test data is limited. However, it is well-established and generally recognized that unconfined creep test results are consistently conservative.

F.2-3 Installation Damage (AASHTO, 1990)

The effect of installation damage on geosynthetic reinforcement shall be determined from the results of full-scale construction damage tests. Values must be substantiated by construction damage tests for the selected geosynthetic material with project-specific, representative, or a more- severe backfill source, placement and compaction techniques, as described in GRI:GG4 (1990, 1991) and GRI:GT7 (1992) Standards of Practice.

A default factor of safety of 3.0 for installation damage shall be used if appropriate testing has not been conducted. In no case shall an installation damage factor, RF_{ID}, less than 1.05 be used. This minimum must be justified by testing conducted with similar -- or more-severe -- soils and similar -- or more-severe -- placement and compaction techniques.

F.2-4 Durability

The effects of chemical and biological exposure on the reinforcement are dependent on material composition, including resin type, resin grade, additives, manufacturing process, and final product physical structure (AASHTO, 1990). A literature review on geosynthetic durability was recently completed and is presented in Elias (1990). An extensive laboratory and field-testing program investigating geosynthetic durability was initiated in 1991 by the FHWA.

Chemical durability of geosynthetics is mainly affected by pH of the surrounding environment. The following are recommendations for acceptable soil fill pH:

- $3 \leq pH \leq 9$ is recommended;
- $12 \leq pH \leq 3$ should not be used; and
- $pH > 9$ should only be used with supporting test data for the specific product.

These recommendations do not differentiate between specific chemical constituents that can create extreme pH environments. The recommended procedure for measuring the soil pH is by the glass electrode-calomel reference electrode pH meter, on a 1:1 weight ratio of soil to water (Elias, 1990).

Soil pH testing should be conducted on all candidate reinforced and retained fill materials. Testing must be conducted on soils of potential concern, as identified in FHWA-RD-89-186 Durability/Corrosion of Soil Reinforced Structures (Elias, 1990). Soils of potential concern are summarized in Table F-1.

In no case shall a durability factor less than 1.1 be used for a combined chemical and biological degradation factor (RF_{CD} x RF_{BD}), per Task Force 27 (AASHTO, 1990) recommendations.

Polymers used for geosynthetics are generally not susceptible to biological degradation by micro organisms such as fungi and bacteria (Elias, 1990). If biological degradation potential is suspect, geosynthetic resistance to biological effects should be measured as recommended in Elias (1990) and the biological durability factor increased accordingly.

TABLE F-1
GENERAL GUIDE TO SOIL ENVIRONMENTS OF POTENTIAL CONCERN
WHEN USING GEOSYNTHETICS
(after Elias, 1990)

SOIL ENVIRONMENT:	
Acid Sulphate Soils -	characterized by low pH and considerable amounts of CL^{-1} and SO_4^{-2} ions (e.g., pyritic soils in the Appalachian region)
Organic Soils -	characterized by high organic contents and susceptibility to microbiological attack (e.g., dredged fills)
Salt-Affected Soils -	in areas of seawater saturation or in dry alkaline areas as the Southwestern U.S.
Ferruginous -	contains Fe_2SO_3
Calcareous -	in dolomitic areas
Modified Soils -	soils subject to deicing salts, cement stabilized or lime stabilized

F.2-5 Joints, Seams, and Connections

The effect of the joint strength must be factored into design strength when separate lengths of geosynthetics are connected together or overlapped in the direction of primary force development. The value of RF_{JNT} should be taken as the ratio of the unjointed specimen strength to the joined specimen strength. Testing should be conducted in accordance with ASTM D 4595 (1986) for mechanically connected joints and GRI:GG5 (1991) or GRI:GT6 (1992) for overlap joints. Sustained tension tests of 1,000-hour minimum duration should also be conducted on mechanically connected joints, according to GRI:GG4 (1990, 1991) and GRI:GT7 (1992). A load level equal to the allowable strength, T_a, is suggested for long-term testing. Limits on number and location of joints and seams in a slope or wall structure should be addressed in the project specifications.

As discussed in Chapter 9, the connection geosynthetic strength to the wall facing element may limit strength and, therefore, control the allowable design tensile strength. Connection strength must be addressed in wall designs.

F.2-6   Default Partial Factor Values

Documented testing on materials is always preferred. However, use of default partial factor values is allowed for some parameters when actual test documentation is not available. Acceptable default values for reinforced-soil structures are presented in Table F-2. Values are conservative to encourage product-specific testing and research. Completing a comprehensive testing program will provide significant material cost savings and increased user confidence.

TABLE F-2
GEOSYNTHETIC DEFAULT VALUES FOR VARIOUS
PARTIAL FACTORS
(after AASHTO, 1990, GRI:GG4a, 1990; GRI:GG4b, 1991; and GRI:GT7, 1992)

	Installation Damage	Creep[1]	Chemical Degradaton[2]	Biological Degradation	Joint/Seam Strength[3]
FS	3.0	5.0	2.0	1.3	2.0
NOTES:					
1. A creep default value may be used only for preliminary design actual test data is required for final design (Task Force 27 Report, 1990).					
2. A chemical default value should not be used for soil conditions listed in Table F-1; actual test data should be used for final design.					
3. A default value of 1.0 may be used if no joints or seams are present.					

F.3   DESIGN STRENGTHS

The procedure summarized in Section F.2 is for computation of long-term allowable strength, T_a. The methods for determining design strength, T_d, for reinforced wall and slope designs are presented in this section. The design strength determination usually varies between wall and slope designs, as discussed below. Design strengths for embankments on soft foundations are presented in Chapter 7.

F.3-1   Reinforced Slopes

There are two common approaches to limit equilibrium design of reinforced slopes, as discussed in Chapter 8. One method uses design charts. The second method uses computerized slope stability analyses, either to design (i.e., layout the geosynthetic) or analyze a trial layout. Typically, both of these approaches use a design strength that is taken as equal to the long-term allowable strength, that is:

$$T_d = T_a$$

However, the safety factor against excess total strain or rupture of the geosynthetic reinforcement is usually addressed differently with these two approaches to design.

With chart designs, a safety factor against slope instability is typically incorporated by factoring the soil shear strength. With computerized analyses, a safety factor is computed and reinforcement is added until the targeted safety factor is achieved. A minimum safety factor against internal failure planes of 1.3 is typical. Minimum safety factors of 1.3, 1.3, and 1.5 against compound, external, and horizontal sliding failures, respectively, are recommended (Berg, 1993).

F.3-2   Retaining Walls

The limit equilibrium design procedures used for reinforced wall design, as summarized in Chapter 9, use a factored geosynthetic tensile strength in the analysis computations. Therefore the design strength is taken as equal to the quotient of the long-term allowable strength by the selected safety factor, that is:

$$T_d = \frac{T_a}{FS}$$

This provides an overall safety factor that accounts for variations from design assumptions and provides a reasonable assurance against wall failure. A minimum factor of safety, FS, of 1.5 has been typically required.

F.4   IMPLEMENTATION

The determination of the partial factors for creep (including extrapolation), chemical durability and biological durability require extensive testing and is product-specific. Testing standards for determination of these partial factors are not fully developed; thus, test procedures and/or result interpretation can vary. Therefore, evaluation of supplier submittals is not an easy process for many agencies.

Detailed lists of items to be supplied by potential geosynthetic reinforcement material suppliers and system suppliers, are presented in Berg (1993). Guidelines for assigning evaluation responsibilities to the various public works agency organizations are also provided in these guidelines. However, these guidelines are focused upon scenarios where agencies evaluate materials and use an approved products or vendor list system.

Implementation of geosynthetic reinforced slope and wall technologies has been significantly hampered by the review required to assess geosynthetic long-term allowable strength. When an approved products or vendor list is not used, many public works agencies find the procedure too laborious and time-consuming for post-bid evaluation of materials. Other agencies are not able to implement a complete review process, amongst their various organizational units, for a variety of reasons, or do not feel that they have the in-house expertise to complete an evaluation. Unfortunately, failure to take advantage of well-documented, cost-effective geosynthetic reinforced slope and wall technologies is costly to owners and taxpayers.

It is recognized that some public works agencies need an easier and/or quicker procedure for implementing long-term allowable strength quantification and geosynthetic product acceptance. Therefore, simplified procedure for determining long-term allowable strength is presented in Section F.5. This alternative procedure is to be used in conjunction with, and to complement, the detailed procedure presented in Section F.2.

F.5   SIMPLIFIED LONG-TERM ALLOWABLE STRENGTH DETERMINATION

An alternative for computing a long-term allowable strength (Section F.2) is presented in this section. This alternative procedure is complementary to the more detailed procedure.

The goal of providing an alternative method is to foster widespread use of geosynthetic reinforced slopes and walls in transportation facilities. Specific objectives include:
- providing an easy-to-use method for determining design strengths;
- providing agencies with a method to generically specify geosynthetic reinforcement with a defined default design strength (through a defined default long-term allowable strength and a design safety factor), that suppliers will be required to use unless a detailed evaluation of long-term allowable strength on their specific products has been completed by the agency;

- providing long-term allowable strengths that are conservative and economical;
- providing long-term allowable strengths that are sufficiently punitive that thorough testing and evaluation of geosynthetic materials by manufacturers and suppliers is still promoted;
- providing an interim method until a sufficient number of materials and/or systems are approved by an agency and an approved list specifications is developed;
- providing a method for use with conservative soil environment parameters; and
- maintaining the current state-of-practice.

These objectives and goals can be achieved using a single default factor to account for creep, installation damage, durability, and connections. At first glance, the product of the partial factors, RF, presented in Table F-2 could possibly be used. However, the product of the partial safety factors presented in Table F-2, exclusive of the seam or connection factor, is 39; and it is 78 with the seam or connection factor included. These numbers are too high and do not meet the objective of providing a reasonable and economical value. The partial factors presented in Table F-2 were developed for use when one or two partial factors are undefined, along with other, defined partial factors.

A reasonable default reduction factor, RF_d, which meets the stated goal and objectives, is presented below. This default reduction factor should be limited to projects where the soil environment meets the following requirements:
- granular soils (sands, gravels);
- $5 \le pH \le 9$;
- biologically inactive environments; and
- maximum backfill particle size of 20 mm.

Other qualifiers on application of this default reduction factor is limiting use to projects where:
- maximum retaining walls height is 10 m;
- face element shall be a nonagressive environment for the geosynthetic;
- maximum reinforced slope height is 15 m;
- geotextile reinforcement meets specification strength requirements for High Survivability for separation applications (Chapter 5); and
- the manufacturer certifies that the supplied geosynthetic is intended for and fit to use as long-term soil reinforcement.

For retaining walls and slopes, a default long-term allowable strength, T_a, should be defined with a reduction factor, RF_d, equal to 7. This value will be economical for some products on most projects, but not economical for all products in all situations. This reduction value has been developed recognizing that most users desire minimal risk when using any material. A reduction factor less than this value would likely discourage thorough testing by some suppliers. A lower number would also likely result in new products being manufactured to achieve as high of an ultimate strength -- at the lowest cost -- as possible. A reduction factor greater than 7 would likely be too prohibitive and uneconomical. Obviously, the value of 7 is not fixed, as it is based upon engineering judgement. This approach therefore, is applicable with a value other than 7.

Use of this alternative allowable strength procedure for structurally-faced, reinforced soil retaining walls does not eliminate the requirement of connection strength testing. Testing shall be conducted to define the ultimate, short-term connection strength, T_{conult}. The default long-term allowable connection strength, T_{ac}, should be defined with a reduction factor, RF_d, equal to 4. The value of 4 is not fixed, as it is based upon engineering judgement. This approach therefore, is applicable with a value other than 4. Thus, the design strength for retaining walls is defined with a safety factor applied to the controlling allowable strength -- the lower of T_a (i.e., $T_{ult}/7$) or T_{ac} (i.e., $T_{conult}/4$).

An additional implementation obstacle is establishing a consistent method for defining ultimate strength. The procedure, ASTM D 4595, Tensile Properties of Geotextiles by the Wide-Width Strip Method, is recommended for geotextiles. The GRI Test Method GG1 Geogrid Rib Tensile Strength is recommended for geogrids.

Again, use of a default reduction factor, RF_d, is for complementary use with the more-detailed procedure. Blanket, long-term use of a default reduction factor will penalize many current suppliers and limit economic benefit of geosynthetic reinforced structures. Those manufacturers and suppliers that have or will conduct the extensive testing to document partial safety factors should be allowed to use those factors, after agency review and approval. Exclusive use of a blanket default value will also severely impede further evolution of this technology.

F.6   SOIL-REINFORCEMENT INTERACTION

Two types of soil-reinforcement interaction coefficients or interface shear strengths must be determined for design: pullout coefficient, and interface friction coefficient (AASHTO, 1990).   Pullout coefficients are used in stability analyses to compute mobilized tensile force at the front and tail of each reinforcement layer in slopes and at the tail in walls. Interface friction coefficients are used to check factors of safety against outward sliding of the entire reinforced mass.

F.6-1   Pullout Resistance

Design of reinforced slopes requires evaluation of the long-term pullout performance with respect to three basic criteria:

    i)   pullout capacity; i.e., the pullout resistance of each reinforcement should be adequate to resist the design working tensile force in the reinforcement with a specified factor of safety (FS_{PO}), where FS_{PO} is typically set at a minimum of 1.5; and

    ii)   allowable displacement; i.e., the relative soil-to-reinforcement displacement required to mobilize the design tensile force should be smaller than the allowable displacement; and

    iii)   long-term displacement; i.e., the pullout load should be smaller than the critical creep load.(Christopher et al., 1990)

The pullout resistance of the reinforcement is mobilized through one or a combination of two basic soil-reinforcement interaction mechanisms. The two mechanisms by which load may be transferred between soil and geosynthetic are: i) interface friction; and ii) passive soil resistance. Geotextile pullout resistance is developed with an interface friction mechanism. Geogrid pullout resistance may be developed by both interface friction and passive soil resistance against transverse elements. The load transfer mechanisms mobilized by a specific geogrid depends primarily upon its structural geometry (i.e., composite reinforcement versus linear or planar elements, thickness of in-plane or out-of-plane transverse elements, and aperture dimension to grain-size ratio). The soil-to-reinforcement relative movement required to mobilize the design tensile force depends mainly upon the load transfer mechanism, the extensibility of the reinforcement material, and the soil type. (Christopher et al., 1990)

The long-term pullout performance (i.e., displacement under constant design load) is predominantly controlled by the creep characteristics of the soil and the reinforcement material. Soil reinforcement systems will generally not be used with cohesive soils susceptible to creep. Therefore, creep is primarily an issue of the reinforcement type. The basic aspects of pullout performance in terms of the major load transfer mechanism, relative soil-to-reinforcement displacement required to fully mobilize the pullout resistance, and creep potential of the reinforcement in granular (and low-cohesive) soils for generic extensible reinforcement types are presented in Table F-3. (Christopher et al., 1990).

TABLE F-3
BASIC ASPECTS OF REINFORCEMENT PULLOUT PERFORMANCE IN GRANULAR AND LOW COHESIVE SOILS FOR MECHANICALLY STABILIZED EARTH
(after Christopher et al., 1990)

Generic Reinforcement Type	Major Load Transfer Mechanism	Displacement to Pullout	Long-Term Performance
geogrids	frictional + passive H.D.	dependent on reinforcement extensibility (25 to 50 mm)	dependent on reinforcement structure and polymer creep
geotextiles	frictional (interlocking) L.D.	dependent on reinforcement extensibility (25 to 100 mm)	dependent on reinforcement structure and polymer creep
NOTE: L.D. - low-dilatency effect; H.D. - high-dilatency effect			

Pullout resistance of geosynthetic reinforcement is defined by the lower value of:
  i) the ultimate tensile load required to generate outward sliding of the reinforcement through the soil mass; or
  ii) the tensile load which produces a 38 mm displacement.

Several approaches and design equations have been developed and are currently being used to estimate pullout resistance by considering frictional resistance, passive resistance, or a combination of both. The design equations use different interaction parameters, therefore it is, difficult to compare the pullout performance of different reinforcements for a specific application (Christopher et al., 1990).

A normalized definition is recommended as presented by Christopher et al. (1990). The ultimate pullout resistance, P_r, of the reinforcement per unit width of reinforcement is given by:

$$P_r = 2F^* \cdot \alpha \cdot \sigma'_v \cdot L_e$$

where:

L_e = the embedment or adherence length in the resisting zone behind the failure surface, (m);

2 = the reinforcement effective unit perimeter; for geogrids and geotextiles, (dimensionless);

$L_e \cdot 2$ = the total surface area per unit width of the reinforcement in the resistance zone behind the failure surface, (m^2);

F^* = the pullout resistance (or friction-bearing-interaction) factor, (dimensionless);

α = a scale effect correction factor, (dimensionless); and

σ'_v = the effective vertical stress at the soil-reinforcement interfaces, (kN/m^2).

The pullout resistance factor, F^*, can be most accurately obtained from pullout tests performed on the specific, or representative, backfill to be used on the project. Refer to GRI Test Method GG5 (1991) and GT6 (1992), as applicable, for pullout test procedures. Note that this test method produces pullout interaction coefficients that are classified as either short-term or long-term. **Design of reinforced soil walls and slopes for permanent applications requires use of long-term interaction coefficients**.

Alternatively, F^* can be derived from empirical or theoretical relationships developed for each soil-reinforcement interaction mechanism or provided from the reinforcement supplier. For any reinforcement, F^* can be estimated using the general equation (Christopher et al., 1990):

$$F^* = \text{Passive Resistance} + \text{Frictional Resistance}$$

or

$$F^* = F_q \cdot \alpha_b + K \cdot \mu^* \cdot \alpha_f$$

where:

F_q = the embedment (or surcharge) bearing capacity factor

α_b = a structural geometric factor for passive resistance

K = a ratio of the actual normal stress to the effective vertical stress; it is influenced by the reinforcement's geometry

μ^* = an apparent friction coefficient for the specific reinforcement

α_f = a structural geometric factor for frictional resistance

Passive resistance is applicable only to geogrids, and not to geotextiles. The passive resistance portion of the above equation assumes that long-term passive resistance will occur across transverse geogrid ribs. This requires sufficient long-term junction strength between the transverse and longitudinal ribs to assure stress transfer. Long-term stress transfer is assured if the geogrid is creep-tested with the through-the-junction method per GRI:GG3a (1991). Long-term pullout interaction coefficients should be quantified for geogrids with either:

i) quick, effective stress pullout tests and through-the-junction creep-testing of the geogrid per GRI:GG3a test method (1991);

ii) quick, effective stress pullout tests of the geogrid with severed transverse ribs;

iii) quick, effective stress pullout tests of the entire geogrid structure if summation of shear strengths of the joints occurring in a 12-inch length of grid sample is equal to or greater than the ultimate strength of the grid element to which they are attached (AASHTO, 1990); or

iv) long-term effective stress pullout tests of the entire geogrid structure.

Long-term pullout interaction coefficients should be quantified for geotextiles with:

v) quick, effective stress pullout tests.

Test Method a) Controlled Strain Rate Method For Short-Term Testing per GRI:GG5 (1991) and GRI:GT6 (1992) is recommended for testing under conditions i), ii), iii), and v) above. Test method d) Constant Stress (Creep) Method For Long-Term Testing per GRI:GG5 (1991) is recommended for condition iv) above. Joint shear strength shall be measured in accordance with GRI:GG2 (1988) and ultimate strength shall be measured with either GRI:GG1 (1988) or ASTM D 4595 (1986) for condition iii) above.

Long-term testing may also be required if cohesive soils are utilized, to define long-term effective stress (drained) pullout resistance. Procedures and results for long-term testing in cohesive soils have been presented by Christopher and Berg (1990). Their method is a combination of GRI:GG5 (1991) and GRI:GT6 (1992) methods c) Incremental Stress Method for Short-Term Testing, and d) Constant Stress (Creep) Method For Long-Term Testing.

F.6-2 Interface Friction

Soil-geosynthetic interface friction should be determined in accordance with ASTM D5321.